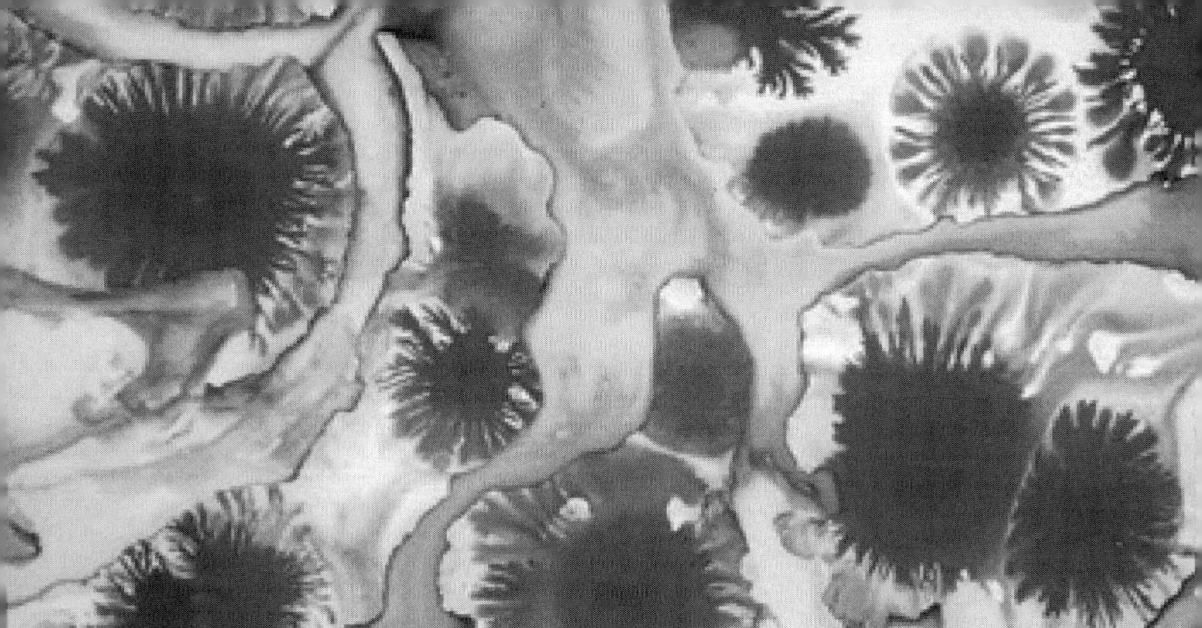

SOIL and FERTILIZER
Concepts and Practices

Dr. Chiu-Chung Young

SOIL and FERTILIZER
Concepts and Practices

ISBN: 978-986-5663-01-8
GPN: 1010301290
DOI: 10.6140/AP.9789865663018
Publishing Date: August, 2014
Price: NT$ 680 / USD$ 30

Author: Chiu-Chung Young
Editor-in-Chief: D. J. Guan
Executive Editor: Chunsheng Huang, Ching-Hua Liao, & Wan-Ting Yeh
Cover & Layout Designer: Mei-Hsiu Lin
Cover Picture: Chiu-Chung Young
Typesetting: Wen-Lin Liu

Publishing Issuer: Der-Tsai Lee
General Manager: Chris Cheng
Manager: Ya-Chu Fan
Publishing Specialist: Kevin Yang
Publisher: National Chung Hsing University
       No. 250, Guoguang Rd., South Dist., Taichung City 402, Taiwan
       Tel: +886-4-2284-0291 / Fax: +886-4-2287-3454
       E-mail: press@nchu.edu.tw

Airiti Press Inc.
       18 F, No. 80, Sec. 1, Chenggong Rd., Yonghe Dist.,
       New Taipei City 234, Taiwan
       Tel: +886-2-2926-6006 / Fax: +886-2-2923-5151
       E-mail: press@airiti.com

© Copyright 2014 Airiti Press Inc.

All rights reserved, this work is copyright and may not be translated or copied in whole or part without the written permission of the publisher (Airiti Press Inc.). Use in connection with any form or by means of electronic or mechanical, including photocopying, recording or any information storage and retrieval system now known or to be invented is forbidden.

About the Author

NAME: Dr. Chiu-Chung Young (楊秋忠)
Email: ccyoung@mail.nchu.edu.tw
Website: http://140.120.200.173/

POSITION:
- National Chair Professor (Lifetime Honor), Ministry of Education, Taiwan
- Chair Professor, Department of Soil and Environment Sciences, National Chung Hsing University, Taiwan
- Executive Director, Advisory Committee, National Chung Hsing University, Taiwan
- Former Vice Presidents, National Chung Hsing University, Taiwan
- Former President, Chinese Sustainable Agriculture Association
- Former President, Chinese Society of Soil and Fertilizer Sciences
- ROC Member of the International Union of Biological Sciences (IUBS)
- ROC Member of the Scientific Committee on Problems of the Environment (SCOPE)

HONOR AND SCHOLARSHIPS:
- Scientific Committee on Problems of the Environment (SCOPE): Life Achievement Awards on Environmental Sciences (2010)
- The Executive Yuan of ROC: Outstanding Achievement Award in Science and Technology (2004)
- Ministry of Education: National Chair (2005-2008, 2010-lifetime honor), Academic Award (1999), Outstanding Teacher Award (1994), Outstanding Education Member Award (1992)
- National Science Council: Outstanding Research Awards (1985-1987, 1987-1989, 1989-1991, 1991-1993, 1993-1995 for 10 years 5 times), Special Research Fellow Awards (1996-1998)(1999-2001), Outstanding Research Fellow Award (2003), Technology Transport Award (2008)
- National Chung Hsing University: Chair Professor (2005 to now), Distinguished Alumnus (2012), Outstanding Research Teacher Award (1999)
- Academic Society: Outstanding Research Awards from Chinese Soil and Fertilizer Society (2006), Agricultural Association of China (1994), Chinese Agricultural Chemical Society (1990)
- University of Hawaii, USA: Distinguished Visiting Scholar of the College of Tropical Agriculture and Human Resources (2008)
- National Cheng Kung University: Outstanding Achievement Award in Science and Technology of Environmental Bioremediation (2005)

MAJOR CURRENT RESEARCH INTERESTS:
- Soil microbiology and biochemistry
- Microbial fertilizer and soil-crop relationships
- Soil organic compounds and allelopathy
- Organic waste treatment, organic fertilizer

PUBLICATIONS:
- Total 614 papers
- Refereed papers: 202
- Conference papers: 210
- Technical and extension education papers: 182
- Books: 20

Soil and Fertilizer: Crop Cultivation and Soil Problems

(A) Normal vegetable production in the field soil.
(B) Normal pineapple production in the field soil.
(C) Alkaline soil causes pear tree's younger leaves tuning yellow.
(D) Excessive nitrogen fertilization causes over shoot growth of pear tree.
(E) Wilt and rot caused by *Fusarium oxysporum* in the soil.
(F) Rotten root caused by soil flooding.

Soil and Fertilizer: Crop Diseases and Soil Problems

(A) Tomato bacterial wilt causes by *Ralstonia solnacearum*.
(B) The symptoms of deficiency calcium in Tomato.
(C) Root-knot is caused by nematodes in cucumber.
(D) Root rot is caused by soilborne disease in *Amaranthus* continuous cropping.
(E) Root rot is caused by soilborne disease in rice paddy.
(F) Poor rooting of pineapple is caused by pesticide residues in the soil.

Soil and Fertilizer: Functions of Rhizobial Inoculation

(A) Root nodules of peanut after rhizobial inoculation.
(B) Rhizobial selection is important for better inoculant for soybean.
(C) Rhizobial inoculation increases peanut production in Yen-Chang field test.
(D) Rhizobial inoculation increases the growth of *Leucaena* plants in the pot test.
(E) Rhizobial inoculation increases the growth of soybean in Hsin-Zu field test.
(F) Rhizobial inoculation increases the growth of soybean in Wi-Pu field test.

Soil and Fertilizer: Cultures and Function of Phosphate Solubilizing Microorganisms

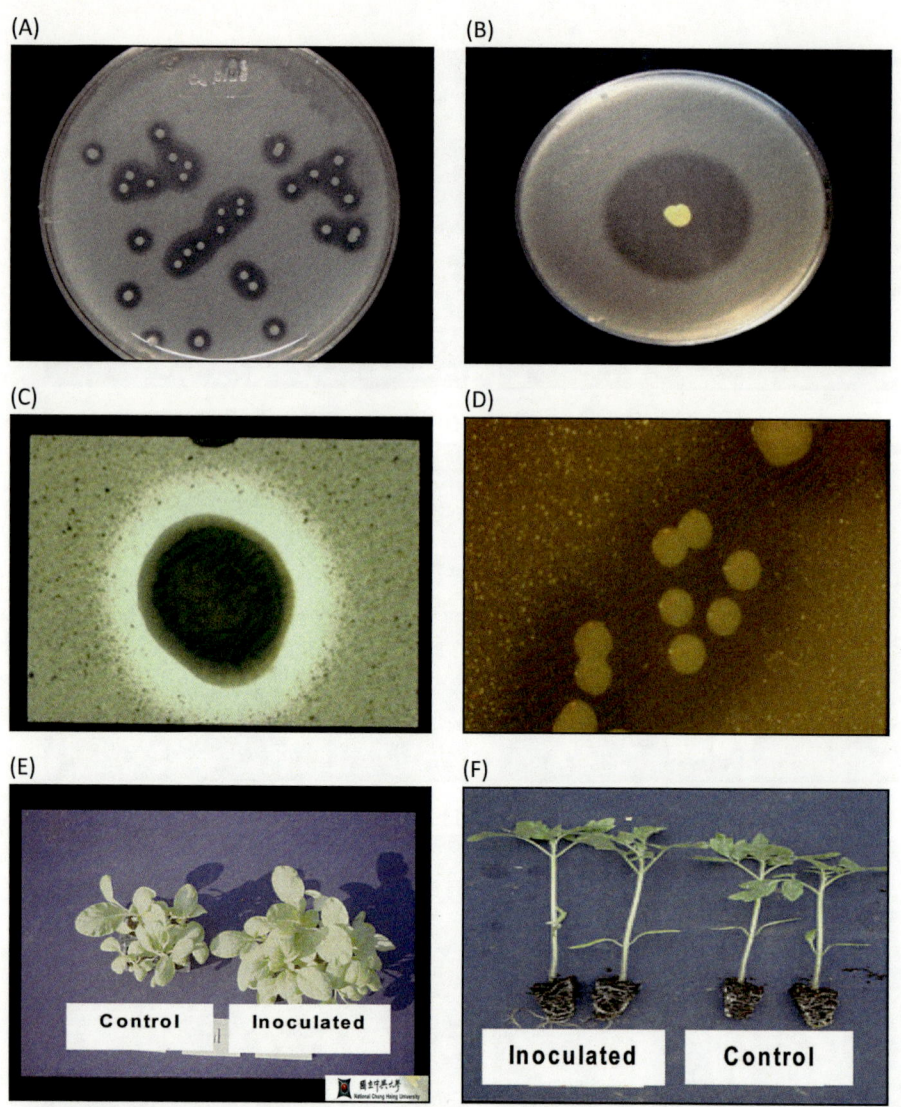

(A) Phosphorus-solubilizing bacteria (PSB) dissolve insoluble phosphates in the plate.
(B) Supper PSB dissolves insoluble calcium phosphates in the plate.
(C) Clear zone around PSB to dissolved calcium phosphate in the microscope observation.
(D) Clear zone around PSB to dissolved iron phosphate in the microscope observation.
(E) PSB inoculation increases the growth of *Brassica* plants.
(F) PSB inoculation increases the growth of tomato plants.

Soil and Fertilizer: Cultures and Functions of Arbuscular Mycorrhizae

(A) Mycorrhizal spores in the inoculants.
(B) Mycorrhizal inoculant production in green house.
(C) Arbuscular mycorrhiza (AM) inoculation increases the growth of maize roots.
(D) Arbuscular mycorrhiza inoculation increases the growth of maple seedlings.
(E) Arbuscular mycorrhiza inoculation increases the growth of melon plant in the field.
(F) Arbuscular mycorrhiza inoculation increases the growth of soybean plants.

Soil and Fertilizer: Organic Wastes and Treatments

(A) Leaching water problems in organic waste.
(B) Big space is need for composting of organic wastes.
(C) Fast treatment (1 h) for oil palm fiber to produce organic fertilizer.
(D) The batch reactor for fast treatment of organic wastes is to produce organic fertilizer.
(E) The continuous reactor for fast treatment of organic waste is to produce organic. fertilizer.
(F) Effect of animal feces after the fast treatment on the growth of *Ipomoea aquatic*.

Soil and Fertilizer: Microbial Fermenters and Researchers

(A)

(B)

(C)

(D)

(E)

(F)

(A) Manual fermenter.
(B) Modern fully automatic fermentation control system.
(C) Modern fermenter tanks.
(D) The handbook for microbial fertilizers.
(E) Group photo of Microbial Fertilizer Measurement Training Course.
(F) Group photo of Dr. Chiu-Chung Young's lab.

Soil and Fertilizer: Author's Family

(A)

(A) Group photo of Dr. Chiu-Chung Young's family.

Preface

Agriculture builds upon crops integration and the environment, with which its yield depends strongly on healthy soil foundation. With that in mind, the knowledge of soil and fertilizer is crucial to maintaining an environment for crop production with optimal nutrients, water and oxygen. Soil is one of human's precious resources. The protection and nurturing of our soil is thus an integral part of sustainable development.

Effective soil management is considered not only to be knowledge of technology, but also of art. In practice, to make a potential use of the land, the management strategies need to cover all the differences, such as characteristics of the soil, unique plant and climate to each geographical location. Such an approach rises to a more important issue nowadays, because the loss of cultivable lands and the need of high quality agricultural products are increasing.

This book is a translation of Soil and Fertilizer (in Chinese, 9th edition, 2010). This is based on my first-hand experiences in research and field application over the past 40 years. In addition, many materials also come from up-to-date textbooks and research articles that I came across during my teaching career. I consider this reference book with the state of art would be helpful for those who are interested in the field of soil and fertilizer. The book emphasizes on practical skills in real-world applications supplemented with theory, which allow readers to put theory into practice efficiently.

I was born and raised in a farmer's family during which I enjoyed hiking in the mountain trails and paddy fields, and raising cattle. Ever since, I have stayed in connection to farmland activities and, particularly, plant and soil. My undergraduate and graduate studies were at the Department of Botany, and Food Institute of National Chung Hsing University, respectively. My master study was at the Institute of Botany, Academia Sinica, and doctoral study was at the Department of Agronomy and Soil Sciences, University of Hawaii. These deeply rooted experiences have always motivated me to contribute my expertise to the farmland where I come from. The soil nourishes hundreds of millions of living organisms. Thus, it is the mother of crops, and the basic of our nature. Every single living organism originates from soil. I hope that every single one from all regions of the world should take thoughts and protect the environment and the soil for our descendents.

This book is dedicated to my parents, family tutor: Mr. Jack Pan; respective teachers, Dr. Duane P. Bartholomew, Dr. S. Sun, Dr. C.H. Chou, Dr. T. C. Juang; my wife Kuei-Yin, and my kids Li-Sen and Li-Hao.

I would also like to thank reviewers Dr. Zueng-Sang Chen and Dr. Ren-Shih Chung, from Department of Agricultural Chemistry, National Taiwan University, and as well as Dr. Shi-Chung Wu and Miss Men-Yu Chen.

Chiu-Chung Young
National Chair Professor
at Department of Soil and Environmental Sciences
National Chung Hsing University, Taichung, Taiwan
July 04, 2014

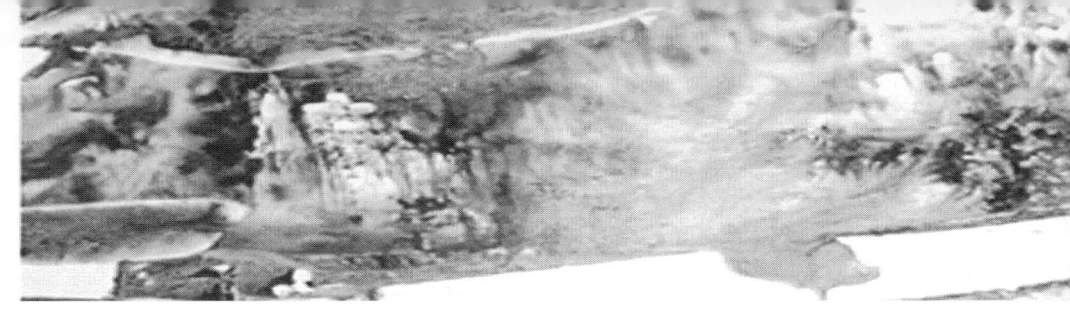

SOIL and FERTILIZER
Concepts and Practices

Contents

Chapter 1: Understand the Human Ancestral Treasure: The Soil 1

Soil Introduction 1
    Protection of the Soil is a Mission for Every Generation 1
    The Constitution and Sources of the Soil 2
    The Development of the Soil 2
    The Soil is also a Living Natural Body 3
    The Soil is a Remediation Factory 4
    The Soil Confers Buffering Capability 4
    The Soil can Get Sick 4
    The Key to Environmental Protection 5

Concerns on Soil Degradation 7
    Factors for Soil Degradation 7
    The Result of Soil Degradation and Methods for Improvement 8

Soil Pollution Management Issues and Concerns 11
    The Severity of Soil Pollution 11
    The Characteristics of Soil Pollution 12
    Preservation of the Soil is the Key Task 12
    The Direction and Goal of Soil Preservation 13

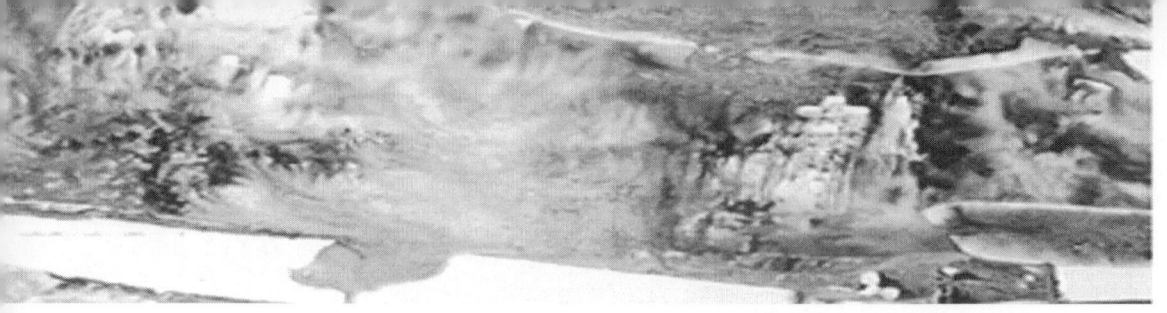

Chapter 2: Essential Concepts for the Improvement of Soil Fertility and Soil Management — 15

Soil Productivity 15
    What is "Fertile Soil"? 15
    Essentials to Improve Soil Fertility 16
    Conclusion 24

Chapter 3: The Series of Soil Fertility Maintenance — 29

The Maintenance Guidelines of Inorganic Fertility 29
    Yield Factors 29
    Guarantee for Soil Complementary 30
    Concepts of Proper Application of Inorganic Fertilizer 30
    The Prevention of Problems and Soil Conservation 31

The Maintenance Guidelines on Organic Fertility 32
    Efficacy of Soil Organic Matter 33
    The Preservation Methods of Soil Organic Fertility 34

The Maintenance Guidelines on Soil Microorganisms 36
    The Functions of Soil Microorganisms 36
    The Occurrence of Problems in Soil Microorganisms 37
    Methods of Soil Microbial Conservation 38

Chapter 4: Diagnosis of Problem Soils — 41

Simple Diagnosis Methods 41
    Simple Diagnosis of Soil Problems 42
    Simple Diagnosis for Cultivating Crops with Symptoms 43

Diagnosis by Application of Chemical Analysis 46
    The Sampling of Problem Plants for Chemical Analysis 48

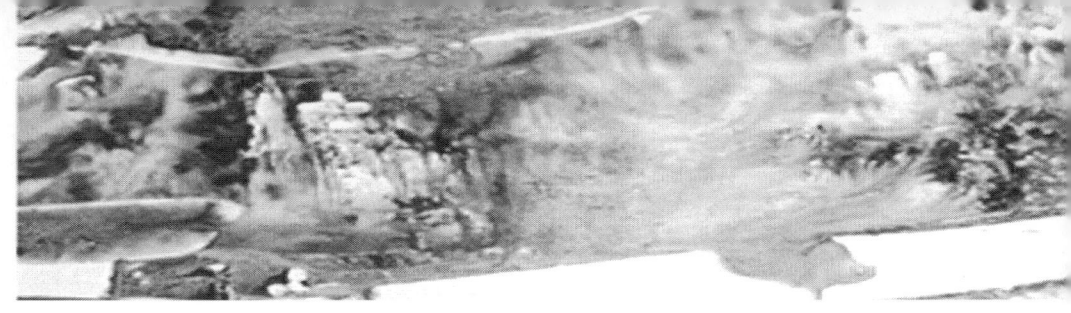

    The Sampling of Problem Soils for Chemical Analysis **49**

    Evaluation the Analysis Data with Symptoms **49**

Problem and Diagnosis of Pollution 51

    Do not Let Soil, Mother of Crops, Cry **51**

    Properties of Soil Pollution **52**

    Soil Polluted by Water **53**

    Air Pollution **54**

    Safe Food is Based on Safe Soil **55**

    The Direction and Goal of Soil Conservation **55**

Chapter 5: Fundamentals of Plant Nutrients and Fertilization 57

The Classification and Application of Fertilizers 57

    The Definition and Classification of Fertilizers **57**

    The Basic Principles of Applying Fertilizers **59**

    Calculation of Fertilizer Application **60**

    Guidelines of Fertilizer Application **61**

Features and Guidelines of Application for Nitrogen 63

    The Relationships between the Soil Nitrogen and Plants **63**

    The Sources and Types of Nitrogen Fertilizer **63**

    The Forms and Transformation of Soil Nitrogen **64**

    The Loss of Nitrogen Fertilizer and Guidelines for Improvement **65**

    Guidelines of Improving the Availability of Nitrogen **66**

Nitrogen Transformation and Soil Microorganisms 68

Features and Guidelines of Application for Phosphorus 70

    The Relationship between the Phosphorus in Soil and Plants **70**

    The Forms and Transformation of Soil Phosphorus **71**

    The Types of Phosphorus Fertilizer **72**

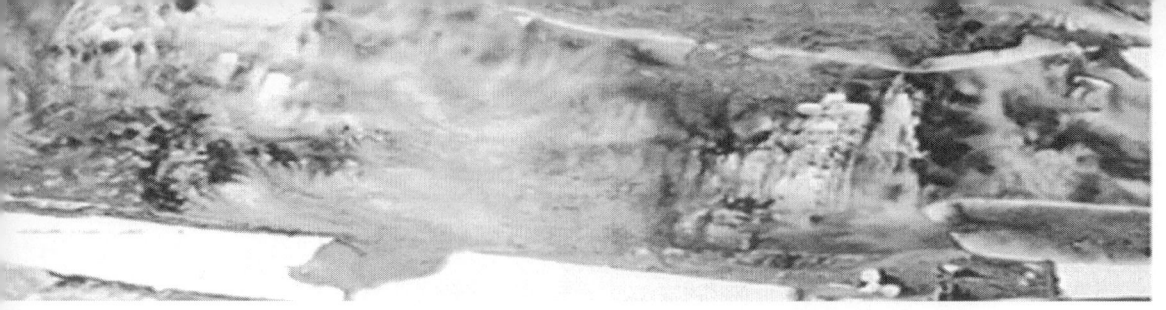

    The Availability and the Fixation of Phosphorus Fertilizers 73

    Guidelines of Enhancing the Availability and Function of Phosphorus Fertilizers 74

Features and Guidelines of Application for Monopotassium and Monocalcium Phosphates 76

    The Features of Monopotassium and Monocalcium Phosphates 76

    The Function of Monopotassium and Monocalcium Phosphates 77

    Guidelines of the Application of Monopotassium and Monocalcium Phosphates 78

    Guidelines for Application of Monopotassium and Monocalcium Phosphates 79

Features and Guidelines of Application for Rock Phosphorus Fertilizers 80

    The Types of Rock Phosphates 81

    The Function of Rock Phosphates in the Soil 82

    The Features of Rock Phosphates 82

    The Quality of Agricultural Rock Phosphates 83

    The Guidelines of Application for Rock Phosphates 83

    The Additives Affect the Solubility of Rock Phosphates 84

Features and Guidelines of Application for Potassium and Magnesium 85

    The Forms and Uptake of Potassium and Magnesium in Soil 85

    The Influence of Potassium and Magnesium on Plants 86

    Types of Potassium and Magnesium Fertilizers 87

    Guidelines of Application of Potassium and Magnesium Fertilizers 87

Features and Guidelines of Application for Calcium and Sulfur 88

    The Function of Calcium and Sulfur on Plants 89

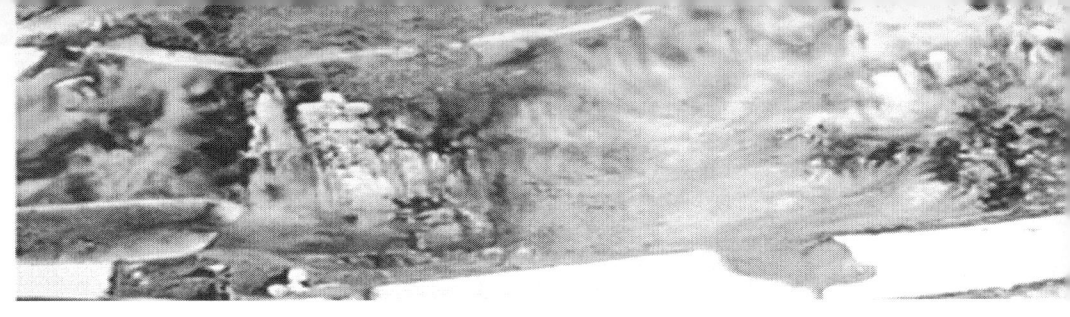

    The Symptoms of Calcium and Sulfur Deficiency **90**

    The Activity of Calcium and Sulfur in the Soil **90**

    The Sources and Types of Calcium and Sulfur Fertilizers **91**

    Guidelines for Applying the Calcium and Sulfur Fertilizers **92**

The Microelements 93

    The Microelements are Indispensable **93**

Features and Guidelines of Application for Boron and Molybdenum 94

    Features and Guidelines of Application of Boron **94**

    Features and Guidelines of Application of Molybdenum **97**

Features and Guidelines of Application for Microelements of Iron, Copper, Zinc and Manganese 98

    The Function on Plants and Symptoms of Microelement Deficiency **99**

    The Factors Affecting Microelement Deficiency **100**

    The pH Affects the Availability of the Microelements **101**

    The Application of Iron, Copper, Zinc and Manganese, and the Tactics for the Deficiency **102**

Features and Guidelines of Application for Silicon Fertilizer 104

    Preface **104**

    The Function of Silicon on Plants **104**

    The Features and Ingredients of Silicon Fertilizer **107**

    The Application and Guidelines of Silicon Fertilizer **107**

    Notices for Application of Silicon Fertilizer **108**

The Siderophores and the Health of Plants 109

    The Necessities for the Plant Health **109**

    Iron is an Indispensable Element for Organisms **109**

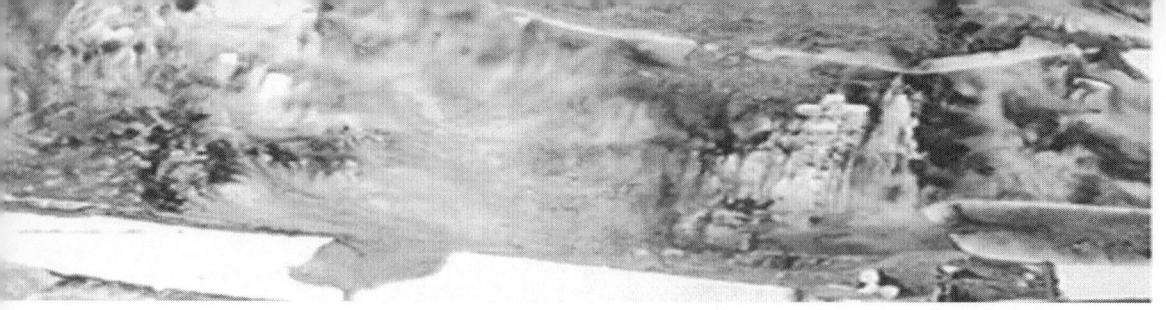

    The Features of Siderophores **109**
    The Function of Siderophores **110**
    The Application of the Siderophores **111**
    Conclusion **111**
Methods and Guidelines of Fertilization for Solid Fertilizers 111
Methods and Guidelines of Fertilization for Liquid Fertilizers 113
    The Development History of Liquid Fertilizers **113**
    The Types of Liquid Fertilizer **113**
    Equipments and Methods of Application for Liquid Fertilizer **114**
    The Advantages and Disadvantages of Liquid Fertilizer **115**
    Methods and Guidelines of Liquid Fertilization **116**
Effects and Guidelines of Application for Slow Release Fertilizers 117
    The Advantages of Slow Release Fertilizers **118**
    Types and Materials of Slow Release Fertilizers **119**
    Methods and Guidelines of Slow Release Fertilization **121**
Fertilization and Quality of Agricultural Products (1): Nitrogen Fertilizers 122
    The Purpose of Fertilization **122**
    The Concept of Quality and Factors Influenced Quality **122**
    Nutrient Supply of Crops and Food Quality **124**
    Relationship between Nitrogen Fertilizer and Quality **124**
Fertilization and Quality of Agricultural Products (2): Phosphorus, Potassium, Calcium, Magnesium, Sulfur and Trace Elements 125
    Relationships between Phosphorus Fertilizer and Quality **125**
    Relationships between Potassium Fertilizer and Quality **126**
    Relationships between Calcium, Magnesium, Sulfur and Quality **127**

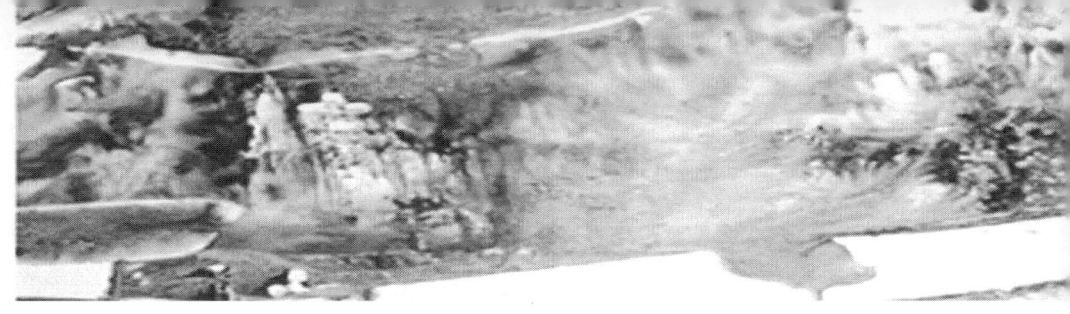

    Relationships between Trace Element and Quality **128**
    Relationships between Fertilization, Food Quality, and Health of Animals and Human **129**
    Conclusion **130**

Advantages and Disadvantages of Vegetables Cultivated by Water Culture 130
    Water Culture has Great Potential, but We should Pay Attention to the Disadvantages **130**
    The Advantages and Disadvantages of Water Culture **131**
    Strengthen Basic Researches on Water Culture **133**

The Application and Problems of Fertilizer 134
    Never can One Fertilizer be Omnipotent **134**
    Reinforcement of Farmer Education on Fertilizer Application **135**
    Concept of Application of Fertilizers and System of the Vendors **135**
    Concern on Conservation of Soil Environment and Biological System **136**

Common Problems and Improvement of Fertilization 137
    The Inappropriate Fertilization will Cause Big Trouble **137**
    Recognizing the Function of Fertilizers **138**
    Excess or Shortage of the Fertilization **138**
    Appropriate Quantity of Fertilization **141**

Problems of Excessive Fertilization and Guidelines for Improvement 141
    Problems Caused by Excess of Fertilization **141**
    Improving Ways of Excessive Fertilization **143**
    Improving Ways for Excessive Fertilization for Perennial Crops **144**

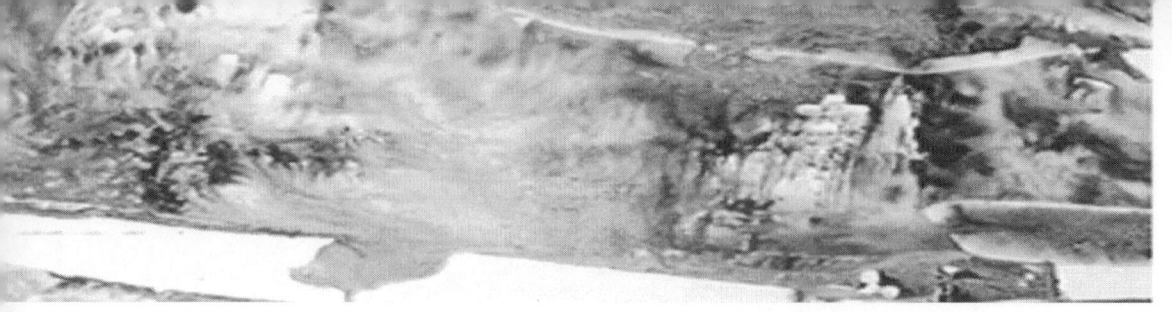

Fertilization and its Quality Influence on the Human Health 147
   The Content of Nutrition and Effects of Deficiency 148
   Assessment According to Medical Index 149
   Correct Fertilization and Healthy Food of Human 149

Chapter 6: The Concept and Application of Foliar Fertilization 157

The Concept and Application of Foliar Fertilization (Part 1) 157
   The Correct Concept of Foliar Fertilization 157
   The Absorption of Foliar Nutrients 158
   Advantages and Disadvantages of Foliar Fertilization 159
   Conditions for Foliar Fertilization 161

The Concept and Application of Foliar Fertilization (Part 2) 162
   The Timing for Foliar Fertilization 162
   The Selected Concentration for Foliar Fertilization 162
   Factors Affecting Foliar Fertilization 164
   Guidelines for Applying Foliar Fertilization 164

Chapter 7: Guidelines of Improvement of Fertilization and Soil Management 169

Guidelines to Improve the Effect of Fertilization 169
   Causes for the Decline of Fertilization Efficiency 169
   How to Reduce the Loss and Volatilization of Fertilizers 170
   Pay Attention to Soil pH Value 172
   Improve the Fertilization Technique 174
   In Accordance with the Need and Characteristics of Crops 176
   It is Easy to Reach the Essence of Matters, if You are Aware of the Important Sequence of Development 177

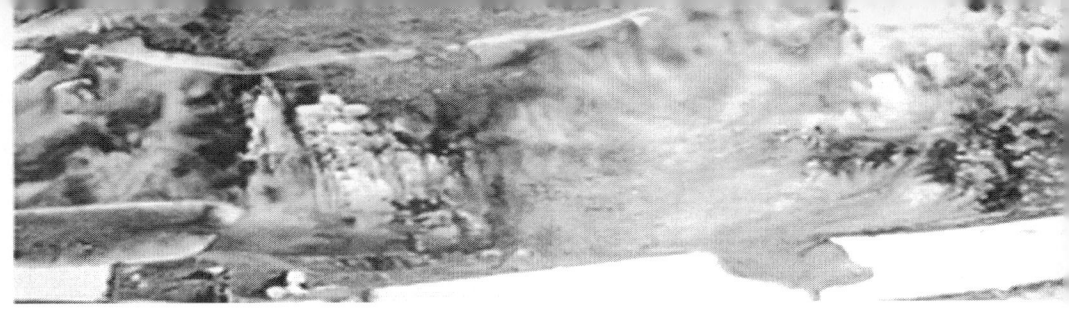

Nutrients and Soil Management for Winter Crops 178
    Characteristics of Winter and its Relationship with Crops 178
    Nutrient Requirement and Absorption Capacity of Winter Crops 179
    The Availability of Soil Nutrients in Winter 179
    Guidelines for the Management of Soil and Fertilizer in Winter 180

Soil Environment and Management in Rainy Season 182
    Soil Environment in Rainy Season 182
    Soil Problems in Rainy Season 184
    The Influence of Rainy Season on Root System 185
    Guidelines of Soil Management in Rainy Season 186

Guidelines for Fertilization under Climate, Disease and Pest Stresses 188
    Climate Stress and Fertilization 188
    Resistance to Diseases and Fertilization 190

Problems of Soil pH and Guidelines of Diagnosis and Management 192
    Understand the Features of Soil pH 192
    Sources of Soil pH 193
    The Source of Acidity in Soils 193
    The Source of Alkalinity in Soil 194
    How do We Know whether the Soil is too Acidic or too Alkaline 195
    The Harmful Effects of Extreme Acidity or Alkalinity of Soils 196
    Guidelines to Neutralize Extreme Acidity or Alkalinity of Soils 196

Problems of Soil Electric Conductivity, Salinity and Guidelines of Diagnosis and Management 200
    Understand Problems of Soil Electric Conductivity and Salinity 200
    The Source of the Soil Electric Conductivity 200

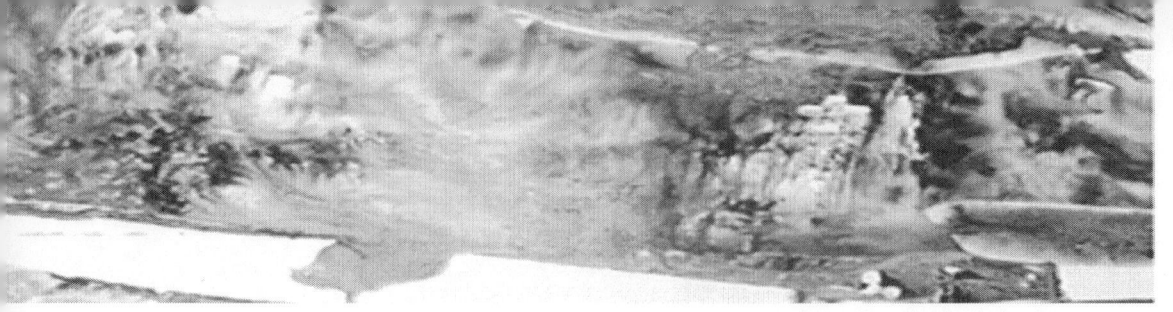

    Problems of Soil EC and Saline Soil Damage **201**

    Methods to Improve Saline Soil Damage **202**

Characteristics of the Saline-Alkaline Soil Formation and Improvement Guidelines 204

    Causes of the Formation of Saline-Alkaline Soil **204**

    Types of Saline-Alkaline Soils **205**

    Guidelines to Improve Saline Alkaline Soils **206**

Guidelines of Soil Fertilization and Management in Slope Lands 208

    Common Improper Ways of Reclamation and Cultivation **209**

    Soil Types and Limiting Factors for Slope Lands' Crops **210**

    Guidelines of Fertilization for Slope Lands and Orchard Management **211**

Conservation of Forest Soils and Use of the Soil Microorganisms 212

    Soil Conservation in Forest Lands **212**

    Strategies to Improve Soil Fertility of Forest Lands **216**

Soil Problems and Solutions of Continuous Cropping 218

    Occurrences for Continuous Cropping **218**

    Problems of Continuous Cropping in Soil **218**

    The Control of Soil Problems Caused by Continuous Cropping **220**

Guidelines for Management and Fertilization of Coarse Soil 223

    Differentiation of Soil Texture **223**

    Characteristics of Coarse Soil **223**

    Guidelines for Cultivation Management and Fertilization of Coarse Soil **224**

Relationships between Management of Soil Fertilization and Crop Diseases 226

    Soil Fertilization is Close to Diseases **226**

    Soil is Related to Plant Diseases **227**

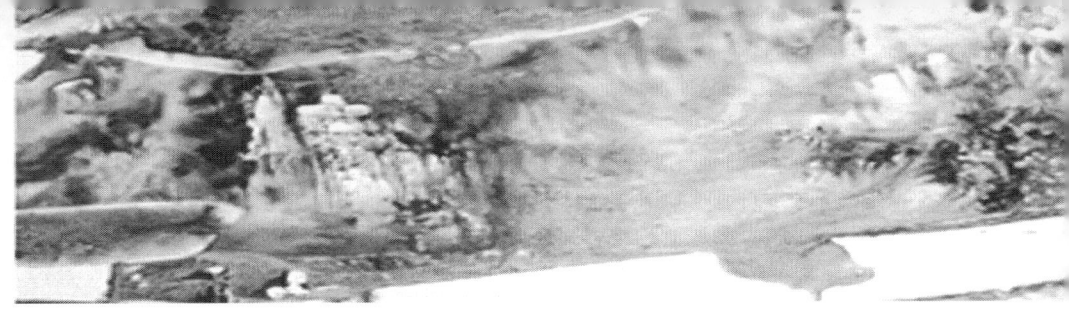

    Relationships between Nutrients and Diseases **228**

    Application of the Soil Microorganisms and Biological Control **230**

    Guidelines of Management of Soil Fertility and Cultivation **233**

Guidelines of Soil Management for Plant Nursery 234

    To Ensure Healthy Seedlings by Plant Germplasm Resources and Environment **234**

    Common Problems in Soil Environment of the Nursery **234**

    Guidelines of Soil Management for a Healthy Nursery **236**

Management of Soil and Fertilizer in Production of Vegetables 238

    Main Concepts of Soil and Fertilizer Management for Vegetables **238**

    Important Roles of Vegetables **239**

    Qualities of Vegetables **239**

    Concepts and Influencing Factors of Vegetable Quality **240**

    Nutrient Supply and Quality of Vegetables **241**

    Fertilization, Food Quality and Human Health **246**

    Conclusion **247**

Conservation and Application of Irrigation Systems in Soil Problems of Facility Cultivation 247

    Soil Problems in Facility Cultivation or Greenhouses **247**

    Guidelines for Soil Conservation of Facility Cultivation or Glasshouse **248**

    The Irrigation Systems in the Facility Cultivation **249**

Guidelines for Fertilization and Management in Golf Course Turf 250

    Proper Management could Reduce the Occurrences of Soil Problems **250**

    Basic Understandings of Turf Growing Environment **251**

    Fertilization Management of Turf: Increase Nutrient Availability and Reduce Fertilizer Loss **253**

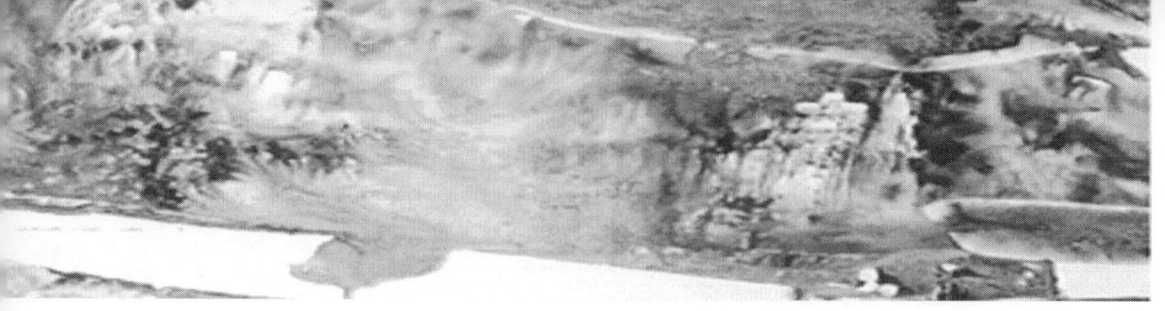

    Diagnosis of the Nutrient Problems **255**

Managements for Fertilizer Conservation, Water Conservation and Water Saving in the Golf Course 257

    To Strength the Fertilizer and Water Holding Capacity of Turf Soil is the Fundamental Method of Soil Management **257**

    Key Methods for the Fertilizer Conservation of Capacity Improvement **258**

    Guidelines and Methods for the Water Holding Capacity Improvement **259**

    Integral Function of Fertilization and Water Spraying Together with Fertilizer Conservation, Water Conservation, and Water Saving **260**

Guidelines of Rational Fertilization to Fruit Trees 262

    Preface **262**

    Understand Characteristics of Crops before Fertilization **262**

    Understand the Characteristics of Fertilizers **263**

    Understand why the Fertilizer Efficiency Decreases **266**

    Guidelines in Rational Fertilization **267**

Soil Medium and Fertilization Management in Rose Cultivation 271

    Soil Adaptability of Rose **272**

    Soil Improvement during the Cultivation **272**

    Approaches to Improve the Poor Soil Properties **272**

    The Medium and its Features for Potted Rose **273**

    Concepts and Methods of Fertilization **274**

    Types and Methods of Fertilization **275**

    Summary **277**

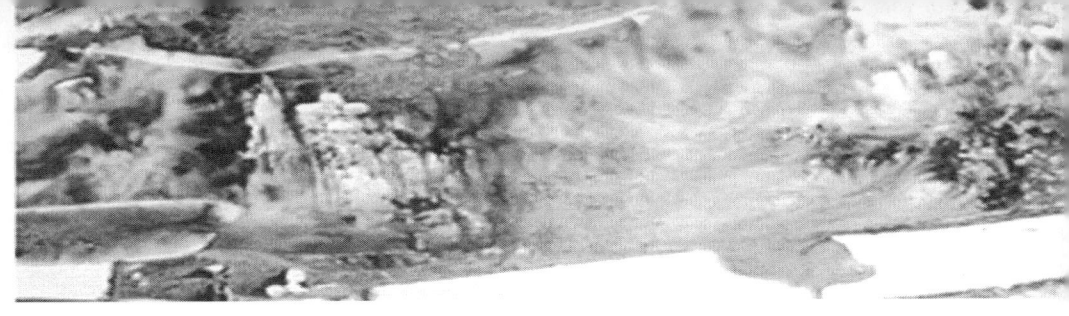

Cover Grass Cultivation and Management of Soil Fertility 278
Functions of Cover Grass Cultivation 278
Effects of Grasses on Soil Physical Property 279
Effects of Grasses on Soil Chemical Property 280
Effects of Grasses on Soil Biological Property 281

Chapter 8: Soil Organic Matter and Organic Fertilizers 283

Introduction to Soil Organic Matter 283
Soil Organic Matter Content is a Standard for Good Agricultural Soils 283
The Best Feature of Soil -- Humic Acid 285

The Characteristics and Applications of Peat 289
Many Agricultural Soils Require Organic Matter 289
The Formation and Characteristics of Peat 289
The Slow Decomposition of Peat is the Best in Organic Fertilizers 290
Applications of Peat and its Advantages in Agriculture 291
The Effects of Peat 292
Fundamentals and Attentions for Peat Application 293

Application Guidelines for Organic Fertilizers 294
Reasons why Organic Matter can be Used as a Fertilizer and a Soil Conditioner 294
The Source and Application Guidelines of Organic Fertilizers 295

Function, Quality and Selection of Organic Fertilizers 301
The Functions of Organic Fertilizers 301
The Quality of Organic Fertilizer 303
Guidelines of Right Organic Fertilizers and Application Selection 305

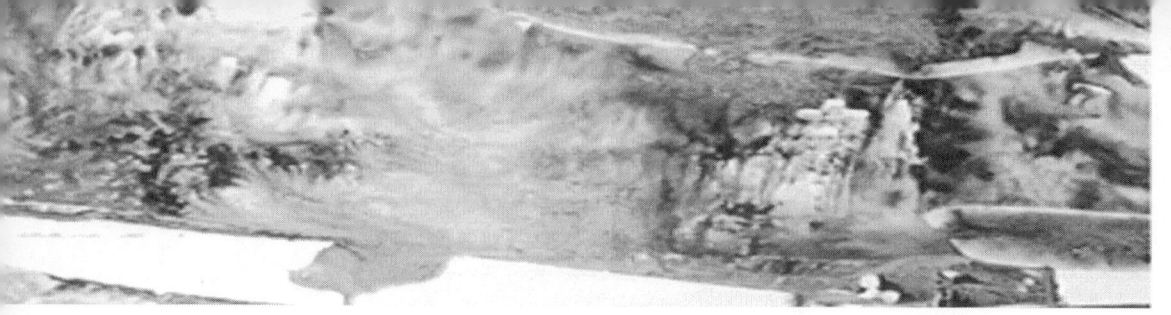

The Application of Amino Acid Fertilizers 307
    The Source and Function of Amino Acid Fertilizers 307
    Guidelines and Attentions for the Application of Amino Acid Fertilizers 308

Applications of Green Manure and its Effects on Soil Conservation 309
    Green Manure is the Moisture Controller of Soil Conservation 309
    The Effects of Green Manure 309
    The Application Method, Amount and Selection of Green Manures 310
    Attentions for Green Manure Application 314
    The Necessity of Green Manure Application -- Planting Green Manures is Better than Fallow 314

The Application and Recycling for Crop Residues 315
    The Significance of Recycling for Crop Residues 315
    The Purposes and Strategies for the Recycling of Crop Residues and Straws 316
    Methods to Recycle Crop Residues and Straws 316

The Application of Organic Wastes 318
    The Necessity for the Development and Application of Organic Wastes 318

The Effects, Production and Changes of Compost 319
    The Effects of Compost 319
    Selection of Raw Materials and Methods for Composting 321
    Properties and Changes During Composting 323
    Conclusion 326

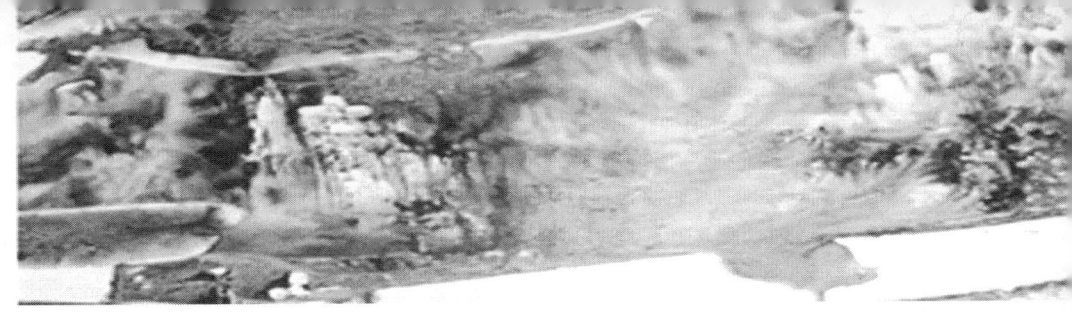

Stabilizing Technologies of Composting and Composting-Free Technology 326
  The Stabilizing Process of Composting 326
  Technologies to Accelerate the Stabilizing Process of Composting 327
  The Stabilizing Process of Organic Wastes by Composting-Free or Fast Treatment Technology 330
  Innovative Concepts for Stabilization of Organic Wastes by Composting-Free or Fast Treatment Technology 330
  Conclusion 332
Composting and Fast Treatment Technology of Kitchen Wastes 332
  Disposing Problems of Kitchen Wastes 332
  Composting Problems of Kitchen Wastes 333
  The Characteristics of Kitchen Wastes' Fast Treatment 334
  Effects of Kitchen Waste Organic Fertilizer Produced from Fast Treatment Technology 335
  Conclusion 336
Guidelines for Characteristics and Application of Bark Compost 337
  The Origins of Tree Bark and its Value 337
  Characteristics of Bark Compost 337
  Good Qualities for Mature Bark Compost 337
  Application Effects of Bark Compost 338
  Application Range of the Bark Compost 338
  Application Rate, Application Time and Attentions for Bark Compost 338
Housefly Propagation by Organic Fertilizer and Improvement Methods 339
  Problems in Housefly Propagation by Organic Fertilizer 339

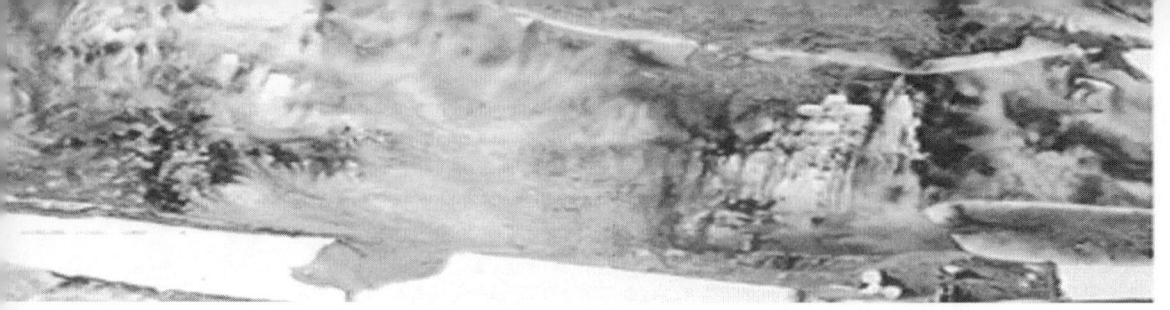

    The Improvement Method for Housefly Propagation in Organic Fertilizer **342**

Production Methods of Liquid Organic Fertilizer 343
    Application Guidelines of Liquid Organic Fertilizer **345**

Application of Cultivation Media 346
    Cultivation Media Expect Further Attention and Development **346**
    Necessary Properties of Good Media **346**
    Characteristics and Application Guidelines of Cultivation Media **350**

Introduction and Outlook of Organic Agriculture 352
    The Development Outlook and Significance of Organic Agriculture **352**
    The Advantages and Disadvantages of Organic Agriculture **353**
    Organic Agriculture Needs Further Experiments and Studies **355**

Organic Agriculture Cultivation and its Specifications Foreword 356
    The Objectives and Basic Production Principles of Organic Agriculture **357**
    The Methods and Strategies of Organic Agriculture **358**
    Predictable Difficulties of Organic Agriculture **358**
    Soil Quality Management in Organic Agriculture **359**
    Pest and Disease Preventives Applicable in Organic Agriculture **361**
    Adjuvant Applicable in Organic Agriculture **364**
    Strategies for Sustainable Agriculture **365**

Microelements and Heavy Metal Elements of Organic Agricultural Soils 368
    Crops Production Should not be Short of Microelements **368**
    Microelements Application in Organic Agriculture **368**
    Heavy Metals and Factors of Organic Agricultural Products **369**

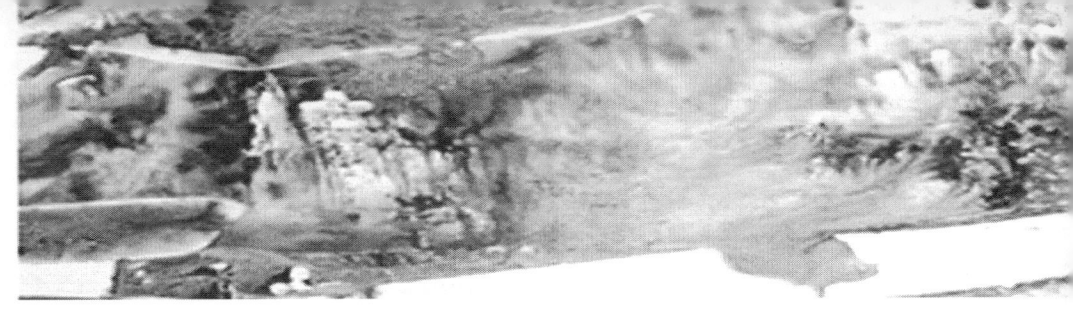

Chapter 9: Soil Microorganisms — 371

Soil Microorganisms and its Requirements for Growth 371
- The Importance of Soil Microorganisms 371
- Growth Requirements for Soil Microorganisms 372
- Classification Based on Carbon Source 372
- Classification Based on Energy Utilization 372
- Classification Based on Oxygen-Dependent 373

Types and Change Problems of Soil Organisms 373
- Types of Soil Organisms 373
- The Inhabitation and Types of Soil Organisms 374
- Importance of Soil Microbial Types to the Soil 377
- Problems of the Existence and Change of Soil Microorganisms 379
- More Researches on Soil Ecology are Needed 381

Soil Microbial Diversity and Its Conservation 382
- Introduction of Soil Microbial Diversity 382
- Resources of Soil Microorganisms 382
- Functional Value of Soil Microbial Diversity 382
- Protection Methods of Soil Microbial Diversity 384
- Prospective 385

Definition and Applying Purpose of Biofertilizer 386
- Definition of Biofertilizer 386
- Application Purposes of Microbial Fertilizer 387
- The Necessity of the Soil Microorganism Applications 388

The Relationships between Plant Nutrient Absorption and Soil Microorganisms 390
- Plant Root System and Soil Microorganisms in the Rhizosphere 390
- The Influence of Microorganism on Plant Nutrient Absorption 391

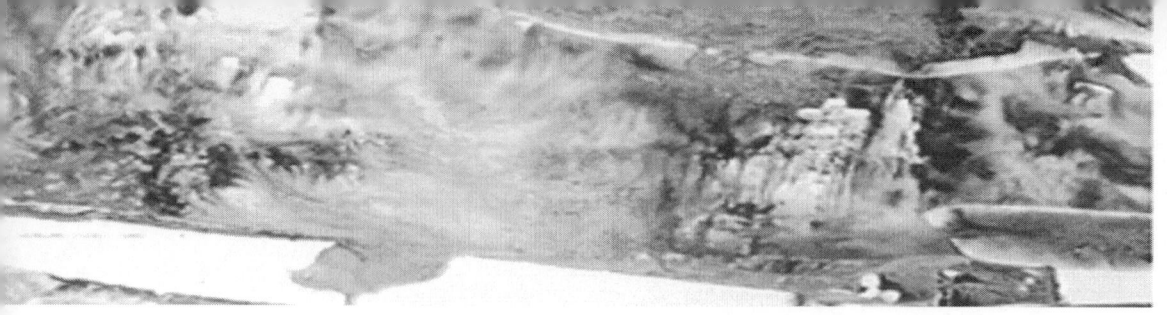

    Reasons of Microorganism's Influence on Plant Nutrient Absorption **393**

    Pay Attention to Soil Microorganism Management **394**

Functions, Types and Characteristics of Nitrogen-Fixing Microorganisms 394

    The Importance of Biological Nitrogen Fixation **394**

    Types of Nitrogen-Fixing Microorganism **395**

    Taxonomy of Rhizobia **395**

    Functions and Characteristics of Rhizobia **397**

Necessity and Development of Rhizobia and Mycorrhizal Fungi 398

    Soil Microorganisms and Crop Production **398**

    Necessary for Inoculating Rhizobia and Mycorrhizal Fungi **399**

    Effect of Biotic Soil Conditioners **401**

Introduction of Endomycorrhizal Fungi 401

    Research History of Mycorrhiza Fungi **401**

    Distribution of Endomycorrhizal Fungi **402**

    Classification of Endomycorrhizal Fungi **402**

    Morphology of Endomycorrhizal Fungi **402**

    Function of Endomycorrhizal Fungi **402**

    Relationship among Endomycorrhizal Fungi, Phosphorus Fertilizer and Environment **403**

Phosphate Solubilizing Microorganisms 404

    Relation between Soil Phosphorus and Plants **404**

    Relationships of Phosphorus-Solubilizing Bacteria among Phosphorus of Soil and Fertilizer **405**

    Types and Characteristics of Phosphate Solubilizing Microorganisms **406**

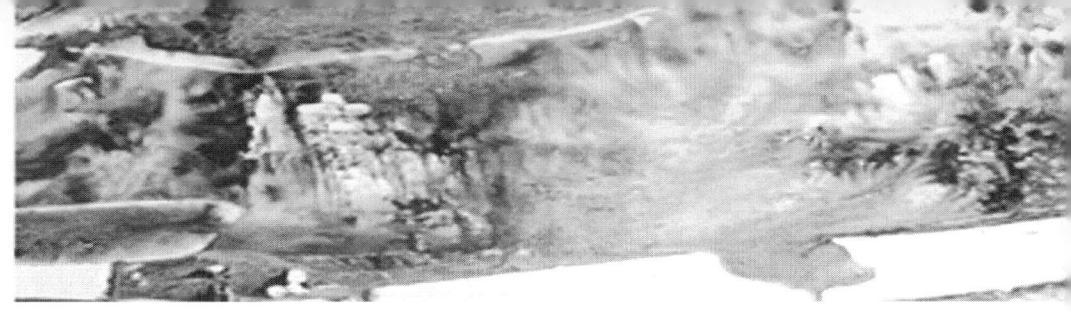

 Benefits of Phosphate Solubilizing Microorganism on Crop Growth **408**
 Effects of Applying Phosphate Solubilizing Microorganisms **409**
 Methods of Phosphate Solubilizing Microorganisms Application **409**

Roles of Microbial Application in Organic Agriculture 410
 Introduction **410**
 Roles and Application of Soil Microorganisms in Organic Agriculture **411**

Types of Microbial Fertilizers and the Quality of its Application 415
 Purposes of Microbial Fertilizer Application **415**
 Types and Functions of Microbial Fertilizers **416**
 Quality Requirements for Microbial Fertilizers **418**
 Preserving Instruction and Noticing Items of Microbial Fertilizers **418**
 Application Methods of Microbial Fertilizers **419**
 Notices of the Inoculation **421**

Application of Microbial Fertilizer in Sustainable Agriculture 421
 Application of Microbial Fertilizer is an Important Factor of the Successful Sustainable Agriculture **421**
 Why could Microbial Fertilizer Replace Part of Chemical Fertilizers? **422**
 Guidelines for Getting the Best Effect of Microbial Fertilizers **424**
 Reasons for the Application of Microbial Fertilizer could Hardly Lead to the Residual Problem in Large Quantity **425**

Interactions among Soil Microorganisms 426
 The Importance of Soil Microorganisms **426**
 Interactions among Soil Microorganisms **427**
 Application and Correlation of Microorganism **432**

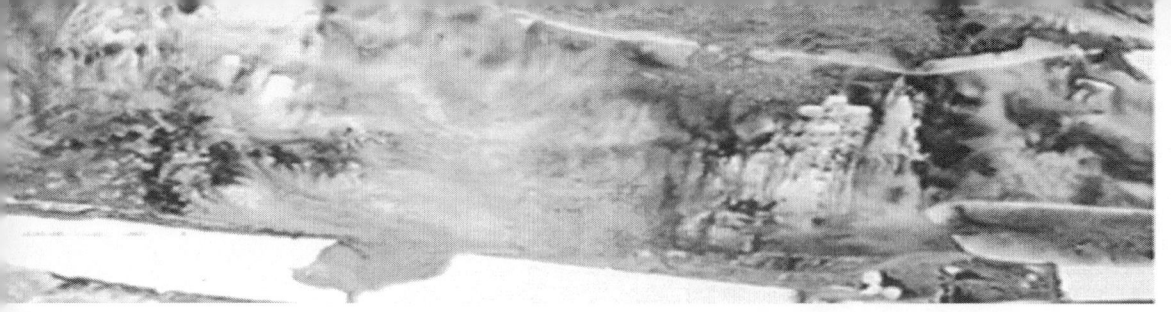

    The Possible Impact of Biotechnology and Transgenic Crops on Agriculture and Environment 432

        21st Century is the Century of Biology and Environment 432

        Uncertain Effects of Agricultural Biotechnologies and Transgenic Crops 433

        Possible Impact of Novel Organisms on Agricultural Ecosystem 434

        The Development of Biotechnology should be Careful for the Possible Crisis 435

    Advantages and Disadvantages of Soil Sterilization and Pesticide and Guidelines for Preventive Management 435

        Major Functions of Soil 435

        Advantages and Disadvantages for the Application of Anti-Pests and Anti-Pathogens in Soils 436

        Notice the Possibilities the Application of Soil Sterilization for Anti-Pests and Anti-Pathogens 438

        Guidelines of Soil Sterilization Management and to Fight against Pests and Pathogens 439

    Conclusion 441

Chapter 10: Q & A: Mailbox for Soil and Fertilizer   443

    Q & A for the Diagnosis of Problem Soils 443
    Q & A for Plant Nutrients and Guidelines of Fertilization 444
    Q & A for How to Improve Fertilization and Management of Soils 452
    Q & A for Organic Matters and Organic Fertilizers 465
    Q & A for Soil Microorganism 475

References   481

Chapter 1

Understand the Human Ancestral Treasure: The Soil

1.1 Soil Introduction

The soil is one of the basic resources as well as an invaluable treasure granted by the Creator to us mankind. Our lives are dependent on the soil as it produces necessary foods and fiber. The human diet includes plants and animals. Plants grow in the soil, absorb the moisture and nutrients from the soil to grow, and convert solar energy into biochemical energy. The animal diet derives directly or indirectly from plants. Basically, human life cannot live without the soil.

Soil is so important to us that we ought to protect it well. However, few industrial-developed countries truly understand how to cherish and protect it. Conversely, people overuse and pollute our soil inadvertently. It is regretful that in many highly developed and civilized regions people emphasize healthy food, beautiful environment, and energy application, yet lacking basic knowledge about soil protection and utilization. Hence, protection of the soil should become public awareness and be considered the most important task nowadays.

1.1.1 Protection of the Soil is a Mission of Every Generation

The soil that our ancestors handed down to us is the most precious treasure for humankind. Cultivated under our ancestor's traditional methods, the soil is handed down from one generation to another to sustain the basic existence and necessity of the humankind. How could this valuable treasure which we have to hand down to our descendents endure our generation's endless devastation? If we do not cherish and protect the soil, how can we be responsible for our ancestors?

1.1.2 The Constitution and Sources of the Soil

The soil is the top natural layer that covers the Earth's crust and consists of substances such as the fragmented mineral substances and the organic matters. Influenced by differences in parent rock, climate, landscape, etc., soils from different regions present different features. The basic material of the soil is constituted of different proportions and mixtures of diverse classes of mineral substances and organic matters. The mineral soil mainly consists of mineral substances (above 70% or 80%), while the histosol mainly consists of organic matters (above 20% or 30%). Of the land surface of the Earth, the largest coverage is the mineral soil.

The interspaces between constituent solids of soil permit the existence of air, moisture, and organisms. These substances greatly influence physical and chemical properties of the soil as well as the development of plants, and thus are important factors in the soil. The mineral substances in the soil originate from rocks of the Earth's curst, which were weathered by various physical, chemical, and biological forces and broke into fine particles, accumulated there or displaced to other places.

The organic matter originates from the residues and excretes of animals, plants, and microorganisms, which were decomposed and polymerized into organic matters of the different structures. Organic matters are often mixed with minerals into complexes which are not prone to be decomposed.

1.1.3 The Development of the Soil

The rocks undergo weathering and turn into minerals. The residues and excretes of animals, plants, and microorganisms are decomposed and polymerized into organic matters. The minerals and organic matters accumulate on the surface of the Earth's curst and develop into soil by the natural forces. This process is called the development or the formation of the soil.

The process of the development of soil is through biochemical weathering and soil horizonation as discussed separately below.

The Biochemical Weathering:

The roots of plants dig into the fragmented mineral particles while the residues of the aerial plant parts accumulate on the soil surface. These residues are turned over and displaced by animals and allow the blending of the organic matters with minerals. Microorganisms participate in this process via their exudates that allow the decomposition and polymerization of the organic matters. These processes further alter the solubility and the decomposability of the minerals. These biochemical forces make minerals produce various unstable substances that are prone to move with water.

The Soil Layer Development:

The leaching of soils is a process by which the water washes from top- to subsoil due to the permeation of rain and/or melting water from ice. Thus, soil can be divided into the leached-out layer (eluviated horizon; E layer) and the washed-in layer (illuviated horizon). These two layers, the A layer (topsoil) and the B layer (subsoil), are often called the soil profile and is used to differentiate from the third layer, which is called the C layer. The C layer is called the parent rock layer, namely the layer that has not been washed. The degree of the development of the soil profile is divided into three periods according to the visibility of the layer development. The relatively younger soils with little or no profile development is one that bears no obvious feature of development; the developed soil is the one that presents clear horizon features of development.

1.1.4 The Soil is also a Living Natural Body

To protect the soil instead of deteriorating it, we need to firstly recognize the soil. The soil is not a disordered random of matters. The soil is "a natural body that originates from the rock layer of the earth that has been weathered and influenced by terrain, climate, organisms, period of time and several other factors over tens of millions of years."

The soil is composed of minerals, organic matters, water, air, and living organisms. The former four are lifeless, while the organic matter comes from the living organisms. The living organisms in the soil include microorganisms, protozoa, insects, etc. If the soil is seen in a magnified view, its dwelling organisms are as active as that of human and other animals on earth. Commonly, every gram of the soil contains about one hundred million microorganisms that can be divided into very complex categories. The different microorganisms have their own niches, and they exert mutually balancing influences on each other. Under natural circumstances, there will not be any big change.

We can also regard the soil as a multi-cellular organism. The integrated soil system usually confers structures and functions similar to an usual living cell. Therefore, we can regard the soil as a kind of living organism; imprudent use of soil will annihilate, diminish, destroy, or impair the soil. An inert soil is just like the rock powder on the surface of the moon, which lacks the ability to yield and cultivate plants. Hence, we humans need to cherish and take good care of the soil.

1.1.5 The Soil is a Remediation Factory

Soils perform five essential functions: physical stability and support, filtering and buffering, nutrient cycling, water relations, and biodiversity and habitat in the global ecosystem. The soil has a lot of lovable characteristics, such as acting as a remediation factory. The microorganisms in the soil possess an immense enzyme system and such that it can decompose organic matters of all kinds with varying degrees of efficiency. There are many animal, plant, and microbial residues or organic wastes and exudates in our living environment. The soil decomposes these organic matters and absorbs them in order to maintain the clarity of earth environment. Each hectare of soil with its satisfying draining capability can purify at least one hundred metric tons of organic matters annually. In order to make this happen, organic matters should be mixed with the soil. Contrarily, when organic matters or waste debris exceed the soil's remediation capability, the soil's remediation ability will be disabled. In this case, stench and toxic gasses will be released and devastate our living environment.

1.1.6 The Soil Confers Buffering Capability

Another characteristic of the soil is its buffering capability, which provides a necessary environment for the dwelling organisms in the soil and allows them to survive for generations. When acidic matters are added, the alkaline characteristics can be neutralized. On the contrary, when alkaline matters are added to the soil, its acid characteristics can also be neutralized by the alkaline. These abilities allow soils to resist large variations in pH and be maintained in a relatively stable state. When too much soluble salts or fertilizers are applied, inorganic and organic matters in the soil would adsorb part of them and prevent the damage that would be caused to plants. When salts are inadequate, the soil can slowly release the necessary salts adsorbed on the soil and provide the nutrients needed for plant growth and development. By its nature, soil is the mother of crop production.

There are clear differentiations of the four seasons: spring, summer, autumn and winter in many places on the Earth, and thus the temperature within one year fluctuates a lot. Only the temperature of top one centimeter of the soil is easily influenced by climate. At places where the annual temperature variations could reach 30°C, the fluctuation of soil temperature decreases when the depth of the soil increases. In 70 to 100 cm below the soil surface, the temperature nearly stays constant. The stable temperature of the soil provides an optimal environment for the growth and development of plant roots and microorganisms in the soil.

1.1.7 The Soil can Get Sick

In addition to the existence of crop pathogens in soil, the soil itself can also get sick. The soil confers buffering capability, but the ability is limited and would be impaired if we consistently discharge abundant industrial waste water, acids, chemical fertilizers, salts, heavy metals, and high-temperature materials in the soil. In this case, the soil will get sick and lose the ability of crop production.

The balance between the microorganisms in the soil has evolved over a long time and thus allows plants to grow healthily. When the soil's nutrient cycles are broken, or when too much pesticides or fertilizers are used, the biological balance between the microorganisms in the soil will be disrupted, which will in turn victimize the plants. Therefore, we should be careful when using these matters.

The "chemical war" between plant species in the soil is attributed to the existence of allelopathic substances generated from plants. They will make soils sick and lead to the decline of soil fertility (productivity). Plant root exudates or residues decompose and often release toxic substances against plants. When they are accumulated in the soil over time, the toxic substances will harm plants. In other words, when two species of plants grow nearby and compete for sunlight, moisture, and nutrients, one of them may release toxic substances to poison the other. This process is called the "chemical war" between plants in the soil.

In agriculture farming systems, the previous crop may release and accumulate toxic substances in the soil. Before these toxic substances are decomposed, they will poison the next crop planted. This is another type of "chemical war between plants." The microorganisms in the soil play an important role in avoiding the chemical war between plants, for they can decompose these toxic substances. Therefore, in order to generate and maintain a healthy and balanced biological ecosystem, the destruction of soil microorganisms by pollutants should be prohibited.

1.1.8 The Key to Environmental Protection

The soil plays an important role in the environment and the ecosystem. The environment encompasses atmosphere, water, soil, and living organisms. Any environmental safety measures and pollution preventions are closely related to these environs. Advances in human civilization are associated with the exploitation of water and soil resources. The application or syntheses of the exploited resources create waste water and toxic chemicals, and when concentrated at sites, it severely threatens the existence of humankind and/or the natural balance of the ecosystem. For example, the exploitation of fossil energy, the use of plastic products, and pollution of massive heavy metals would produce unprecedented problems to the air, water, and soil. The air and water pollutions will ultimately result in the pollution of the soil. Conversely, soil pollution will also cause pollution to the air and the water.

The two powerful and efficient weapons for increasing crop production are fertilizers and pesticides. However, they may lead to immense soil pollution if overused or misused. The abundant or misused fertilizers will enter the water and cause problems, such as the case with excessive nitrate pollution in our drinking water that caused the Blue Baby Syndrome. When abundant phosphorus flows into water reservoirs or rivers, it will lead to the massive reproduction of algae, exhaust soluble oxygen, and in turn cause large-scale death of fishes. This is another problem that deserves our attention. What is more, pollutions of toxic and heavy metals, radioactive, and hard-to-decompose chemicals, when unabsorbed by the soil, will enter the food chain via water or plant absorption, accumulate in our body, and seriously threaten our health. Therefore, it is necessary to protect our soil environment for the sake of the health of humankind.

The prevention of the soil erosion is one of the environmental protection tasks. Soil erosion is irretrievable; even worse, when soil is flushed into the rivers or water reservoirs, it raises the level of the river bed and congests the reservoirs, polluting the water source, and changing the water ecosystem. The damage could be beyond estimation. If environmental protection in the slope land and mountainous region is poor, it will cause severe soil erosion. In order to protect our soil and reduce soil erosion, deforestation should be limited, covering crops should be planted in fruit orchards, and blind digging should be forbidden. Forest protection could not only decrease the magnitude of soil erosion, but is also in close relationship with the volume of water conservation. Forest soil can retain rain water and prevent it from eroding into the river; thus, it can prevent the occurrence of flash flood and mud slides. Therefore, preventing soil erosion can protect water reservoirs and our living safety.

Soil is closely related to environmental protection. Because contamination of the soil may cause a lot of problems to the environment, our knowledge of the soil will help environment protection. The quality of life is highly valued nowadays; protecting the soil for us humankind's survival is necessary to improve our life quality in the future.

As the predecessor said: "The significance of life is to create a continuing new life of the universe; the goal of our living is to improve the whole human living." The soil nourishes hundreds of millions of living organisms; therefore, it is the mother of crops and the base of nature. Every single living organism originates from the soil. I hope that every person from all regions of the world can be concerned with the protection of the environment and the soil for our descendents.

1.2 Concerns on Soil Degradation

The soil is one of human's basic resources and our lives rely on it as it offers us necessary food and fiber. The human diet includes plants and animals. Plants grow on the soil, absorb moisture and nutrients in the soil to grow, and converts solar energy into biochemical energy. Animal diet derives directly or indirectly from plants. Basically, human life cannot live without the soil.

The soil is the mother of agriculture; it not only produces our food but also functions as a state of security, a reservoir of water resource, a purifier of water and air, the mediator of climate, and the nurturer of the ecosystem and landscape. Especially in places with high population, the soil is considered a priceless treasure.

1.2.1 Factors for Soil Degradation

There are several causes of soil degradation. The factors differ according to either the natural conditions or the artificial farming management methods. The natural conditions include parent rock, locality, climate, elevation, terrain, groundwater level, water quality, etc. The artificial farming managing methods include cultivating methods, water quality of irrigation, fertilizers and pesticides, etc. Detailed explanations are as following:

Natural Factors:

Parent Materials:
Long-term cultivation of soils that contain rocks with strong acidic, alkaline, or salt materials will dissolve more acidic, alkaline, or salt materials and result in soil degradation.

Climatic Factors:
The climatic factors include wind, rain, water, snow, temperature, climate, and seasonal conditions such as drought and flood, all of which would cause soil degradation to some extent.

Locality:
The locality is a generalized result based on a prolonged period. For example, soils of an arid region present a high alkaline feature, while the rainy soil shows a high acid feature. The soils of deserts and sandy riverbeds show a weak water-holding capacity.

Elevation:
At lower elevations, the temperature is higher, which would decrease the amount of organic matters in the soil.

Topography:
Steep topography speeds up soil erosion. Lowlands are prone to water accumulation and poor draining.

Groundwater:

The overuse of groundwater will deficit the water resource and lower its level, resulting in drought of farm lands.

Water Quality:

Soil degradation occurs when the farm lands are irrigated by water of poor quality (such as high salt, high calcium, or high iron).

All the natural factors mentioned above that result in the decline of soil is mainly attributed to the leaching and erosion of rain.

Artificial Factors:

Application of Agrochemicals:

In agriculture, fertilization is necessary for providing nutrients for the growth of crop. However, long-term overuse or misuse of agrochemicals (chemical fertilizers and pesticides) will cause soil acidification, unbalance of nutrients, decrease in soil organic matter content, and decline of soil physicochemical and biological properties. There are incidences of animal and poultry manures overuse that result in the organic eutrophication of the soil. Overuse of lignin-high sawdust and tree barks can also cause white fungi rot and subsequent development of root rot in crops.

Poor Irrigation Water Quality:

The quality of the irrigation water differs according to their localities. The drawing of the inappreciable water from wells or rivers will lead to the acidification, alkalization, and salinization of the soil. The use of the industriously polluted water or municipal waste water (such as cleanser) that are especially high in heavy metals, organic wastes, and the pathogenic bacteria content will also cause serious soil pollution.

Cultivation Methods:

The facility or green house cultivation makes soils degrade easily. Excessive use of fertilizers will lead to the soil degradation in the house cultivation without rain washing. Continuous cropping is also an important factor that influences the biological, physical and chemical decline of the soil.

Soil degradation problems caused by the artificial factors mentioned above are very common.

1.2.2 The Result of Soil Degradation and Methods for Improvement

Deteriorations in soil physical, chemical, and biological properties are common types of soil degradation. Amongst, soil degradation of the physical properties includes soil compaction, formation of soil top crust, soil erosion (by water or wind), and poor drainage. Soil degradation of the chemical properties includes soil acidification, alkalinization, loss of fertility, unbalance of nutrients, and the presence of toxic substances. Soil degradation of the biological properties includes decrease of organic matter content, extinction of macro- and micro-organisms, and the decline of pest ecosystem in the soil. Detailed explanations are listed as following:

The Decline of Physical Properties of the Soil and Methods for Improvement:

Soil Erosion:
At sloping lands, steep slopes and terrains, or places with weak geological structures, soil erosion is more prone to occur due to the impact of water and wind. In seismically active fracture zones, soil erosion is especially prone to happen when abundant water is washed into these zones to an unbearable force (such as higher than 100 tons/ha/year of soil erosion) that eventually results in mud slides. Other incidences such as over cultivation of land also cause to soil erosion. When the topsoil losses cover plants and declines in water permeability, leaching of water will increase and erosion canals will be formed, which will not only make the already deficient groundwater resource unable to be replenished, but will also cause to issues such as the descending of the stratums along the seashore. Soil erosion is the main reason for the decrease of the topsoil and the cause of soil degradation. It is therefore important to enforce the conservation of soil and water.

Soil Compaction:
Soil compaction occurs when the weight of heavy machinery and animal activities are imposed on the soil, especially after raining, for example, when bulk density in the soil compaction layer is higher than 1.80 g/cm3. Under these circumstances, soil porosity will be disrupted and reduced. It can be solved by deep plowing and turning over the soil.

Crust Formation of Topsoil:
When rain drops hit the soil surface, the soil surface glues together. It is especially for silt loamy soil. Application of organic matters can improve this problem.

Poor Drainage:
Heavy rainfall or flood is the main reason for poor drainage, particular in clay soil. The sinking of water banks or badly designed seacoasts are also important reasons. Improvements of drainage systems should be undertaken in regions using water draining facilities.

The Decline of Chemical Properties of the Soil and Methods for Improvement:

Acidification and Alkalinization of the Soil:
The main acidification of the soil can be attributed to acid rain, acid parent materials, absorption of abundant cation nutrients by plants, excessive application of acid-forming fertilizers, and excessive amount of organic acids. The main alkalinization of the soil can be attributed to the intrusion of sea water into the soil layers, water evaporation from the soil exceeds rainfall, and excessive use of fertilizers in arid areas. When the soil gets

too acidic (pH < 5.5) or too alkaline (pH > 7.5), neutralizing soil conditioner should be applied. For example, apply calcium carbonate and peat to neutralize acidic and alkaline soil, respectively.

Decrease in Organic Matter Contents in the Soil:

In rainy, hot, and humid environments, especially in dry farmlands or slope lands, soil organic matters decompose easily and become deficient; its amount generally less than 10 g/kg. This should be dealt with by increasing organic matter contents in the soil by different kinds of organic fertilizers.

Unbalance of Nutrients in the Soil:

Disordered use or overdose of chemical fertilizers will antagonize the availability of different nutrients to plants. In some cases, plant absorption of nutrient elements will be impeded and result in metabolic problems. No essential elements, secondary elements, or trace elements should be over-applied. Regular monitoring of the soil condition in regards to the unbalance of elements and nutrients is required for farmlands that constantly apply chemical fertilizers.

The Accumulation and Pollution of Salts and Heavy Metals:

Polluted water, inadequate irrigation, or excessive use of chemical fertilizers will cause the accumulation of salts or heavy metals in the soil. The soil polluted by heavy metals merits attention. Intensive plantation on a certain crop land will lead to the excessive accumulation of salts from fertilizers. This can be alleviated by rotation between dry and wet crops. The paddy field cropping can wash out considerable amounts of soluble salts or alternatively select drought-resistant crops.

The Decline of Soil Biological Properties:

The Death and the Extinction of Soil Organisms:

Farming practices nowadays commonly apply excessive amounts of pesticides and fertilizers. Pesticides contain toxic substances that kill pests, bacteria, and weeds, and can be extremely harmful to soil organisms. For example, the macro fauna, such as the earthworms, have been observed to decrease in number or eliminated to extinction in the soil. The reduction and extinction of the earthworms can be a good and visible indicator for soil bios stem and worth considerations. Therefore, the excessive and disordered use of chemical matters should be avoided.

The Occurrence of Soil Pests and Diseases:

The occurrence of soil pests and diseases could be attributed to many factors, mainly the decline of the soil environment (soil acidification and nutrient unbalance) and the increase in the population of pests and pathogens. Common problems of continuous cropping and soil pest and disease include root rot, stump rot, bacterial wilt, soft rot, root knot diseases, late blight, nematode diseases, etc. Excessive amount and varieties of pathogenic microorganisms in the soil, in addition to poor soils, such as soils with rough soil texture and the acidic soil, will increase the incidence and magnitude of the disease.

Excessive pathogens in the soil are related to inappropriate soil management such as continuous cropping, long-term upland farming, and soil acidification. The prevention and maintenance of the soil's microbial system is far more important than the remediation process for problems. The key to preserving soil microorganisms is to apply long-term approaches for the maintenance of soil's physical, chemical, and biological environment.

Soil degradation should not only be the cultivator's concern; it must also become a concern for every civilian to significantly reduce pollution. This especially accounts for countries with limited farm land. Japan had implemented the "Soil Productivity Enhancement Law" and had paid special attention to soil degradation issues, which is worth learning from as an example to ensure our safety and health in regards to "production, living and ecology."

1.3 Soil Pollution Management Issues and Concerns

1.3.1 The Severity of Soil Pollution

Soil is the center of our survival, our living, and our life. The soil is our Mother Earth; it nurtures all types of organisms and serves both as producer and purifier. Since ancient times, the soil has been bearing a great responsibility for the regeneration of each living creature on the earth. Nowadays, the rapid development of the industry pollutes our soils, which, in turn, threatens our personal health and degrades the environmental quality. Therefore, the conservation of our soils has to be an urgent issue. In the future, we should not benefit from the current interests in sacrifice of our descendents' welfare. The concept of "no see, no trouble" should be revised; otherwise, all of us may be the direct victim of soil pollution.

To cope with soil pollution issues, a complete survey report is needed in order to avoid the repetitive occurrence of similar pollutions. The key to the issue concentrates on people and regulatory controls. As such, in each society there should be an intact set of laws for coping and management of soil pollution. Unlike short-term water and air pollutions that can be alleviated by natural forces such as rain and wind, once the soil is polluted, it is a long-term disaster. The soil is the origin of our food chain. Just like the eaten foods cannot be taken out after eating, soil pollution, once occurred, is hard to be alleviated. Therefore, the implement of the rules and regulations for the prevention of soil and groundwater pollution is related with the safety and health of all the civilians.

1.3.2 The Characteristics of Soil Pollution

The severity and danger of soil pollution equal to that of water and air pollutions. Prevention of soil pollution is far more important than remediation of soil pollution, and the reasons are as following.

Soil Pollution is Difficult to Be Removed and is Persistent in Soils:

Both water and air pollution are likely to be alleviated by natural forces such as rain and wind. However, it is very important to know the characteristics of soil pollutants -- once the soil is contaminated, the materials that cannot be decomposed such as heavy metals will adhere to the soil and will be persistent and difficult to be removed.

Soil Pollution is Latent and Difficult to Notice:

Water and air pollution are easy to be recognized and responded by animals, plants or people. However, when the soil is polluted, it will accumulate unnoticed and result in severe problems if unchecked.

Soil Pollution Will Result in Secondary Pollution:

Water and air pollution usually round up in the soil or in the ocean. Similarly, the mobile parts of soil pollutants will enter water and air, and in turn pollute the water resource, the groundwater, the sea, and the air.

Pollution Remediation is Costly and Has a Weak Recoverability:

It is of great difficulties to deal with soil pollution, and the remediation process is costly. The property of the soil will be destroyed completely even after the remediation process, and will result in soil degradation. Therefore, prevention of soil pollution is far more important than remediation.

1.3.3 Preservation of the Soil is the Key Task

In order to improve the competitiveness and acceptance of crop products by consumers, the products should reach three standards: freshness, safety, and quality. Therefore, soil preservation should be an important agricultural strategy in the future to ensure the confidence and reliance of the consumers.

Among the three standards, freshness is the characteristic of local products because which do not need to be transported long distances. Besides good management, the compatibility between the crop and the soil is also important to the production of safe and high-quality products. The soil is the mother of crop production. The essential key to a safe and high-quality crop product is healthy soil. The key to maintain healthy soil is to preserve it; otherwise, there will be severe outcomes as listed below:

The Decline of the Physical and Chemical Properties of the Soil:

Including soil compaction, acidification, salinization, toxication of the soil, decrease in organic matter contents, unbalance of nutrients, reduction productivity, reduced efficiency of fertilizers, increased fertilizer application, increase in cost, increase in the incidence of disease, increased pesticide application, and the decline of the soil biological ecosystem; all in sequential facts. Furthermore, decline of the physical and chemical properties of the soil will deteriorate the water-holding capacity; thus, water source will decrease and result in shortage of groundwater and sinking of the land.

The Decline of the Biological Property of the Soil:

The disordered balance of the biological ecosystem will increase the amount of pests and pathogens, the application of pesticides, and the capacity of the soil decreases, and weaken the remediation ability, all of which will result in the decline of the environment and biodiversity, and the death or extinction of species.

Therefore, the preservation of soil is closely tied to a healthy ecosystem. In the near term, soil preservation is related to our healthy diet and the quality of our lives. In the long term, it will influence the health and survival of our descendents. A sustainable agriculture relies on fresh, safe, and high-quality crop products. It is a key and also a direction to improve agricultural competitiveness. Thus, soil preservation is the principle task.

1.3.4 The Direction and Goal of Soil Preservation

Soil preservation is the basis of agriculture. Only good soil preservation can bring hope to agriculture, as well as healthy products and the well-beings of all civilians. Therefore, soil preservation is the basis of the society, the nation, and the people. The preservation of the soil should be divided into the following two directions:

Soil Protection:

The goal of soil protection is to decrease the destruction and decline of the soil. The reduction of soil pollution and soil erosion should be borne in mind when cultivating the soil. Rules and regulations should be implemented and enforced for the prevention of underground water pollution. Active promotion of soil protection alongside the Water-Soil Protection Act is the key. The farmers are the vanguards of soil protection; therefore, improving the education and training of farmers is crucial to the whole task and needs the support of the government personnel at all levels. In addition, soil protection practices should also include a reasonable fertilizing concept. Soil protection should be listed in the country's construction plans. The "Soil Productivity Enhancement Law" is a good example to illustrate that the construction should not only be related with that of the hardware but should also include the improving of life quality in which the significance of the construction really lies.

Soil Care:

The goal of soil care is to enhance the function of the soil. The soil's functions include productivity, remediation capacity, water and fertilizer holding capacity, biological balancing, and buffering capacity. Soil care is a more positive concept than soil protection. Once cultivated, the soil will surely be deteriorated. If we hesitate to cure the soil, it will make crop production more and more difficult, there will be increased uses in fertilizer application and disease incidences, and more and more pesticides and fertilizers will be used, which will all in turn increase the production cost, make the quality of the products worse and the environment enter a vicious circle. Soil care should include a reasonable fertilizing concept, elimination of the soil limiting factors, and careful application of organic fertilizers to increase the organic content in the soil. The government should enhance long-term inspections on the condition of the soil and the fertilizer, enforce rules and regulations for soil care, and educate more fertilizing technicians. Carrying out concrete actions to improve the soil productivity and listing soil preservation as an item of the government's performance appraisal are fundamental to safeguarding the health of all the people.

Chapter 2

Essential Concepts for the Improvement of Soil Fertility and Soil Management

2.1 Soil Productivity

The core of agricultural production is agronomical, horticultural, and forage crops. Factors that influence the yield and quality of crops include soil status, crop cultivar, environment and climate, cultivation management, and disease and pest control. It is important to consider these five factors for improvement of soil productivity at any area. Economically speaking, because only investment can bring income, we need to choose the best crops for the most suitable soil. Soil is the mother of crop production. To enhance soil fertility, we must understand the soil and recognize the constituents of a fertile soil. Only after this can we understand the disadvantages of our soil and the limiting factors that restrain productivity, and improve and eliminate those negative factors, turning infertile soils into fertile. Therefore, understanding the advantages and disadvantages of our soil, enforcing soil conservation, and improving soil fertility are the guidelines for high productivity and quality of crops.

2.1.1 What is "Fertile Soil"?

Soil is the medium for crop cultivation. "Fertile soil" refers to a soil that can meet the need for crop growth, that is, enough supply of "nutrients" and "water."

Such nutrients include elements like nitrogen, phosphorus, potassium, calcium, magnesium, sulfur, manganese, iron, boron, zinc, copper, molybdenum, chlorine, etc. Supplementary elements such as sodium, vanadium, cobalt, selenium, silicon, and other trace elements are required differently by different crops.

Apart from the moisture and a full set of nutrients, it is also important that fertile soil provides enough "oxygen," which is extremely important for upland crops. All the root system of plants requires oxygen for respiration. Therefore, good air permeability of soil is a significant factor for managing uplands.

Fertile soil needs to create both appropriate biological and non-biological environments for the root system to play its role, and offer porosity and a buffered environment (proper pH, salinity, and temperature) for the root system to extend.

In conclusion, "fertile soil" is not difficult to obtain. Under better management, poor soil can be transformed into fertile soil with the ability to fully supply the three important elements, i.e., nutrients, water, and oxygen. The protection of the physical, chemical, and biological properties of the soil is important for the maintenance and preservation of soil fertility. Above all, high crop productivity is easy to be achieved with sufficient soil fertility, appropriate selection of crops according to the climate, as well as a good cultivation and management approaches.

2.1.2 Essentials to Improve Soil Fertility

Unlike temperate regions where production is only once a year, subtropical and tropical areas are hot, rainy, and humid, in which farmlands are cultivated intensively. In these regions, the degradation of soil fertility, soil fatigue, and other soil problems occur very frequently, especially with uplands and continuous vegetable cropping. These soil issues should be taken seriously. While in dry regions, soil issues are often related to salt accumulation and soil alkalization.

Soils in different areas are quite different due to the differences in climate, environment, cultivating system, and parental rock materials. In addition, different fertilization programs and management systems will also lead to different soil characteristics. In order to improve soil fertility, we need to firstly understand the soil problems and limiting factors that affect land productivity. In the following sections, we will discuss several frequent soil problems and diagnosis and resolution of which.

Inappropriate Soil pH and the Fact of Soil Acidification:

How does It Happen:

Acidic (pH < 5.0) or alkaline (pH > 7.5) soil is prone to cause the availability of mineral nutrients to plants. Especially in acidic soil, crops are difficult to adjust to this adversity. Soil acidification is very common when there is frequent leaching by heavy rain,

CH2 Essential Concepts for the Improvement of Soil Fertility and Soil Management

acidic parental rock materials, excessive absorption of cations by plants, high organic acid contents, and poor buffering capability.

How to Diagnose It:

The determination method includes using a pH electrode or a colorimetric method with litmus test paper. The pH of soil should be determined by thoroughly mixing fresh soil with water. Under dry soil condition, it is to be avoided directly inserting the meter to the soil for the pH measurement, because the pH diagnosis is to determine the concentration of hydrogen-ion dissolved in water.

How to Solve It?

- Use neutralizer. Acidic soil often uses lime and alkaline materials, such as agricultural lime, magnesia lime, oyster shell powder, and dolomite powders to neutralize. These neutralizer materials are better applied together with organic matters. On the other hand, alkaline soil uses acidic materials such as sulfur powder, diluted sulfuric acid, and acidic peat to neutralize. Neutralization of acidic or alkaline soils should be progressed through several years and avoid overuse.

- Apply organic matter on acidic or alkaline soils to enhance the effectiveness of nutrients.

- Instant application of foliar fertilizers can temporarily improve acidic or alkaline soils. This can only cure the symptoms, while the fundamental solution is to improve the soil quality.

Deficiency of Soil Organic Materials:

How does It Happen:

In rainy, hot, or humid environment, soil organic matter degrades rather fast. Uplands and slope lands are prone to be deficient in organic matters. Excessive cultivation of farmland often results in the deficiency of soil organic matters. As a result, the productivity of uplands and fruit trees will decrease, and farmers will resort to more chemical fertilizer application, which turns out to be a waste rather than a good way to improve productivity. Furthermore, soil diseases and environmental pollution issues will increase. Focusing on the enrichment of organic materials is essential to improving such problem soils. In order to increase the productivity of uplands, we must pay more attention to the maintenance of soil organic matter content, because organic matters have several functions, including:

- Improve soil physical properties -- improve soil aggregate structure, soften the soil, and improve the aeration and drainage capability.

- Enhance the soil water holding capacity.

- Slowly release the nutrients that are necessary for plants.

- Chelate micro-nutrients and improve the solubility of plant nutrients.
- Enhance the buffering ability of the soil to mitigate the acid-base reaction of the soil.
- Adsorb and exchange plant nutrients and increase the adsorption of slow-released fertilizers.
- Improve beneficial microbial activities and antagonize pathogenic blooms.
- Reduce the effects from phytotoxins or toxic materials created by human or from the natural environment.
- Partial constituents of organic matters can improve plant growth.
- Dark color of organic matters is useful to absorb heat and improve early spring planting.

To sum up, the organic matter content of the soil must be improved to achieve the above ten results.

How to Diagnose Soil Organic Matter Content:

Apart from sending soil samples to agriculture related organizations to analyze the organic matter content of soil, one can use his/her eyes to observe. For example, if the soil after drying appears hard, red or yellow in color, and without any soil aggregate structure, it means that the soil may be deficient in organic matters. The more organic matters in the soil, the darker the soil appears.

How to Solve It:

We need to apply organic matters or organic fertilizers to increase soil organic matter content. We can also grow green manure crops, such as sunn hemp (*Crotalaria juncea*), sesbania, milk vetch, beans, alfalfa, clover, rape, etc. Adopt the cultivation system by rotating with legumes is the best way to improve soil organic matters. As for management, we need to increase covering plants to reduce soil erosion and topsoil erosion to protect soil organic matters. As for upland cultivation, we should reduce tillage and plant upland crops to preserve soil organic matters. The application of peat and humic acid in recent years has helped to increase and stabilize soil organic matters.

Poor Physical Properties of the Soil:

How does It Happen:

Soils of the slope fields are usually red or brown in color, either acidic or alkaline in pH, and deficient in organic matter. These soils usually have poor physical properties such as poor soil aggregate structure and poor water holding capacity that make them become dry and compact easily. These are typical barren lands. When the soil texture is of larger size or high clay content, it will also become problem soil. If the soil texture is too large the

CH2 Essential Concepts for the Improvement of Soil Fertility and Soil Management

size, the water holding capacity will be poor; if it is of high clay content, its drainage will be poor and inhibit root growth. Poor drainage or high groundwater level of a region will make the soil oxygen-deficient. If the root system is shallow, the plant will lodge easily. In some regions such as slope lands, droughts are often and may occur earlier in dry seasons. In consequence, crops and fruit trees are weak in absorbing nutrients and will also cause problems.

How to Diagnose It:

Measuring soil texture (sand, silt, and clay content) often applies sedimentation machinery, flotation-weigh device (hydrometer method), or hand touch to judge whether the soil is too sticky or too sandy, and measure soil porosity and determine whether it is too compact. When the method of hand touch is adopted, we can grab a hand full of soil samples and add water to moist the soils. After choked in our hands, if the soil sample is high-content in sand particles, it is sandy soil; if it is slippery, it is silt soil; if it is slimy with weight and can be pinched into a strip, then it is clay soil. The soil porosity, structure, and aggregation, which are closely related to the aeration and drainage ability of the soil, can be easily observed through our eyes. If grey spots appear in the soil profile, we can infer that the soil has poor drainage ability. The closer the grey spots are to the surface of soil profile, the more severe the problem is.

How to Solve It:

- Poor Water Drainage:

    A rather economical way to solve the problem of regional-inhibited water drainage is to choose water-tolerant crops or varieties, or to use high ridges cultivation. The drainage system can be applied with a simple open pipe drainage style or an underground pipe drainage system. The construction of the drainage system should be economical and allow efficient drainage and long-term usage.

- Poor Soil Structure:

    Poor soil porosity and aggregate structure can normally be solved by increasing the organic matter contents accompanied with lime. For alkaline soil, the problem can be improved by just increasing the organic matters. If we want to improve the quality of the subsoil, we can deep plow or use either vapor-pressure, hydraulic or pneumatic type to loosen the deep soil or apply deep fertilization.

- The Soil is too Sandy or too Sticky:

    Apart from choosing the types of crops that can grow on such soils, we can use transported top soils from other places and apply organic materials to improve the soil. Both water holding capacity and nutrient holding capacity are very weak in sandy soil; to improve it, one can use organic matters. Sticky soil implies that the soil porosity is poor. This kind of soil is like mud when wet, and like rock when dried. Applying organic matters can improve the formation of the aggregate structure of the soil and increase soil porosity that allows better water inflow and outflow. We should take out the top soils from the land when building a house to avoid local soils being buried under the building.

- The Soil is too Dense and Compact:
  We can use sufficient fertilizers along with deep plow to improve it. Under the circumstances, we should use more coarse organic fertilizers, grow crops that are able to break the compacted layer, or apply some earthworms in the soil.

Nutrient Unbalance of the Soil:

How does It Happen:

Unbalance or overdose of chemical fertilizers may cause antagonistic effects between nutrients, decrease the availability of nutrients, or interrupt with plant metabolism. Regardless of whether they are macro or micro elements, they all can never be used in excess. For example, when nitrogen is used in excess, the crops tend to grow tall and bushy, liable for diseases, hard to blossom, fruits dropping, and difficult to regulate the production period. When we use too much phosphate, the crops will grow slowly and result in the insufficient absorption of trace elements. When we use too much potassium, the crops tend to be deficient in magnesium, calcium, and other elements. When we use too much trace elements, the crops will get poisoned and the growth will be inhibited. One should be aware of the unbalance of certain soil nutrients, when fertilizers have been applied for a long time in order to avoid these issues.

How to Diagnose It:

Nutrient unbalance of the soil cannot be observed through our eyes. We can identify symptoms on plants through its appearance or through chemical analysis of the plant and the soil. The soil, equally important as our body, needs to be examined at regular intervals. Regular diagnosis is especially important for the intensive cultivation system, which needs at least to be examined every 3 to 4 years. We should use the soil and plants with typical problems as samples for analysis in comparison with the healthy soil taken nearby. This will allow us to find the problem in a timely manner. However, problem soil sample and healthy soil sample need to be packed separately before they are sent to related institutions for analysis.

How to Solve It:

- Understand the Characteristics of Crop Varieties:
  Because different crops require different nutrients from the soil, from soil or crop diagnosis we can take different solutions. When crops lack certain nutrients, we can firstly supply them with foliar fertilizers and later improve the soil fertility by soil fertilization. For example, the crops growing in the

unsuitable acidic and alkaline soil condition can be improved through this way to ensure the balanced absorption of nutrients. If the unbalance is caused by over nutrient, then we need to use a longer period of time to improve it. We can use the organic matters that have well decomposed, such as composts, humus, and peat, to absorb the excessive nutrients, reduce its fertilizer injury, and finally control it to reach a balance.

- Match up with the Crop Rotation System:
  When nutrients of the soil are unbalanced, we can apply the crop rotation system with various plants to absorb the excess of soil nutrients and reduce fertilizer injury of crops. We can also use the rotation system between upland and paddy field to reach a balanced condition of the nutrients. Besides, the proper use of fertilizers can improve the unbalance of the soil nutrients.

- Use Fertilizers to Contend against the Nutrients Unbalance:
  If the slope or hilly land cannot be rotated with rice paddy cultivation, we may not depend on the rain to rush away the redundant nutrients. Apart from using organic matters with well decomposition, we can utilize the competitive elements as a remedy. For example, the deficiency of trace elements caused by the overuse of phosphor can be recovered by deep irrigative fertilization and foliage dressing. The deficiency of magnesium (Mg) caused by the overuse of the potassium (K) can be replenished by the use of magnesium sulfate or magnesium silicate. Both elements can obviously reduce the fertilizer injury caused by some over application of nutrients.

Problems and Pollution Caused by the Accumulation of Salt and Heavy Metals:

How does It Happen:

Pollution and irrigation by improper water, the overuse and improper use of chemical fertilizers, and organic fertilizers with high content of salts will all lead to the accumulation of salt and heavy metals. Because soil has the ability of absorption, a lot of salts and heavy metals are accumulated in it; only a small amount of which will be drained away by the water, and some of the rests will be absorbed by the plants. Problems will undoubtedly occur as days and months go by. This is a problem that needs to be paid close attention to. The salt accumulation will be more problematic in areas absent of rain and high-temperature environments.

How to Diagnose It:

The most precise way is to measure the content of salt and heavy metal in the soil through chemical analysis. We can also use electric conductivity meter to measure it. When the conductivity is greater than 4 dS/m, most plants have the problem in growth. The accumulation of salts can be observed by our eyes. We can observe the dry surface and may find some white powders or crystalline salts. A large accumulation of salts will produce white powder, which is harmful to the growth of many plants. Besides, we need to pay attention to the source of irrigation water and see whether it has been polluted by the factories.

How to Solve It:

The accumulation of salts and heavy metals in the soil, which is caused by the overuse of the fertilizer of salts type in the intensive farming, can be dealt with by the crop rotation of uplands and paddy fields. And, a large amount of soluble salts will be washed away in the paddy field. Also, we can choose the salt-resistant crops because the crops differ in the salt-resistant degree. The use of organic materials and acid-producing microorganisms is also beneficial to the wash of salts, because organic acids which are produced by the decomposition of organic substance and microorganisms can enhance the solubility of salts.

However, the pollution caused by the accumulation of heavy metal is not easy to be washed away or wiped off. A small amount of the heavy metal pollution can be adsorbed by the organic materials, and the absorption of the plants can be reduced through the dilution effect. If the soil is heavily polluted, we should not grow edible crops or forage crops to prevent the pollution from entering the human food chain. If we must use the soil, we should carefully choose some non-edible plants to grow, such as forest, fiber, and some other special crops.

The transportation of soils from some other places can also be applied to solve the accumulation of salts and the pollution of heavy metals. It can dilute the soil to ease the injury of salts with high concentration and heavy metals. For example, when the surface soil is polluted and we have no choice but to use the subsoil, we can use the soil transported from other areas and increase the content of organic substances to greatly improve the quality of the subsoil. The quality of the soil can be enhanced after the use of appropriate fertilizers.

The Problem Caused by the Erosion of Topsoil:

How does It Happen:

The erosion of soil has a great impact on soil fertility, especially the soil of slope lands. It is especially important to conserve the water and soil, when the hilly land is cultivated into an orchard. The surface soil is valuable because it endures years of natural improvement. When surface soils are lack of vegetative protection, the valuable surface soil will be washed away even though it is rather compact. This will cause not only the loss of the soil fertility, but also the pollution and the sedimentation of water reservoir, rivers, and the water resources. The erosion of the surface soil is often accompanied by the loss of fertilizer, which cannot be neglected.

CH2 Essential Concepts for the Improvement of Soil Fertility and Soil Management

How to Diagnose It:

This problem is very easy to be diagnosed. We can know whether the topsoil is losing by noticing whether the runoff water during the rain is muddy.

How to Solve It:

We can use covers or grow grasses to conserve the surface soil. Especially, topsoil of the orchard can be well protected by growing grasses. And, we don't need to extirpate weed in the rainy seasons. Ostensibly, the grasses compete to absorb fertilizers, but after the grass cutting, mulching, and plowing, the fertilizer absorbed in the grass returns to the soil, which is rather beneficial to the fertility of the topsoil. Besides, the cultivation of grass can help the rain water leach into the deep soil layer for plant absorption, which is useful to increase the organic substances from plant residues and keep the fertility of the soil.

The Problems of Soil Pest and Pathogen:

How does It Happen:

There are many factors that can cause pest and disease problems of the soil, including the unhealthy physical property, chemical property, and biological property. The roots of the crops may have problems because of pathogen bloom and low resistance of plants. The occurrence of pests and diseases of the soil has something to do with the climate and the resources of irrigation water. And, the infectious diseases of seedlings can never be ignored, because the healthy seedlings are very crucial.

How to Diagnose It:

In order to know root diseases of crops and the types of diseases, we need to adopt some special methods. The cause of the root death may be biological or non-biological, or both interact as cause and effect. The problem of nematode can be easily observed, because crops with nematode often have some expanded tumors in their roots.

How to Solve It:

In order to solve the problems of pest and disease, we can use the chemical agents, physical treatment (such as high temperature treatment), and biological treatment. In addition, we can use the soil conditioner to improve the soil quality through chemical, physical, and biological methods, and to inhibit the proliferation of pests and pathogens, or to improve the resistant ability of crops. For example, we can apply lime to regulate soil acidity, which is useful to control some pests and diseases. We can use the organic materials to improve the soil quality, which can control some diseases. However, we need to pay attention to the fact that unprocessed animal feces can easily cause the infection of various types of pathogens and pests. Such materials must be buried under the soil to reduce the bad smell and pest problems.

Agrochemicals (pesticides and fertilizers) are essential in modern agricultural production, and there are no big problems when it is properly used. The beneficial soil microorganisms are useful to resist the threat of pathogens. And, such biological balance is necessary. The overuse and improper use of agrochemicals is harmful to the beneficial microorganisms in the soil and rhizosphere.

The Soil Problems Caused by Continuous Cropping:

How does It Happen:

Many uplands are confronted with soil problems because of continuous cropping. The problems are caused by continuous cultivation, management, and application of pesticides and fertilizers on the crops. And, problems are mainly related to the physical property, chemical property, and biological property of the soil. These problems are often caused by too many pathogens and pests, toxic substances, salt accumulation, and nutrient unbalance.

How to Diagnose It:

Long-time cropping of the same crop or crops of the similar type can result in poor plant growth, which cannot be totally improved by pesticides or fertilization. Under such condition, the seedling will be withered, the root will be rotted, or the new leaves will grow poorly or even stop growing.

How to Solve It:

Crops cannot grow healthily when cropped continuously. We should stop continuous cropping. There are many advantages of rotation cropping, and it's best to adopt the rotation cropping with upland and paddy crops. The soil problems caused by continuous cropping can be solved through the choice of other crops. And, they can also be solved by the application of lime materials in acidic soils to improve the soil acidity and the microbial environment as a whole. We can also use organic fertilizers or plant the green manures for improvement. Organic materials can adsorb toxic substances, increase the decomposition of toxic substances, have the effect of counterbalancing microorganisms of the soil, and reduce the nutrient unbalance by supplementing nutrients. We can adopt the application of beneficial microorganisms to counterbalance the biological property of the soil. Besides, we can change the environment to prevent some continuous cropping problems, such as flooding and drying the soil by turning it over.

Under continuous cropping, crops tend to have many roots and shoots, whose residues may also be the source of the problems. Therefore, getting rid of the crop residues can reduce the injury of toxic substance and pest problems.

2.1.3 Conclusion

The above eight problems do not exist in all soils. Therefore, you should consider from your own farming land, think of the possible factors that may contain the limiting factors of the soil, and solve the bottlenecks that reduce the soil fertility accordingly. And then the soil quality will definitely be improved. Only with proper preservation of the soil can we have healthy soils and improve the soil's production fertility. For the current agricultural production, regulation and change of the production seasons and high quality should be top priorities. In doing these, the coordination of soil is essential. By doing more observations, more tests, more recordings, our agriculture will have a better tomorrow with more "testing spirits."

Table 2-1. Common Soil Problems and Their Explanations

Soil Problems	Explanation
1. Strong acidic soil	The pH of the soil is less than 5.5, deficiency of calcium, magnesium, molybdenum, phosphorous, and other nutrients.
2. Strong alkaline soil	The pH of the soil is greater than 8.0, sometimes deficiency of manganese, boron, and zinc. Nitrogen is easy to volatilize, and phosphorous is easy to be fixed.
3. Sandy soil	Weak water holding capacity, deficiency of nutrients, suffering easily from the nematodiasis.
4. Coastal monsoon disaster area	Monsoon damage.
5. Salt accumulative soil	Electric conductivity is greater than 4 dS/m, high content of salts, not suitable for non-salt-resistant crops.
6. Red soil	Little organic materials; deficiency of certain nutrients; too sticky; subsoil is compacted.
7. Soil with poor water drainage	Inhibit the growth of crops, especially the uplands.
8. The subsoil is too compact and soil with plow pan	Apart from the sandy soil, the red soil and clay also have such problems. Inhibit the extension of roots and the nutrient absorption from deep horizon, coupled with the poor drainage and permeating system, which decrease the crop production.
9. Soil mineral crust	This may happen to all types of soil. The deficiency of organic matters and poor soil structure will make it difficult for crops to germinate, which will reduce the crop production by less number of plants.
10. Shallow soil and gravelly soil	It is difficult for seeds to germinate, which will reduce the crop production by less number of plants. The soil depth is less than 40 cm; it mainly happens in the river banks and lower reaches of the stream, delta areas; failure to develop.
11. Other problems	The year-round cultivation and improper management of fertilizer application make all types of soil have the possibility to have the problem of salt accumulation, nutrient unbalance, continuous cropping problems, toxic substances, erosion of topsoil, loss of the beneficial microorganisms, pathogens of the soil, and pollution of heavy metals.

Table 2-2. Methods and the Goal to Improve Different Soil Problems

Soil Problems	Methods to Improve the Soil	The Goal
1. Strong acidic soil	Use lime, coal cinder, dolomite fines, and other alkali materials; use it year by year (2 to 3 years), coupled with the use of organic materials.	Increase the pH value to 5.6 ~ 6.0.
2. Strong acidic soil with strong soluble ability (degraded red soil)	Use the silicate slag, and increase the use the trace elements like manganese (Mn).	Free iron in soil: 2300 > Fe > 1,400 mg/kg; Reduction Mn: 350 mg/kg > Mn > 20 mg/kg; Fe in rice leaf at tillering stage: 300 mg/kg > Fe > 70 mg/kg; Mn in whole plant: 2,500 mg/kg > Mn > 20 mg/kg.
3. Soils deficient in silicon	Use silicate slag.	Increase the content of SiO_2 in rice straw to 27%.
4. The soil that needs to add the organic materials	Use the organic fertilizers including the composts, coarse organic fertilizers, manures, rice husk; pay attention to the water holding capacity and improving soil structure to ensure the increase of the organic substances of the soil and plant green manures.	Use 2 to 10 tons of organic fertilizers per hectare per year.
5. Soils deficient in nutrients	Regulate the soil condition by supplementing the absent elements according to the soil analysis.	Keep improving until the symptoms disappear.
6. Strong alkaline soil	Use the organic materials, sulfur powder and peat.	Regulate the pH value of the soil less than 7.9.
7. Sandy soil	Improve the soil according to the properties of the soil a. Use more organic materials to improve the soil structure and enhance the water holding capacity. b. Use lime to improve the acid soil.	The same goal with those mentioned above.

Table 2-2. Methods and the Goal to Improve Different Soil Problems (Continue)

Soil Problems	Methods to Improve the Soil	The Goal
	c. Use the sulfur powder to improve the strong alkaline soil and reduce the pH value. d. If the soil is deficient trace elements, and the content of trace elements in the organic fertilizers is little, implement some trace elements. e. If the subsoil is finer than sandy topsoil, turn over the subsoil, mix it, and use some phosphorous fertilizers; if the layer of sandy topsoil is too compact, and it's not economical to mix the subsoil, then set up a closed conduit 40 cm beneath the surface soil in case of rain water flooding. f. The water holding capacity needs to be enhanced to improve the sandy soil in the monsoon zone. Use organic fertilizers to reform the structure of the topsoil; reduce the acidity or the alkalinity or implement the trace elements; plant tall grass to resist strong winds; plant trees along the field balks; set plastic wind shield; increase irrigation works.	
8. Salt accumulated soil or the soil with salt accumulation	Use concealed conduit to drain off water or find new water resources; plant rice at least once a year to wash off salts and cultivate the salt-resistant crops instead.	Conductivity of the soil is less than 2 dS/m within 1 meter deep.
9. Red soil	Use organic materials, lime, or silicate materials; choose phosphorus fertilizer; break the subsoil; deep fertilization.	

Table 2-2. Methods and the Goal to Improve Different Soil Problems (Continue)

Soil Problems	Methods to Improve the Soil	The Goal
10. Poor drainage soil	Use drainage or improve the quality of soil texture in soil profile including simple channel and deep ditch.	Transient oxygen deficiency is not allowed within 100 cm of the soil profile (< 5% O_2).
11. Compact rhizosphere soil or soil with plow pan	Use chemical or physical methods to break soil compaction and use deep fertilization.	
12. Shallow soil and gravel soil	Transported soil removed from other place to improve the soil, discarding gravels, application of organic materials, enhancement of water, and fertilizer holding capacity.	Effective soil layer of topsoil depth >15 cm.

Chapter 3

The Series of Soil Fertility Maintenance

3.1 The Maintenance Guidelines of Inorganic Fertility

3.1.1 Yield Factors

Agronomic, horticultural, and forage crops are cores of agriculture; factors affecting their production are soil, cultivation management, variety, climate and environment, pathogen and pest control, etc. All these factors should work in coordination with each other to achieve the goal of increasing crop production; if any of which becomes limiting factor, it will result in crop failure. Rely on one single factor cannot increase the production. For example, good new species alone cannot increase production; it still needs proper season and climate, together with the management of soil and cultivation, and proper prevention of diseases and pests. Among these factors, soil and cultivation managements are the most important, for the soil is fundamental to agricultural production, and only healthy soil can produce healthy crops.

In some areas, agricultural system is intensive with plants growing all year around on the land, which has taken up a large amount of nutrients. What's more, the fallow period is brief. Therefore, it is necessary to exert special protection of the land to avoid soil problems and decline that are easily triggered.

Remember, farmlands have been passed down generation after generation to produce the food that we need! We still need to pass the land to our descendants. In modern times, with the improvement of industrialized agricultural science and technology, we should prevent soil destruction, exhaustion of soil nutrients, and soil pollution, as well to preserve the soil and sustain its fertility. Especially under the circumstance of intense continuous cropping system, soil is under a rather heavy burden. To keep long-term fertility of the soil, this series put forward maintenance essentials and methods on inorganic soil fertility, organic fertility, and soil microorganisms in these three aspects.

3.1.2 Guarantee for Soil Complementary

In order to protect and preserve soil fertility, we should start by getting to know soil property. The compositions of soil include inorganic matter (water, air, and minerals), organic matter, and organisms that include physical, chemical, and biological properties in nature. Any soil composition or property should not be neglected.

The best productivity of a piece of land relies on the complementary of these items; we cannot excessively consume the soil organic and inorganic nutrients, and must never break the balance between the microorganisms in the soil. For example, soil acidification caused by excessive consumption of soil nutrients, which is also common in the intensive farming, would surely lead to nutrient absorption problems and is the main cause of diseases.

3.1.3 Concepts of Proper Application of Inorganic Fertilizer

Application of inorganic fertilizer is the most effective tool in modern agricultural production, and fertilization is an art work. In general, improper or excessive fertilization is the most fatal to soil preservation which is as dangerous as people taking in high dose supplements of nutrients. The guidelines of inorganic fertilization are no more than proper time, proper amount, proper method, proper location, and proper kind of fertilizer according to the needs of crops. For example, long-term application of ammonium sulfate and single superphosphate would lead to soil acidification. Likewise, if too much nitrogen fertilizer is used during the period of paddy rice seeding, the leaves would turn deep green, which may easily trigger serious rice blast. Excessive or improper application of chemical fertilizers will not only directly affects the production and quality of cops, but also indirectly influences the health of human beings. For example, we should be careful to prevent vegetables from taking into the overly applied nitrate or other nitrogen which is as dangerous as nitrate in salted meats. Speaking of the quality of delicate agriculture production, soil plays an essential role; we cannot neglect the research of crop quality and soil fertility and management.

The concepts of using chemical fertilizer in balance are of vital importance. There exists a certain ratio of cations (nutrient elements) taken up by crops that include potassium, sodium, magnesium, calcium, and some microelements. For example, applying potassium fertilizer can increase the content of potassium in leaves, but can reduce the content of other cations such as magnesium. If the soil is slightly deficient in magnesium, large amount of potassium fertilizer would cause the deficiency of magnesium, which is the phenomenon of ion antagonism, commonly occurred in fruit trees. Therefore, the concepts of elemental absorption competition and balanced fertilization are indispensable.

3.1.4 The Prevention of Problems and Soil Conservation

The Problem of Soil Acidification:

The main reasons for soil acidification are:

- Improper application of chemical fertilizer.
- Crop absorption of a large amount of cations.
- Crops or soil microorganisms release a large amount of organic acids which is beyond the soil buffering capacity. Soil acidification would directly affect the metabolism within the crop cell, indirectly influence the effectiveness and ability of nutrient absorption of the crop, and cause the deficiency of microelements, which is most explicitly exemplified in the decrease of availability of phosphorus in acidified soil. Moreover, the occurrence of disease is especially common along with the appearance of soil acidification.

Lime can be used to deal with soil acidification, and its amount should be based on the degree of acidification, soil property, and crop type. The amount of lime applied cannot exceed what is needed. Especially for sandy soil, excessive amount of lime may cause the deficiency of micronutrients. Liming materials are better to be mixed with the soil during soil cultivation; limes run away relatively slowly in uplands compared with paddy fields, and the interval of lime application can be several years.

To avoid soil acidification, we should not use too much acid-forming fertilizers, but use well-fermented organic fertilizers to enhance the soil buffer capacity. The adoption of crop rotation can also help prevent this kind of problem.

The Problem of Salt Accumulation:

Instead of heavy fertilization, we need to find out the reason for poor plant growth. Recently, there are salt accumulation in many farmlands resulted from excessive fertilization. With white salt crystal on soil top layer, crop growth is inhibited. What's more, long-term irrigation with water or groundwater, which has high content of salts, can also lead to the salt accumulation problem of farmland.

The most effective solution for the soil salt accumulation problem is leaching. We can transform the upland into a paddy field to dissolve and drain salts, or adopt strip cultivation, using rain water or spray water supply to partly reduce salt damage. The cost of subsurface pipe drainage or soil turning, however, is relatively high. We can also plant resistant crops of high salt to adsorb large amounts of salts, or plant crops with high salt tolerance to guarantee the harvest before salts are cleaned out.

In the prevention of soil's salt accumulation, apart from not abusing chemical fertilizer, we should also pay attention to the salt content of irrigation water. It is safer when electrical conductivity is less than 0.05 dS/m. When it exceeds 2.25 dS/m, salt accumulation can be rather serious.

Problems of Deficiency in Inorganic Nutrients:

The occurrence of nutrient deficiency varies with the difference of crops and soils. Apart from that the soil which lacks certain elements itself, the decrease of availability may also lead to the deficiency of some nutrient elements. For example, alkaline soil would have the symptom of yellow leaf for peanuts; but for other plants, this symptom is not obvious. After improvements to reduce the soil alkalinity, the symptom of yellow leaf of peanuts can be eliminated, which demonstrates that peanuts are highly sensitive to alkaline soil, and that nutrients' absorption and availability are relevant to the pH value of soil.

When dealing with nutrient-deficient problems, we need to first of all know which element is insufficient. And, the easiest way is to use the foliage dressing. Respectively spray different element fertilizer, observe which kind of nutrient has the correcting effect, and apply that kind to the soil. We can also apply element analysis on the leaf to detect what element is insufficient. According to the examples listed above, deficiency problems caused by pH degree can only be eradicated by adjusting the pH value of soil.

3.2 The Maintenance Guidelines on Organic Fertility

Among soil preservation and sustaining long-term productivity, maintenance of organic fertility is the most important. Therefore, the content of organic matter in the soil is often used to distinguish the soil productivity and judge whether the soil is fair. Organic matter in the soil is like spinal cord in human body, small in volume yet vital in function. During the conservation of soil, the preservation of organic fertility has been granted the highest priority.

The climate in subtropical and tropical zone is of high temperature and humidity, which promotes the decomposition and consumption of organic matter in the soil. When encountering this kind of farmland of lower productivity and lacking organic matter, farmers would surely increase the amount of chemical fertilizers, which would lead to fertilization waste, and even decrease crop productivity. This may even increase the probability of pathogen and environment pollution. We should apply an effective cure of this kind of problem to yield twice the result with half the effort. To improve the productivity of upland farming, special attention should be paid to the organic matter content of soil. Especially for upland farming, decomposition of organic matter is accelerated. In order to sustain the long-term productivity and fertility of soil, preservation of soil organic fertility admits of no delay and should be paid special attention to.

3.2.1 Efficiency of Soil Organic Matter

The content of soil organic matter consists of two categories: the easily decomposition and the resist-decomposition. Humus and humic substances take up more than 50% and belongs to the resist-decomposition; others, such as protein, fat, amino acid, and cellulose, are easy for decomposition. Humic substance is the most widely distributed major organic matter on the soil surface. Soil humic substance originates from the residues of plants, animals, and microorganisms. The residue is decomposed, synthesized, or polymerized as humic substance.

Soil organic matter has plenty of efficiencies which are listed as follows:

Improve the Soil Physical Property:
- Enhance soil aggregates, soften and stabilize soil, and promote soil aeration and drainage.
- Improve the soil water retention capacity.
- Soil black color can help intake heat and benefit early spring planting.

Improve Soil Chemical Property:
- Slow the process of releasing the nutrient elements that are needed by plants.
- Chelate micro nutrient elements and assist solubility of nutrient elements.
- Improve the buffering capacity of soil and soften the acid-base reaction of soil.
- Adsorb and exchange nutrient elements and enhance the slow released fertilizers.
- Reduce man-made or natural poisonous substance and its effect.
- Some constituents are beneficial to plant growth.

Improve Soil Biological Property:
- Provide soil that are good for microbial activities and increase the check and balance between microorganisms in the soil.
- Protect the diversity of microorganisms in the soil and their ecological function.

All the efficiencies above are meant to tell us that to achieve these advantages of soil we should increase the content of organic matter in the soil.

3.2.2 The Preservation Methods of Soil Organic Fertility

Methods of preserving organic fertility can be divided into the following two categories: to increase soil organic matter content and to prevent the decrease of soil organic matter content. Methods and matters needing attention of soil organic fertility preservation are listed as follows:

Application of Organic Matter or Organic Fertilizer:

The most effective method to maintain the long-term soil productivity is to apply organic matter or organic fertilizer directly, which include various kinds of composts barnyard manures, feces wastes, oil meals, and well fermented organic fertilizers on market. Though general application of organic fertilizers is beneficial to soil, the selection of proper organic fertilizers also needs attention. You should take into consideration which item of soil is intended to be improved when applying organic fertilizers. Meanwhile, the application should be combined with the crops and farming time to achieve the economical and efficient effect.

For example, to improve soil aggregate structure and water holding capability, if the easily decomposing organic matter is applied, the frequency of application should be increased; if resist-decomposition organic matter is applied, the effect would be heightened with the method of deep placement. Resist-decomposition organic matter has high content of lignin which is the main constituent for the transformation of soil putrescibility.

Raw or easily decomposing organic matter or compost can be used. They are instantly decomposed into the soil; therefore, seedlings cannot be planted shortly after the application in case that N immobilization or ammonia producing cause leaf yellowing and inhibit their growth.

Paddy fields of poor drainage are not suitable for excessive application of easily decomposed organic matter or emplacement of too many straws because this is apt for anaerobic decomposition and produce toxic substance, and H_2S to cause rice roots to stop growth or die.

Planting Green Manure Crops:

The growth period of green manure crops can prevent leaching of soil nutrients and reduce soil erosion. After green manure crops are ploughed under soil, the easily decomposed part is decomposed and nutrients are released. The other part is resist-decomposition and can increase the organic matter content of soil. To realize the long-term soil preservation and maintain its productivity, it is an indispensable method to plant green manure crops, both economical and effective, especially for upland.

Many crops can be planted as green manure: some containing less lignin are easily decomposed, while some containing much lignin are resist-decomposition. Generally, they are divided into leguminous green manure crops and non-leguminous green manure crops. In the choice of green manure crops, we need to take into consideration which kind of crops would be planted and whether the land is upland or paddy field, such as the choice of green manure for ordinary paddy field before growing rice. If the leguminous green manure crops contains too much nitrogen and the period of green manure releasing nitrogen coincides with the rice tillering stage, too much nutrients would be supplied, causing rice excessive or unavailable tillering, and resulting in rice production decrease; or, too much nitrogen would make rice seedlings too green, which is susceptible to the serious rice blast. Therefore, green manure for paddy field is usually non-leguminous green manures, such as rape, turnip, etc. However, as for gramineous, upland farming crops (such as maize, broomcorn, sugarcane, etc), leguminous green manures with nitrogen fixation which include *Crotalaria juncea*, sesbania, *Mucuna pruriens*, *Astragalus smicus*, whitetip clover, and Chinese trumpet creeper are chosen. Green manure is not only good for the growth of crops, but also can maintain the long-term productivity of soil. It is also reported that green manure can overcome the pathogens, red rot in particular. Hence, it is truly of various efficiencies.

Adoption of Rotation System:

Crop rotation is a kind of farming mode that changes several kinds of crops at regular time to effectively maintain soil productivity. Constantly planting the same kind of crop would lead to the relatively bigger loss of soil organic matter and soil exhaust, whereas crop rotation has many benefits. And, if leguminous or non-leguminous green manures are adopted in the rotation, they can enhance the effect of soil protection and preservation.

Reduce Plowing and Increase Covering the Surface Soil:

In the better drainage soil, the more plowing, the more soil organic matter is consumed. Reducing soil cultivation and plowing during farming can effectively prevent the deduction of soil organic matter. For example, planting maize without soil cultivation after rice harvest does not reduce its production, but can contribute to increase production; that is to say, no tillage or zero tillage farming is fairly helpful to organic matter preservation.

Rain water would wash out topsoil which contains relatively more organic matter. Therefore, to prevent topsoil from being washed away is of vital importance to sustain soil organic fertility. Increasing covering, planting grasses and residues of crops on the farmland can help improve the lack of organic matter and prevent the erosion of fertile top soil; thus, they are helpful to the preservation of soil organic fertility.

3.3 The Maintenance Guidelines on Soil Microorganisms

Though microorganisms in the soil are too small for our eyes to see, they can be as many as tens of thousands under the observation of microscope. In general, there are more than 1×10^8 microorganisms in per gram of soil, which has played fairly an important role in agriculture. The kinds of soil microorganisms are multifarious and they are divided mainly into bacteria, fungus, actinomyces, algae, and protozoa. Each category of microorganisms takes a certain place on soil particles for the population to grow and balance each other. Through the ages they have coexisted and balanced each other. Under natural environment, the ecological systems of microorganisms do not change dramatically, whereas artificial land farming may change and cause waxing and waning affections according to the tillage types. In regard of their advantages and disadvantages for crops, microorganisms can be roughly divided into beneficial microorganisms and non-beneficial organisms. The purposes of preserving soil organisms referred to in this chapter are to protect the normal living of beneficial microorganisms and to inhibit the development of non-beneficial organisms.

3.3.1 The Functions of Soil Microorganisms

To preserve soil microorganisms, we need first of all to understand the functions and risks of soil microorganisms on plants. The preserving work varies with the changes of time and location, and it is a kind of technique and an art as well. The illustrations below are divided into effects of beneficial and pathogenic microorganisms.

Beneficial Microorganisms:

Beneficial microorganisms in the farmland, which include nitrogen-fixing bacteria, mycorrhizal fungi, nitrobacters, catabolism bacteria, etc., directly or indirectly affect the growth of crops. They are various in kind and broad in efficacy.

The functions of beneficial microorganisms in the soil include broadening the source of soil nitrogen, enhancing the availability of nutrients (especially the phosphorus fertilizers which is difficult to move in the soil), releasing plant hormones, boosting the growth of plant root system, decomposing organic matter to release nutrients, decomposing toxic substance in the soil, contending against pathogens, polymerizing and forming soil humic substances, etc. Each kind of soil microorganisms has played a different role. There must be a certain reason that these microorganisms still exist through the ages.

Especially, antagonizing pathogens and other microorganisms that protect plant root system are even more important. Like the function of some microorganisms in human skin pores, these microorganisms prevent pathogenic microorganisms from invasion. There is some farmlands discovered which has very strong antagonizing pathogens and is not apt to soil disease. This kind of soil is called disease suppressive soil, because of the fact that it is not suitable for pathogens to grow. However, most soil has relatively weak resistance for pathogen invasion. Pathogens massive propagation leads to crop sickness and soil borne deceases. This kind of soil incapable of pathogen antagonism is called disease-conducive soil.

Pathogenic Microorganisms:

Crop disease does not all originate from the infection of stem or leaf on the ground. It is also common that the underground part of the plant gets infected. There are various kinds of pathogens among soil microorganisms, such as *Fusarium*, soft rot bacteria, root rot fungus, root rot nematode, etc. which cause crops to get disease. There are also soil microorganism producing toxic substance and suppressing crop growth, which are called marasmins. These pathogenic microorganisms are most likely to lead to some pathogen massive propagation and toxin accumulation in continuous cropping. Net-house vegetable fields can only exclude some pest and pest-borne diseases and cannot avoid pathogens from soils. Net-house cultivation is pervading in many places. Therefore, soil preservation in net-house cultivation needs especially to be reinforced to avoid soil problems caused by continuous cropping of vegetables such as not growing or seedling root rotting, which would surely lead to yield reduction or poor soil.

3.3.2 The Occurrence of Problems in Soil Microorganisms

The most direct relationship between the crop and the soil is crop root. When problems occur in soil microorganisms, it is plant root that bears the brunt. However, it is not easy to detect root problems because it is buried underground. It is not until stem and leaf show symptoms that the problems can be discovered. For many problems of soil microorganisms, we are incapable of remedying them; or, when the plants have secondary infections, we are not able to detect the original reason.

Therefore, preservation and prevention of soil microorganisms are more important than treatment.

In the following, there are several phenomena for soil microorganism problems in farmlands and orchards.

Excessive Populations of Soil Pathogens:

When pathogens in the soil are too much in population and too many in species, together with poor soils, the disease rate would increase and soil problem may become more serious. Occurrence of soil pathogens is usually relevant to continuous cropping, long-term upland farming, soil acidification, or other improper soil managements.

Unbalanced Populations between Microorganisms:

Soil microorganisms are various in kind; and, unless soil environment has been changed (such as the adding of organic matter and the change of aeration), healthy soil has a fairly balanced ecological system. Though some microorganisms may massively propagate, the balance can be regained after a period of time. Unbalance of microorganisms often leads to the blocking of nutrient supply and crop deficiency of some kind of nutrients. If this situation prolongs, crops would surely be harmed, and some toxic substances in long-term production would damage the root system.

Too Low in Population of Soil Microorganisms:

Too low in population of beneficial soil microorganisms are usually related to too less organic matter and improper application of agricultural pesticides. As the main living habitat for microorganisms, organic substance supplies sources of carbon, nutrients, and energy. Too less organic matter causes the decrease of microorganism amount, poor soil aggregate structure, poor capacity of nutrients and water retention, and poor soil fertility. Improper application of agricultural pesticides or pouring improper pesticides to the plant root would cause massive decrease of beneficial microorganism within rhizosphere and do harm to the roots. Therefore, we should read the usage instructions before applying pesticides.

3.3.3 Methods of Soil Microbial Conservation

The essential to soil microbial conservation is long-term preservation at ordinary times, and the methods of which are shown as follows:

Application of Suitable of Organic Matter:

Supply organic matter is the most effective way to maintain soil microorganisms; yet, we should pay attention that it requires suitable organic fertilizer. In recent years, excessive application of the easily decomposing organic matter such as animal dung frequently occurs to vegetable fields. If this application of excessive amount of organic matter last two to three years, it would lead to soil excessive-eutrophication and salinization, and over microbial propagation, which would do harm to plant roots and give rise to pathogens. We cannot trade long-term productivity for temporary high yielding, or the loss would outweigh the gain.

Inoculation of Beneficial Microorganisms:

Originally, there are many beneficial microorganisms in the soil. Because of long-term soil utilization, the number of some beneficial microorganisms decreases, and for some beneficial microorganisms, such as nitrogen fixing rhizobia, they are originally very little or even none in soil. Application of beneficial microorganisms can enhance the amount and availability of nutrients. For example, nitrogen-fixing bacteria can increase the source of soil nitrogen; bacteria for organic matter decomposition can help quicken organic decomposition to release nutrients; mycorrhizal fungi can increase the phosphorous absorption; and, bacteria releasing antibiotics can inhibit pathogen propagation. The function of inoculation of microorganisms is often affected by soil environment, and we need to pay attention to inoculation methods, and the season and time of microbial application.

Rotation Cropping System:

Cultivation of several kinds of crops in turn can restrain the massive pathogen propagation and decrease disease problems. Continuous cultivation is common in net-house vegetable planting. Therefore, dead soil, which is called so by farmers, is common and cannot be planted with the same kind of vegetable anymore. The most effective treatment for this kind of problem is to adopt rotation cropping and use the upland-paddy rotation to avoid continuous cropping.

Adjusting Soil pH Value:

Soil pH value not only influences crop growth, but also is relevant to the growth of soil microorganisms. Adaptability of each kind of microorganisms has a certain range of pH value. When the soil is acidized, pathogenic microorganisms in the soil would massively propagate together with plant problems, which would worsen the diseases. Adjusting soil pH value can prevent pathogenic microorganisms. Generally speaking, suitable pH value range is between 6.0 and 7.0. Some kinds of crops of higher acid-tolerance capacity, such as tea, strawberry, tobacco, pineapple, etc., can adapt to acidic soil.

Avoid Excessive Use of Pesticides:

Pesticide is the essential prescription of modern agricultural production, and it would not cause major problems with safe usage. To sustain the long-term control and balance function of soil microorganisms, improper or excessive application of pesticide or pouring the remaining of spray liquid agent to roots are harmful to beneficial soil microorganisms and to those in rhizosphere. Soil antiseptic or fumigant is used when damage by disease and pest has occurred to soil, that is to say, they are used as treatment. As for prevention, because they have changed the control and balance of soil microorganisms, the invasion of pathogens would lead t to irremediable consequences.

Changing the Environment of Problem Soil:

As to soils of much pathogen, apart from the above-mentioned methods, we can still apply the method of changing soil environment for preservation. For example, water treatment or planting aquatic crops (such as rice, lotus, taro, etc.) in turn can make the soil be flooded and inhibit the growth of pathogens. We can also use the method of exposing to sunlight by turning the soil, or burning the covered grass to control the number of pathogens. For another instance, soil dressing can remove pathogen microorganisms, and introduce new soil and new microorganisms. There is another method that turns up the subsoil to change the soil environment. Since subsoil has low content of organic matter, organic matter and beneficial microorganism can be applied to improve it. We cannot forget to calculate the cost for the method of changing the soil environment.

Chapter 4

Diagnosis of Problem Soils

4.1 Simple Diagnosis Methods

Soil conservation is a long-term task. Originally, soil has strong buffering capacity and cannot be easily destructed with proper use; improper soil conservation or use may result in "problem soil," "soil sickness," or the so-called "soil fatigue." Once the soil is destructed, a large amount of time and money is needed to recover its health. In order to discover as soon as possible the problem of arable land or the soil which already has problems, we need to diagnose the soil problems. In this series, I intend to introduce the diagnosis of problem soil, which can serve as reference for cultivation.

Diagnosis of soil problems is like seeing a doctor. Generally, we need to firstly take a look, listen, and inquire to know the problem and the history and symptoms of the problem; then, according to the basic information of the problem make the initiatory diagnosis. For those of unknown symptoms, we need to take further analysis and authentication to find out possible problems or causes and finally suit the remediation to the case.

As for diagnosis of soil problems, we need to know the soil condition, management of cultivation, property of crops, as well as the factor of climate. After analysis, research, and discrimination, we might find the possible reason. The reason for each occurrence of problem soil is not necessarily caused by one single factor. Therefore, some problems are easy to be diagnosed, while some are relatively difficult. For convenience, this series would be divided into simple diagnosis and analytical diagnosis, according to the difficulty level.

The task of simple diagnosis is to adopt data of relatively easier detection and research, and predict possible cause according to sciences of soil, fertilizers, and crops. The reason for analysis and authentication diagnosis is that some problems are difficult for simple diagnosis. Therefore, we need to research further into the soil and plant and diagnose according to the analysis result. Simple diagnosis and analysis diagnosis are combined in application and cannot be completely separated. This dissertation would introduce the essentials and methods of simple diagnosis at first.

4.1.1 Simple Diagnosis of Soil Problems

Whether the Soil pH Value is Right for Crop Cultivation:

The initiatory diagnosis of problem soil regards soil pH value as the first target for detection, which is like feeling one's pulse while seeing a doctor. Soil pH value is the index to test whether the soil has any problems, because soil pH value is related to the availability of soil nutrients and the growth and decline of microorganism or pathogens. Though there are differences of various degrees in the optimum range of soil pH value for each kind of crop, it is apt to trigger soil problems when soil pH value is too high (above pH 8.0) or too low (below pH 4.5). In utilization of hilly land, top soil is usually pushed to mountain pit and subsoil is pushed out to the soil surface, which is improper. With the improvement over years and accumulation of organic matter, the top soil is fairly valuable and should not be dumped into the mountain's pit. When developing the hilly land, we should put top soils aside temporarily and push the subsoil into uneven mountain pits or gullies, and then push top soils back for utilization. Especially for red soil of hilly land, its acidity is rather high which has poor effects on crops and we should pay special attention to that, for the improvement of acidic soil cost a lot of time and money.

Whether the Soil Physical Property is Poor:

Generally speaking, we can observe with our eyes to see whether the physical property of soil is good or not; for example, by checking the soil drainage problem, groundwater level, density of soil, aggregate structure, texture, etc. It may vary with the different need of crops. Upland farming, with flooding as its deadliest natural disaster, requires good drainage and a relatively low underwater level, or high beds may be needed. When flooding-intolerant crops are planted at fields of poor drainage in rainy a season, the respiration of the crops' root systems would be terminated. Together with the occurrence of disease, crops are prone to have problems when the sun shines again after rain. This kind of example is most common among vegetable planting.

Whether the Fertilization is at the Right Time and of the Right Amount:

Fertilization can be viewed as a kind of technology and an art as well. Everyone is able to apply fertilizers, but it is more important to apply the right amount of fertilizers at the right time and to the right land according the growth of different kinds of crops. Especially, excessive use of fertilizers is not rare in actual application. Excessive use of chemical fertilizers may cause root rot or fertilizer injury, and would also influence the absorption of other nutrients.

Whether There is Over-Accumulation of Salts:

Pollution and improper irrigation water, or undue amount of salts in the application of chemical fertilizers would cause salt accumulation. In the diagnosis, we can observe whether there are white salts on soil surface after the top soil has been dried. There would be too much salt accumulation while white matters are shown on the soil surface. For other problems, we can still use electric conductivity meter to measure; when electric conductivity is > 4 dS/m, the growth of most crops would be inhibited.

Whether the Problem Soil is Caused by Continuous Cropping:

A lot of uplands have problem soil because of continuous cropping, and the reasons are no more than that something is wrong about the physical, chemical, and biological properties of soil. We need to pay attention to whether the same kind of crop is planted under the condition of continuous cropping or whether its cropping interval is too short. Crops like asparagus, melons, ginger, bean, and Solanaceae are inadvisable for continuous cropping; when planted under the condition of continuous cropping, they are prone to have pathogens triggered by diseases and insects, poor plant growth, and even the death of seedlings. In the diagnosis process, we need to know the cultivation system and concern the fact that the previous crops may influence the growth and production of crops. Adding excessive fertilizers to the crops that already have to the root rot problem caused by continuous cropping would cause more serious withering. We need to nurse crops back to health by foliage dressing.

4.1.2 Simple Diagnosis for Cultivating Crops with Symptoms

When symptoms are shown on crops because of soil problem, it is easiest for observation and detection. We can diagnose these symptoms on time according to their color and form. According to the symptoms of crops' appearance together with the diagnosis of soil, we can easily find out the reason for why problem soil occurs. The relatively easy symptom for problem soil is deficiency in nutrients. The symptom of element deficiency is relevant to that the content is too less in the soil, whereas some deficiency symptoms are caused by incapability of absorption or the abnormal condition of roots, rather than that the soil lacks nutrients. At this time, we need to further clear and diagnose the specific reason. In the following, simple diagnosis of crop with symptoms is introduced.

Withering of Seedlings:

Resistance of seedling roots is the weakest, and is the most vulnerable to soil stress; leaf roll and seedling withering are the obvious features caused by the lack of water absorption by plant roots for leaf transpiration. As for reasons for seedling withering, apart from the possible cause by pathogen, we need to pay attention to injuries by chemical fertilizers, soil being too acidic or too alkaline, subsoil being turned out, too quick decomposition of large amount of organic matters, salts accumulation, poor drainage, injuries by continuous cropping, and the existence of organic or inorganic toxic substances. During diagnosis, you need to take into consideration the soil and cultivation management to find out the possible reason(s).

Abnormal or Withering of Growing Point (Meristem):

Temperature too high or too low would inhibit the growing point of plants. And, if boron or calcium is insufficient in the problem soil, the growing point would be withered and the younger leaves would be deformed. For other crops lacking iron, manganese, copper, zinc, or silicon, usually the top of their stem would not wither. Organic plant toxic substances would also lead to the abnormal growing point; this kind of phenomenon is more serious during drought season.

Change of the Color and Form of Leaves:

The change of leaf color is the reflection of nutrients in plant; deficiency in nutrients is shown not only by the growing rate, but also by the leaf color, which is an important diagnosis index. Moreover, agriculture today gives emphasis to crop quality. Leaf color alone cannot serve as the index for diagnosis of plant health; the thickness and luster of leaves are also important. The color diagnosis of leaves takes place on the above-ground parts of crops, and the degree of the movement of each element in plant is different. The symptoms of element deficiency of each element are shown in the chart in the next figure. For example, old leaves turn yellow when nitrogen is deficient, for nitrogen deficiency is extremely similar to potassium deficiency. But for potassium deficiency, it is usually followed by old leaves' withering from leaf's margin and top, and is presented as scorching form, which differentiates itself from other element deficiency. The diagnosis process of crop's nutrient inhibition is shown in the following Figures 4-1 and 4-2.

CH4 Diagnosis of Problem Soils 45

Younger Leaf and Stem

Older Leaf

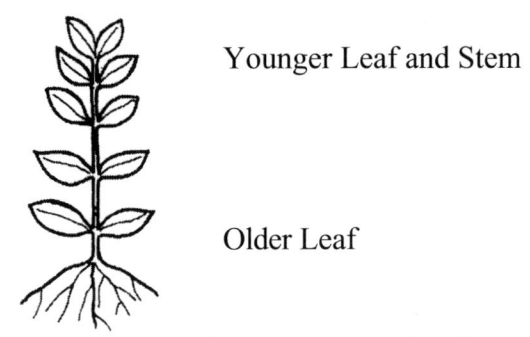

Figure 4-1. Simple Graph of Occurrence Part for the Deficiency of Nutrient Elements in the Plant.

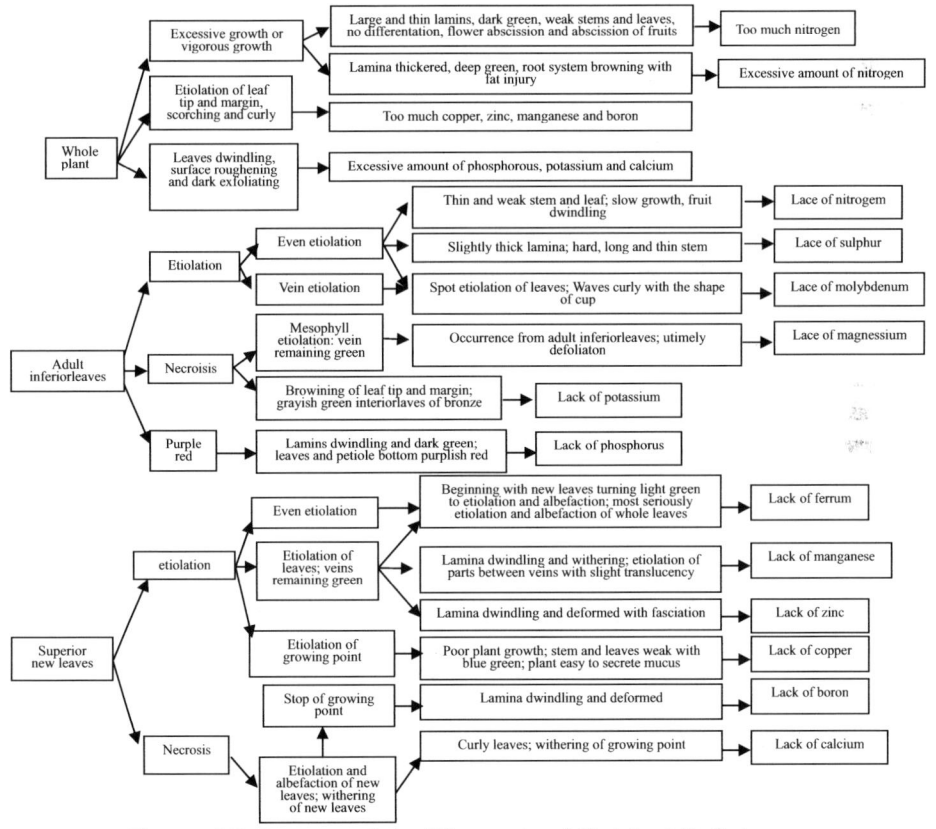

```
                          ┌─ Excessive growth or ──┬─ Large and thin lamins, dark green, weak stems and leaves,
                          │  vigorous growth       │  no differentiation, flower abscission and abscission of fruits  → Too much nitrogem
                          │                        └─ Lamina thickened, deep green, root system browning with
                          │                           fat injury                                                      → Excessive amount of nitrogen
                          ├─ Etiolation of leaf
              Whole       │  tip and margin,       → Too much copper, zinc, manganese and boron
              plant       │  scorching and curly
                          └─ Leaves dwindling,
                             surface roughening    → Excessive amount of phosphorous, potassium and calcium
                             and dark exfoliating
                                                   ┌─ Thin and weak stem and leaf; slow growth, fruit dwindling       → Lace of nitrogem
                          ┌─ Even etiolation ──────┼─ Slightly thick lamina; hard, long and thin stem                 → Lace of sulphur
              ┌─ Etiolation┤                       └─ Spot etiolation of leaves; Waves curly with the shape of cup    → Lace of molybdenum
              │           ├─ Vein etiolation
              │           └─ Mesophyll
              │              etiolation: vein      → Occurrence from adult inferiorleaves; utimely defoliaton          → Lace of magnessium
  Adult       │              remaining green
  inferiorleaves┤─ Necroisis ─ Browning of leaf tip and margin;
              │                grayish green interiorlaves of bronze  → Lack of potassium
              └─ Purple ─ Lamins dwindling and dark green;
                 red       leaves and petiole bottom purplish red     → Lack of phosphorus

                                                   ┌─ Beginning with new leaves turning light green
                          ┌─ Even etiolation ──────┤  to etiolation and albefaction; most seriously                   → Lack of ferrum
                          │                        │  etiolation and albefaction of whole leaves
              ┌─ etiolation┤─ Etiolation of
              │           │  leaves; veins         → Lamina dwindling and withering; etiolation of
              │           │  remaining green         parts between veins with slight translucency                     → Lack of manganese
              │           │                        ┌─ Lamina dwindling and deformed with fasciation                   → Lack of zinc
              │           ├─ Etiolation of         ├─ Poor plant growth; stem and leaves weak with
  Superior    │              growing point         │  blue green; plant easy to secrete mucus                         → Lack of copper
  new leaves  │           └─ Stop of growing
              │              point                 → Lamina dwindling and deformed                                    → Lack of boron
              └─ Necrosis ─┬─ Curly leaves; withering of growing point                                                → Lack of calcium
                           └─ Etiolation and
                              albefaction of new
                              leaves; withering
                              of new leaves
```

Figure 4-2. Process of the Diagnosis of Nutrient Deficiency.

Elements Difficult for Remigration in Plant:
- Transformation of Younger Leaves and Withering of Stem Top:
  Deficiency in calcium and boron.

- Top of Stem Not Withering:
  Deficiency in iron, manganese, copper, zinc, and selenium.

Elements Easy for Remigration in Plant:
- Older Leaf Gradually Withering and Stem Being Thin:
  Deficiency in nitrogen, phosphorus, Sulfur, and molybdenum.

- Older Leaf Turning Yellow and Vein Green:
  Deficiency in magnesium.

- Older Leaf Turning Yellow and Brown Spot:
  Deficiency in potassium.

Flowers and Fruits Dropping and Poor Maturity:

Conservation of soil fertility and long-term supply of right amount of nutrient depend on the combination of management and technology of fertilization. The diagnosis of problems probably depends on the management and fertilization during the growing period. Apart from poor climate that causes flowers and fruit dropping, improper or too early blossom by artificial forcing may also require our reflection. Insufficient or unbalanced supply of soil nutrients would be the source of problems. During blossom differentiation and blossom periods, if there are too much nitrogen fertilizers or the periods coincide with massively nitrogen nutrient released from organic fertilizers, the supply of phosphorus and potassium fertilizers would be insufficient, which may result in fewer blossoms or no fruiting.

The above-mentioned simple diagnosis of problem soil requires careful ordinary observations and regular record-taking. With experiences, we can diagnose the crux of problems.

4.2 Diagnosis by Application of Chemical Analysis

For the diagnosis of problem soil, we can use the method of initiatory diagnosis, together with the management of cultivation and climate factor, to find out the possible reason. However, the occurrence of problems is not decided by a certain single factor. In order to know the complete occurrence of soil problems, we need to combine chemical analyses of plant and soil, thus to obtain the method appropriate to the problem thoroughly. Chemical analyses of plant and soil should be carried by institutions with special equipments so that analyses of plant and soil samples can

Table 4-1. Symptoms and Conditions of Nutrient Element Deficiency Occurrence

Nutrients	Symptoms of the Nutrient Deficiency	Occurrence Conditions of Deficiency of the Nutrient
Nitrogen (N)	1. Plants grow slowly; stems and leaves are tiny; and, fruits become smaller. 2. Beginning from old leaves, the leaves turn yellow green and then yellow and most seriously wither.	Sandy soil; soil of low organic matter.
Phosphorus (P)	1. Can be detected during beginning period of growth and old leaves have the symptoms first. 2. Leaves turn deep green and slightly purplish red, has no luster and most seriously sears and withers.	Most likely to occur to strongly acidic and strong alkaline soil.
Potassium (K)	1. Symptoms begin with old leaves. 2. Inferior leaves turn grayish green with bronze; leaf apex and leaf margin turn brown.	Soil with excessive amount of magnesium or calcium.
Magnesium (Mg)	1. Etiolation and labefaction start with parts between veins but veins remain green. 2. It is more common among leaves near fruits. 3. Defoliation is early in fruit trees.	Acidic soil; easy to occur to soil with too much potassium and calcium during rainy season.
Calcium (Ca)	1. Stem top or younger leaves turn light green, whereas old leaves remain green and most seriously the growing point would be dead. 2. Lignification occurs in younger stem. 3. Inhibition of root growth. 4. Be apt to dropping of fruits, rain injury and decayed fruit.	Acid soil; acid soil with excessive application of magnesium and potassium.
Sulfur (S)	1. Etiolation occurs from inferior leaf to the whole leaf. 2. Mesophyll becomes slight thicker; the stem becomes hard and long and thin; excessive growth.	Soil parent material lacks Sulfur or failure of sulfate fertilizer application.
Boron (B)	1. Younger sprout becomes deformed; terminal bud withers; and growing point stops. 2. Bark has ruck and crack; center part blackens and decays to cavity. 3. Resin and hardening is shown from fruits, and small fruits become deformed.	Neutral soil or alkaline soil; Alkaline soil with calcium fertilizer.

Table 4-1. Symptoms and Conditions of Nutrient Element Deficiency Occurrence (Continue)

Nutrients	Symptoms of the Nutrient Deficiency	Occurrence Conditions of Deficiency of the Nutrient
Zinc (Zn)	1. New leaves have yellow spot; leaflet is small with wither. 2. Obvious etiolation between veins but veins remaining green. 3. Midrib of leaves askew, thus cause both sides of leaf surface to be asymmetric.	Neutral soil or alkaline soil.
Iron (Fe)	1. Etiolation and labefaction between veins.	Neutral soil or alkaline soil.
Copper (Cu)	1. Etiolation occurs to new leaves and growing point; the growth is impeded. 2. Stem leaf becomes weak and turn turquoise; tree trunk and fruits secrete mucus.	Alkaline soil; newly developed soil with plenty of organic matter.
Molybdenum (Mo)	1. Etiolation occurs to veins of old leaves; leaf surface has spot etiolation. 2. Leaf surface is uneven and curly; leaves take the shape of cups.	Strong acidic soil; soil in deficiency in phosphorus.

be analyzed. Hence, the focus of this chapter is not the methods and technology of chemical analysis but the introduction of how to take samples for analysis and the application of analysis. Correct sampling and analysis are the foundation of correct analytical data, which is the key to the analyses of plant and soil and cannot be overlooked.

4.2.1 The Sampling of Problem Plants for Chemical Analysis

Samples for chemical analysis are usually taken from leaves or the above-ground part of plants; and, analysis of leaf is most common. Leaf consists of blade and petiole. For crops in the grass family, we usually take blade rather than petiole as the subject of analysis. Sampling location for different kinds of crops requires consistent criteria.

Sampling for chemical analysis of problem plants is different from sampling aimed at recommended amount of fertilization: blade analysis for problem plants means to detect problems, whereas blade analysis for the purpose of obtaining recommended amount of fertilization focuses on representative blade analysis within the unit area. Blade sample for problem plant should be able to represent the existing problem. Therefore, we need to pay attention to the sampling location on plant, sampling amount, and whether the method is representative; otherwise, it would still be difficult to diagnose with the analyzed data.

In sampling, blades of problems should be adopted, and remember that the chosen blades cannot be stored with soil in the same bag, in case they get polluted by soil and affect the analysis results. As for the visible symptoms for the occurring problems of plants, they are divided into overall occurrence and partial occurrence in the plant. But when taking samples, we need not only take the problem blades, but also take same age and healthy blades near the part, which can serve as the control data of problem blade, and are helpful for contrasted diagnosis. Therefore, during blade picking do not forget to divide as problem blades and healthy blades for their respective chemical analysis.

4.2.2 The Sampling of Problem Soils for Chemical Analysis

Sampling of problem soil also requires us to take representative samples to obtain correct diagnosis. The bigger area problem soil covers, the more attention is needed for the representative sample. Sampling should be directed to the soil which grows problem plants, for only 500 grams of soil is taken for every hectare. Representative soil needs careful sampling; otherwise, it may cause wrong diagnosis.

It is better to take soil around roots as sample for problem soil; yet, soil within rhizosphere is not right for analytical samples aimed for the recommended amount of fertilization, because it contains relatively low content of available nutrients which has been taken up by roots. For problem field, we take soil within rhizosphere, mix several multipoint samples, and take the amount of about 500 grams as soil sample for analysis. For problem soil of orchards, we take soil samples around the trees. Especially for sampling of hilly land of uneven soil, we need to pay special attention to representative and avoid newly fertilized soil or soil at fertilization point; otherwise, the analyzed data is apt cause diagnosis deviations.

According to growth of roots, crops can be divided into short-term shallow root crops and perennial deep root crops; consequently, the depth of soil sampling is also different. For short-term shallow root crops, the sampling soil depth is the top soil of effective root growth, that is, about $15 \sim 20$ cm; for perennial deep root crops, the depth depends on the depth of effective root system, and stratified sampling is adopted because of the depth of roots. Usually, depth of about $15 \sim 20$ cm is viewed as a layer and sampling by soil auger is more appropriate.

4.2.3 Evaluation the Analysis Data with Symptoms

Analytical items for plant or leaf include the content of macro elements and microelements, such as nitrogen, phosphorus, potassium, calcium, magnesium, boron, copper, iron, manganese, zinc, etc. The content of these elements varies with the types of crops. Therefore, during diagnosis, it is helpful to diagnose and acknowledge the problem to compare the analytical data of problem plant with that of healthy plant.

The results of the previously mentioned plant analysis refer to the various amounts of nutrients taken up by plants, which can serve as the basis of recommended fertilizers and the amount in general. However, there is exception: Though the result of plant analysis is deficiency of a certain element, the soil, in fact, does not lack that element. Therefore, diagnosis should combine soil analytical data and other properties. The possible reasons for plant deficiency while the soil does not lack the elements are listed below:

- Sicknesses of plant root system, root rot caused by abnormal factors in the soil, or fertilizer injury would all affect the absorption capacity of root system, and then cause the phenomenon of deficiency in plant.
- Poor physical property of soil, such as poor drainage, deficiency in oxygen, inhibition of root respiration that affects its absorption capacity, weakened oxidation, and ferrous ion that is apt to invade root system and cause toxic effect.
- Unbalance among Nutrients in the Soil:
  For example, excessive potassium would restrain the absorption of magnesium; excessive phosphorus to upland fields would easily cause the deficiency of zinc and iron; when iron and aluminum are very active, the applied phosphorus fertilizers are easily fixed and phosphorus deficiency is apt to occur.
- Unfavorable Climate:
  Temperature influences the growth and activity of plant root system. Most obviously, low temperature influences the absorption of phosphorus fertilizers. For another example, deficiency of magnesium usually occurs in pluvial seasons, whereas deficiency of calcium and boron occurs easily during dry seasons.

Concluding the above reasons, we can see that the diagnosis of plant analytical result needs to be examined in accordance with the analysis of soil and climate factors to for evaluation.

The usual items for soil analysis include pH value, content of organic matter, available phosphate, potassium, calcium, magnesium, silicate, zinc, and salts, which are the most direct proof for diagnosis. With the combination of the data of plant analysis and cultivation management, the reliability of the remedy appropriate to the situation can be even higher.

To conclude, simple diagnosis proposed in the previous chapter and analytical diagnosis in this chapter are both used to evaluate reasons that cause problem soil, and these two methods are supplementing each other. Through correct research and judgment we can solve the problems and attain effects of treatment. During evaluation, we should get started from the relatively easy method. Seek the solution after simple diagnosis based on the crop's appearance; if it fails, we need to further carry on the analytical diagnosis, because analytical diagnosis costs a lot of time and may delay the prime time of treatment. This should be concerned especially for short-term crops, which need more urgent diagnosis and quick treatment. In order to diagnose quickly and correctly, we should take actions for treatment right after simple diagnosis and meanwhile take the samples to carry on analytical diagnoses of plant and soil. By doing so, we can attain the effects of quick remedy and yield twice the result with half the effort.

4.3 Problem and Diagnosis of Pollution

4.3.1 Do not Let Soil, Mother of Crops, Cry

There are many reasons for problem soil, including congenital content defects of soil parent material, improper management, and pollution. Rocks on the Earth's surface take thousands of years to develop into soils by weathering, and the content of rock is the main deciding factor for whether the soil is good or not. If the rock contains relatively more calcium, naturally the formed soil contains also more calcium; likewise, if the rock contains more sulfur, the formed soil would contain more sulfur and become relatively acidic soil. Through the ages, thanks to the activities of living things on and within the soil, plenty of organic matters are accumulated on the surface of soil particles, which enhances the good properties of the buffering capacity of farmland. Pollution not only brings toxic substances to the soil and greatly endangers the health of human beings, but also seriously harms the assets over generation -- the soil. In particular, the harms to beneficial microorganisms and soil's buffering capacity are difficult to be recovered. Beneficial microorganisms can clean soils and contend against pathogens, and can also increase the nutrients and their availability for crops. Soil buffering capacity can restrain the pH value of soil from dramatic changes and supply a good and stable growth condition for crops. Therefore, for the health of all human beings and our descendants, we should work together to protect and care for the soil that we live by.

The rapid development of industry in recent years unavoidably leads to the soil pollution. The prevention of pollution needs to be reinforced especially for soils near factories or downstream waters. The sources of soil pollution include pollution of irrigation water, air pollution, and pollution of agricultural applied things. Generally, the types of pollutants are no more than organic matter and inorganic matter; according to the different sources of pollutants, there are the problem of pollution load and the problem of contaminator, containing the categories of liquid, gas, and solid, which are too numerous to mention one by one. Some pollutants with foul smell or color can be detected; pollutants without foul smell or color and are difficult to be detected are most dangerous.

Each kind of pollutant has a different kind of toxicity or injury to crops, animals, and human beings: some can be toxic and harmful to all the living things and some probably can harm only one kind. Whether the pollutant is harmful to crops or animals, we all need to prevent them from polluting the soil. If the soil is polluted and cannot grow crops, this is our loss. Most dangerously, if the polluted soil can still grow crops, their productions are pollutants with toxic substance. If these vegetables, grains, or fruits are eaten, they will become invisible "killer."

The pollutants in the soil consist of the decomposable and the indecomposable. The decomposable pollutants can further be divided into the quickly decomposable and the slowly decomposable. Many kinds of organic matter are decomposable, and heavy metals are indecomposable. Heavy metals remain heavy metal everywhere and can only transform between inorganic heavy metals and organic heavy metals. For example, mercury includes inorganic mercury and organic mercury; soil microorganisms can transform inorganic mercury into methyl mercury (this is organic mercury). Toxicity and mobility of inorganic and organic matter are different. The problem worth attention is that we cannot overlook dissoluble pollutants. Though some pollutants can be decomposed, before they are decomposed in the irrigation water, the toxic substances might have been untaken by crops and eaten by people.

Apart from massive pollution, the pollutants usually accumulate little by little in the soil and gradually lead to soil pollution. Though the amount of pollution is slight, as time goes by, a large amount of pollutants would accumulate in the soil. Therefore, we need to inspect the safety of the soil's irrigation water to guarantee the safety of grains, vegetables, and fruits.

4.3.2 Properties of Soil Pollution

The severity and harm of soil pollution is no less than that of water and air pollution. Particularly, prevention of soil pollution is far more important than treatment, because the properties of soil pollution are listed as follows:

Soil Pollution is Difficult to Eradicate and is Persistent:

After the soil is polluted, the soil pollution problem depends on the property of pollutant. If indecomposable pollutants such as heavy metals are adsorbed to the soil, they would be difficult to remove and cause fairly long-term harms. However, water and air pollutions can easily be diluted and washed away by rain water or winds of natural force.

Soil Pollution is not Easy to be Detected and has Latency:

If the soil is polluted, without detection, the pollution would accumulate without being realized and cause big problems. However, water and air pollutions can easily influence human, animals, and plants, thus be perceived.

Soil Pollution has Secondary Pollution and is of Recontamination:

The movable parts in soil pollutants would enter the water or the air and pollute water sources, groundwater, oceans, and airs. And, generally, water and air pollutions would assemble in soils or oceans.

The Cost of Soil Pollution Treatment is High and its Recovery is Poor:

The remediation methods of soil pollution are difficult, and the cost for soil recovery treatment is extremely high. If the soil is recovered after processing, the property of soil would be seriously destroyed, and would cause the phenomenon of soil deterioration. Hence, the prevention of soil pollution is more important than soil remediation for the contaminated soil.

4.3.3 Soil Polluted by Water

Types of Water Pollutants:

There are plenty kinds of water pollutants; any matter that pollutes soil, causes decrease in crop production, and harm humans and animals can be viewed as pollutants, including inorganic matter and organic matter. Any salt among inorganic matters would cause the soil problem of salts accumulation. The pollution of heavy matter is most serious and of most toxicity. Therefore, fields and gardens around factories or using waste water should pay attention to whether the irrigation water is polluted in case that they produce unsafe foods. Organic pollutants have many types, such as toxic substances and the wastewater of live stocks. Some organic matters are difficult to dissolve, and we cannot take it lightly. Too acidic (pH value less than 4.5) or too alkaline (pH value more than 8.0) polluted water would seriously affect the soil's buffering capacity, kill beneficial microorganisms, directly injury crops, and cause deficiency of nutrients.

Symptoms of Eater Pollution:

When the soil is polluted by water, some pollutants can be discovered by the symptoms of crops, yet some are difficult to be identified from crops, which also depends on the density of pollutants. Because of the differences of crops and the influence of climate, the degree of symptom occurrence also varies. For example, excessive amount of nutrient elements or deficiency has many symptoms. Generally speaking, the commonly visible symptoms are the inhibition of crop growth and abnormal growth such as above-ground stem and leaves being minus green or discoloring and necrosis; all can be divided into whole occurrence, part occurrence, or spot occurrence of symptoms. Serious polluted crops would have rotten root or the withering of whole plant.

Diagnosis of Soils which were Polluted by Water:

Simple Diagnosis:

Observe the distribution and features of crops with the occurrence of abnormality in the field. The first contact position of water pollution is the water inlet. Therefore, it is most likely for soils at the water inlet to accumulate pollutants, and the symptoms it presents are also the most serious. If the amount of pollutants is fairly large or the pollution has maintained for a long time, the entire field would have the same symptoms. Though some pollutants have no color or smell, there still are many pollutants with color or foul smell which can be detected. Another simple diagnosis of water pollution is to measure the pH value or the value of electrical conductivity: irrigation water with pH value below 4.5 or above 8.0 and the value of electrical conductivity above 4 dS/m could be all polluted.

Chemical Analysis Diagnosis:

We can take the samples of polluted soil or water to related institutes to carry out soil analysis and water analysis. We can also take the sample of injured plants to do plant analysis. Because there are various kinds of pollution and various instruments used for chemical analysis, in order to determine the result as soon as possible, it is better for those analysts to know what kind of factory the pollution source is from. This can serve as a reference for them and be helpful to select correct analysis instruments to detect the problems.

4.3.4 Air Pollution

Air Pollutants:

There are a great number of different kinds of air pollutants, which mainly include the following three categories: sulfur dioxide, fluorides, and chlorides. Some pollutants in the air are transformed into acid, such as sulfate acid, hydrogen nitrate, hydrochloric acid, etc, which falls onto the ground through rain and results in acidic rain. Long-term acid rain would acidify the soil and endanger crops.

Symptoms of Crops of Air Pollution:

Direct influences of air pollutants to crops are relatively obvious. In terms of pollution to soil, acid rain falling onto soil surface is relatively serious. The damage caused to crops by air pollutants is the most serious, especially to leaves during the growing season. Commonly seen symptoms are the color of leaf margin, leaf apex, and the part between veins turning brown or yellow green. What's worse, crops that are heavily-polluted may wither. The speed of symptom presentation also varies with the kind and density of pollutants; generally speaking, they are divided into acute and chronic and can cause growth inhibition, production failure, and aging.

Diagnosis of Air Pollution:

Simple Diagnosis:

Air pollution mainly originates from factories, and crops at downwind location of soot emission are prone to show symptoms. Therefore, crops downwind or upwind can be detected, unless the wind direction is mixing up and the comparison cannot show the pollution. Some air pollutants have foul smell. Usually, air pollutants of sulfur or nitrogen chemical combination stink; other organic matters containing alcohol group, aldehyde group, carbonyl group, etc., also has foul smell and can be detected.

Diagnosis by Chemical Analysis:

We can collect air to carry out the analysis of air pollutant to determine the pollutants. However, because usually we cannot collect polluted air in time, plant analysis can also be used to detect some pollutants. Soil analysis cannot be used to detect air pollution; acid rain is the easiest method for air pollution detection. Collect rain water for several times during the rain, such as collect rain water at the time of rain falling and at intervals to detect pH value and matter contained in the water.

4.3.5 Safe Food is Based on Safe Soil

In spite of water and air pollutions previously mentioned, inappropriately applied agricultural chemicals, such as pesticides and fertilizers, may cause soil pollution. Some soil pollutions are easily detected. But the most terrible soil pollution is undetectable, because crops do not response to these pollutants and do not show obvious responses. It is still dangerous to eat foods that are slightly polluted by pollutants with lower concentration. Since human body has the function of accumulation, people eat a small amount of slightly polluted food everyday and will get sick after a long time. Moreover, these dangerous foods can neither be fed to livestock, because those livestock will eventually be eaten by us and endanger our health. Therefore, vegetable with green color may not be safe. Concerning food security and environmental ethics, regular healthy tests of the soil by soil analysis is very necessary.

4.3.6 The Direction and Goal of Soil Conservation

Soil protection and conservation are the essence of agriculture and environmental ecology in the future; good soil protection and conservation is the hope of agriculture and environmental ecology. In that case, it is then possible to have healthy agricultural products and ecological environment and a healthy nation and living conditions. Therefore, soil protection and conservation is the foundation of society, country, and nation. Soil protection and conservation should emphasize two major directions, soil protection and soil cultivation:

Soil Protection:

The goal of soil protection is to reduce soil destruction and decline. During soil utilization, we should reduce soil pollution and soil losses. The prevention and treatment is the key to soil pollution. However, now, attention is only paid to prevention and treatment of soil pollution, and the concept of prevention has not popularized yet. Soil protection, along with soil and water conservation, is in urgent need of being put in practice actively. To reinforce the work of farmer organization and education is the key to soil protection, which requires implementation and promotion of dedicated units of governments at all levels and personnel. Soil protection should be listed as a national important construction, because construction includes not only the life hardware for the present, but also the improvement of soil environment that is closely related to the quality of life; it is the deeper significance and value of the construction.

Soil Conservation:

The goal of soil conservation is to preserve and increase the functions of soil which include productivity, ability of remediation, holding capacity of water and fertilizer, ecological control and balance force, buffering capacity, etc. Soil conservation is the more positive way compared with soil protection. After soil utilization, soils will have the phenomenon of decline. If we do not carry out soil conservation, agricultural production would become more and more difficult. With the increase of fertilization and pathogens, the amount of pesticides and fertilizers will increase, which will lead to the deterioration of agriculture quality and the rise of agricultural cost, and eventually form a vicious circle of the environment. For soil protection and conservation, we should pay attention to the right amount of fertilization, removal of soil limiting factors, and the application of organic and microbial fertilizers to increase soil organic matter content and soil activity. The government should reinforce the detection of soil and fertilizer, draw up soil conservation policies or regulations to put soil productivity enhancement into practice, and include soil conservation performance in the work of local governments, which is the foundation of the protection of the whole nation's health. In order to ensure the function of soil, Japan has enacted "The Improvement of Soil Fertility Act," which is the practical work of valuing soil assets and has profound value of protection and conservation.

Chapter 5

Fundamentals of Plant Nutrients and Fertilization

5.1 The Classification and Application of Fertilizers

5.1.1 The Definition and Classification of Fertilizers

In the broad sense, fertilizer refers to any materials that can increase soil fertility and crop production or quality. Generally, fertilizer can be classified into three classes: organic fertilizer, chemical fertilizer, and biofertilizer. In the narrow sense, fertilizer refers to substances that can supply plants with nutrient elements, namely carbon (C), hydrogen (H), oxygen (O), nitrogen (N), phosphorus (P), sulfur (S), potassium (K), calcium (Ca), magnesium (Mg), iron (Fe), manganese (Mn), copper (Cu), zinc (Zn), boron (B), molybdenum (Mo), chlorine (Cl), etc. These elements are called the "essential elements of plants." Among those elements, only carbon, hydrogen, and oxygen are from carbon dioxide (CO_2) and water (H_2O); the rest of them are from the nutrient elements of soil. In addition, some other elements are not essential for the plants' growth, but beneficial to it. They are called "beneficial elements." For instance, sodium (Na) is beneficial to the halophytes for adjusting cell osmotic pressure; silicon (Si) can strengthen cell walls and improve plant strength, health, productivity, and disease resistance of hydrophilous and grass family plants; cobalt (Co) can improve the nitrogen fixation of the legumes plants root nodule; nickel (Ni) is essential for urea to hydrolyze enzyme; and vanadium (V) can promote plant growth, etc. The three classification of fertilizer will be illustrated as followings:

Organic Fertilizer:

Through the ages people know that using organic fertilizer is an old way. Ancient people used compost and feces, but nowadays the organic fertilizers vary a lot. The sources of materials and complexity of decomposition are very different, and the source can be classified into two sorts: one is farm organic fertilizer, including farmyard manure, green manure, and discarded resultant organic; the other is commercial organic fertilizer, rooted in the wrack or waste of microorganism, animals, and plants, turf, extraction, or enriched organic fertilizer. The commercial products are compounded according to a certain standard, so the ingredients are comparatively fixed.

There are big differences in the speed of decomposition of organic materials. Generally, green manure, feces, meals, compost, and residues of animals and plants decompose fast, while those organic materials with lignin and high humus, such as peat, tree bark compost, etc., decompose slowly.

Chemical Fertilizer:

Chemical fertilizer refers to the fertilizer made by chemical synthesis or chemical reactions, whose ingredients contain a certain quantity of nutrients for the plants, such as nitrogen, phosphorus, potassium, magnesium, calcium, sulfur, boron, copper, zinc, manganese, iron and molybdenum, and so on, including simple-substances and compound fertilizers. Chemical fertilizer always differs from the natural substances such as rock phosphate. After acid treatment (such as sulfuric acid), it will become chemical phosphorus fertilizer, such as calcium superphosphate. The chemical fertilizer could be fast- or slow-released fertilizer. From the features of solubility, it can also be classified into soluble and insoluble fertilizers. Those soluble ones can be applied not only to soil, but also to those fertilizers for foliage dressing.

Biofertilizer:

Biofertilizer refers to microbial fertilizers that utilize active microorganisms to increase the nutrients or their availability in soils, enhance absorption, or protect the roots. It is usually called microbial fertilizer or microbial inoculants, because of the most frequent use of microorganism. The products of biofertilizer are modern fertilizer with new science and technology, and the classification varies a lot. It is mainly classified according to the function of microorganism. For example, the application of nitrogen fixing microorganisms can increase the source of nitrogen in soil. Nitrogen fixing microorganisms can also be classified into symbiotic nitrogen fixing microorganisms (such as rhizobia) and non-symbiotic nitrogen fixing microorganisms (such as *Azosperilum*). Mycorrhizal fungi can mainly help plants to dissolve phosphorus fertilizer. There are various phosphate-solubilizing microorganisms, and the insoluble phosphate can be dissolved into soluble phosphate. The microbial fertilizers have a low environmental impact. Nutrients in the soil environment are limited, and microbial fertilizers will not destroy the ecosystem after applied to soil; therefore, microbial application should not be the concerns about the contamination.

5.1.2 The Basic Principles of Applying Fertilizers

Soil Amelioration before Fertilization:

Before fertilization, it is necessary to know whether the soil has limiting factors. Take the problem of the soil physical, chemical, and biological properties for example, as it has been mentioned in Chapter 3 regarding the diagnosis of content and the fundamentals to the solution. Generally, the most common way of soil amelioration is to neutralize strong acidic soils and apply lime materials, or to neutralize strong alkaline soils and apply acid peat or sulfur powder. Organic materials can also be used to improve fertility and features of the soil.

To Determine the Type of Fertilizer and Amount of Fertilizer Application:

The following two principles should be noted when determining the amount of fertilizer application.

Law of Minimum Nutrient:

For plant growth, more than 17 kinds of nutrient elements are needed. However, the plant growth is limited by the "most limiting element," which is to say, the growth of plants cannot be fully enhanced without these "most limiting elements," even on the condition that the other nutrients are sufficient. Thus, it is important to know which element is the most needed during fertilization; only by fertilizing according to the "most limiting element" and determining the right type of fertilizer can the effects of fertilization be ensured.

Law of Diminishing Returns:

If there is a deficiency of a certain element in soil, the increment of plant growth will be increased with the augment of the amount of application of this element. When the amount of fertilization tops at one point, the growth of plants stops increasing, and thus the amount of crop production decreases with the increase of the fertilization. This phenomenon is called the law of diminishing returns. Therefore, the more fertilization does not always mean the better. Don't simply increase the amount of fertilization when the crop does not grow well. Sometimes, over-fertilization might lead to a drop in production. Generally, the most economical amount of application is the best choice, which could avoid greater loss and some environmental problems. Therefore, fertilization in proper ways to gain the high yields and the best quality is highly proposed. In addition, a handbook of fertilization can be referred to for fertilizing the common crops.

Precision Fertilization (Formula Fertilization after Soil Testing Method):

Determining the amount of fertilization by soil testing and analyzing the soil fertility or plant is a good way to precisely decide the type and amount of fertilizers. Usually, soil testing and plant nutrient analysis can be done after the harvest of crops to decide the fertilization for crops in the next period.

To Decide the Time of Fertilization:

Deciding the time of fertilization is to match-up the "fertilizer-needed period" of the crop, and to provide proper the type of fertilizer and amount of fertilization. And, reducing the loss of fertilizers should be taken into account, so that the best effect of production and quality, as well as the economic benefits can be achieved.

Different types of crop demand fertilization at different times; except for basal fertilizer, the time for applying additional fertilizer and the corresponding type and amount of fertilization are also very important. For instance, the fruit tree and fruit-vegetables demand much phosphorus and potassium fertilizer in their fruiting period; thus, sometimes it is necessary to apply additional phosphorus and potassium fertilizers.

To Decide the Method and Placement of Fertilization:

Different crops need different ways of fertilization in the field. Generally, there are broadcast distribution, furrow application, hill application, etc. They are different in distribution mode, gradation of top dressing, and deep placement. Before soil cultivation, basal fertilizer should be used, and then start plowing into the soil. Because the furrow or hole application are comparatively intensive, fertilization at a distance from the plants or seeds is recommended in order to avoid the harm from the fertilizer; otherwise, the slow-released fertilizers are needed. For the perennial fruit trees, usually, it is better to apply the basal fertilizer with digging furrows, making soil and fertilizer mixed up, and avoiding the loss of fertilizer due to the broadcast distribution.

5.1.3 Calculation of Fertilizer Application

The content of fertilizer is usually presented by the percentage of N-P_2O_5-K_2O-MgO among which the content of P_2O_5 is 2.29 times as much as P [P_2O_5 (%) = P(%) × 2.29]; K_2O is 1.205 times as much as K [K_2O(%) = K(%) × 1.205]; MgO is 1.667 times as much as Mg [MgO (%) = Mg(%) × 1.667]. The amount of single fertilizer (kg/ha) is calculated based on the percentage of elements in the fertilizer; here is the formula:

Amount of fertilizer application (kg/ha) = Amount of nutrient elements needed (kg/ha) × 100 / Percentage of the element in that fertilizer (%)

For example, a certain crop of 1 ha requires the nitrogen fertilizer of 80 kg/ha, when applying urea (with 46% nitrogen), the amount of fertilizer application is 80 × 100 / 46= 173.9 kg/ha, namely the urea of 173.9 kg/ha should be applied.

5.1.4 Guidelines of Fertilizer Application

Good fertilizer application demands the accumulation of different ways of application for a long period of time. It is important to notice the demand of the soil and crops; too much or too little are not recommended. In addition, the mixing and allocation of fertilizers are also important. Therefore, fertilizer application is not only a technique, but also an art. The followings are guidelines for application of different sorts of fertilizers:

Chemical Fertilizer Attentions:

While applying the chemical fertilizer, it is important to fertilize according to the demands of crops at proper time, with proper amount, in proper way, at proper place, and with proper sort of fertilizers. For instance, it often occurs that the soil becomes acidic because ammonium sulfate and single superphosphate have been used for a long period. Apart from that, if the paddy rice is applied too much nitrogen, the leaves will become dark green, and serious rice blast might occur. Overdose or improper application of chemical fertilizers would not only affect the production and quality of crops directly, but also have an indirect impact on people health. If the vegetable absorb too much nitrate, it is as harmful to people as bacons that contain nitrate, so it should be carefully concerned. When it comes to the quality of exquisite farming crops, soil plays a very important role; thus, the research on the relationship between crop quality and soil nutrients cannot be ignored.

It is important to apply fertilizers in a balanced way. There are a proportion of cations nutrient elements that crops absorb from the soil (such as K^+, Na^+, Mg^{++}, Ca^{++}, and some other microelements). For example, potassium fertilizer could increase the content of potassium in the leaves and reduce the amount of other cations. If applying too much potassium fertilizer, it could lead to a deficiency of Mg or Ca for the crops. This is called the ion antagonism. This henomenon can usually been seen in the fruit trees. As a result, the concept of balance between element absorption's synergism and antagonism should be kept in mind. The synergism and antagonism of different nutrient elements and the impact on the crops' absorption are shown below as Figure 5-1.

The amount of chemical fertilizer applied should be concerned. Some short-term crops are cultivated very intensively, so fertilizers might be over-applied, which would cause the problem of salt deposit.

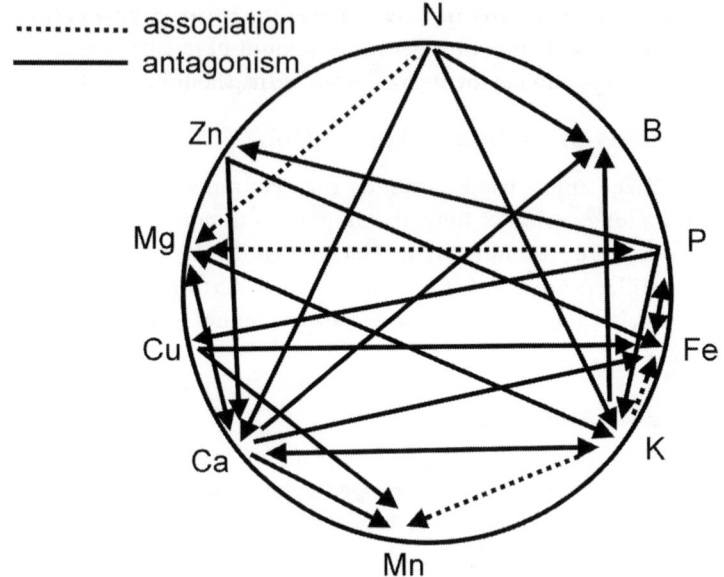

Figure 5-1. The Interaction of Nutrients Elements (Mulder's Chart)

Organic Fertilizer (Illustrated in Details in Chapter 8):

For soils that are deficient of organic matters (less than 2%), only by strengthening some hard-to-decompose organic matters (such as peat) when applying organic fertilizer can the content of organic matters in the soil be increased year after year. However, while the decomposable organic matters mainly supply nutrients that are released, the organic matters that have been accumulated over the years are very few. From this point, it is very necessary to add the decomposable and not easily decomposable organic matters when applying the organic fertilizer. It will not only provide long-period nutrients, but also maintain the function of soil, which is obviously a win-win method for the soil.

Microbial Fertilizer (Illustrated in Details in Chapter 9):

As for applying the microbial fertilizer, it is fundamental that the function of the microbial fertilizer can be realized only when the plant roots or seeds are fully reached by the microorganisms from the microbial fertilizer, and that it is effective only if the organic matters touch the organic decomposing microorganisms. When applying the microbial fertilizer to the soil, the effect will be seen much sooner if it is mixed with the organic and inorganic nutrients or the attached substance (such as peat).

5.2 Features and Guidelines of Application for Nitrogen

5.2.1 The Relationships between the Soil Nitrogen and Plants

Generally, surface soil contains 0.1 ~ 0.3% nitrogen, and some even contain less than 0.1%. In the plant nutrients, nitrogen, phosphates, and potassium are the most demanded nutrients, which are called the "three key nutrients of fertilizers" in fertilization. Different plants weigh differently in plant dry matters according to the difference of growth and position of plant (about 2 ~ 5% nitrogen). Among the fertilizer of different elements, nitrogen fertilizer is the easiest to observe by the changes to the color and shape of the plant. The influence of nitrogen fertilizer is listed as follows:

- Amplify the leaf area, and thus improve the photosynthesis and increase the growth of roots and production.
- Enhance the amount of protein in plants and fruits.
- Enhance the leaf color.
- Lessen the thickness of cell wall and increase the body of plants cell.
- Increase the amount of water in cell cytoplasm and ratio between stem and roots.

The farmers always over-apply the nitrogen fertilizer, which will cause the following results:

- Weaken the cell wall, which is apt to cause pests and diseases, drought injury, and freeze injury.
- Postpone the harvest of plants and fruits.
- With the augmented leaf area, during water shortage, the transpiration will increase, plants will be apt to wilt, and production will be reduced.
- If the stalks of the cereal plants are weakened, the lodging of crops may occur.
- The fruit trees might grow excessively, which would lead to decrease in flowering and fruit weight, or in the quality of sugar plants.

5.2.2 The Sources and Types of Nitrogen Fertilizer

There are various sources of nitrogen fertilizer, including organic and inorganic ones. The nitrogen fixation of legumes or rhizosphere microorganisms could synthesize amino acid through fixing N_2 gas (cannot be utilized by the plants directly) into NH_3 (can be utilized by the organisms) by symbiosis or non-symbiosis. This can be also seen as one of the sources of nitrogen fertilizer.

The organic and inorganic sources of nitrogen fertilizer are explained as follows:

Sources of Organic Nitrogen Fertilizer:

All kinds of organic matters contain nitrogen; especially, the protein which consists of amino acids contains high amounts of nitrogen. Generally, the animal meats (fish meal, meat meal) and the meal of the legumes contain the highest organic nitrogen fertilizer. The organic nitrogen fertilizer can only be taken up and utilized by plants after organic nitrogen is decomposed into small molecules or inorganic nitrogen. The organic nitrogen fertilizer made up of amino acids will be decomposed into inorganic nitrogen after applied to the soil.

Sources of Inorganic Nitrogen Fertilizer:

Only a few of inorganic nitrogen fertilizers are gained from the mineral substance accumulated for a long time, such as Chile saltpeter (with sodium nitrate), while most are the outcomes from urea or ammonium sulfate after nitrogen from air N_2 is synthesized industrially into NH_3 through electricity, high temperature and pressure, or $CaCN_2$. Therefore, inorganic nitrogen fertilizer can be classified into ammonium, nitrate, calcium cyanamide, and urea, etc. On the combination among these, and there are even some organic slow release nitrogen fertilizer, such as formaldehyde urea, isobutylidene diurea, etc, which can postpone to be uptake by plant and utilized after decomposed by microorganisms. The sulfur coated urea is one kind of physically slow release nitrogen fertilizer in the commercial market.

5.2.3 The Forms and Transformation of Soil Nitrogen

The soil nitrogen can be divided into several forms in different soil positions:

- Organic nitrogen.
- Mineral nitrogen in the soil solution or exchange positions.
- Nitrogen in residues.
- Ammonium fixation on the clay.
- Gaseous nitrogen.

Therefore, the nitrogen in soil could be classified into organic and inorganic nitrogen. It holds different proportion in different soil. Generally, the content of organic nitrogen in soil is much higher, with the proportion of about > 95%, however the inorganic of < 5%.

The Organic Nitrogen:

The organic nitrogen is mainly in $-NH_2$ form, among which the amino acid holds about 20 ~ 50%, hexosamine holds about 5 ~ 10%, and the structures of half of which have not been identified yet. The organic nitrogen can be classified into easily mineralization and stable forms. As for the mineralized nitrogen, unless it is applying with a great deal of organic fertilizer, it generally occupies about one third of the organic nitrogen. After the organic matters are decomposed by microorganisms and NH_4^+ is produced, the upland will produce NO_2^- through nitrification, and finally the NO_3^- will be formed.

The Inorganic Nitrogen:

NH_4^+ and NO_3^- are the most common forms of inorganic nitrogen; there will be a little NO_2^- in soils with a high pH value. The rests are the N_2O and the N_2 in the air after denitrification. The denitrification occurs under the conditions of paddy and poor aeration soils when nitrate is transformed to gas forms.

In the organic nitrogen, only some of the small molecules can be mobilized in the soil; most of the rest nitrogen of polymers cannot easily be mobilized. And, the inorganic nitrogen- NO_3^- is the most movable form in the soil. However, NH_4^+ is easy to be fixed and adsorbed by the soil.

5.2.4 The Loss of Nitrogen Fertilizer and Guidelines for Improvement

The nitrogen fertilizer is the easiest to be lost and ran off; therefore, it is necessary to know the ways of loss. The ways of nitrogen loss can be classified into: (a) the loss by leaching; (b) the loss by ammonia evaporation; and (c) the loss by gaseous nitrogen from anaerobic denitrification and nitrification. They are illustrated as follows:

The Loss by Leaching:

Phenomenon that may Occur:
Rain water and irrigation runoff transfer the soluble nitrogen fertilizer out of soil or into the groundwater. The NO_3^- is the most movable, so it is the easiest to be lost by leaching.

Guidelines for Improvement:
To avoid the leaching problem of nitrogen in the rainy seasons, application of the nitrogen fertilizer should be reduced before raining; and, don't spread it on the hillside fields. Slow release and organic nitrogen fertilizers could be used to decrease the nitrogen loss resulted from its high solubility in water.

The Loss by Ammonia Evaporation:

Phenomenon that may Occur:
When the urea and ammonium are applied to the soil with pH over 7.5, NH_4^+ would be transformed into NH_3, and the gas would be emitted; this would be more serious especially in the seasons of higher temperature and strong wind.

66 Soil and Fertilizer

Guidelines for Improvement:
The ammonium (such as ammonium sulfate) should be avoided being applied to strongly alkaline soils. Sulfur powder or peat could be used to neutralize strongly alkaline soils, and nitrate fertilizers could be a choice. It is better to be applied to subsoil than to topsoil. And, slow release of nitrogen fertilizers, such as sulfur coated urea, could be used to prevent fast loss. Other substances (such as urease inhibitors, hydrocarbons or herbicides) are also functional, but the use of chemicals to the fields should be avoided as much as possible.

The Loss by Gaseous Nitrogen Loss:

Phenomenon that May Occur:
- Anaerobic Denitrification:

When the soil is under the conditions of poor drainage or cultivation in paddy field, there will be oxygen deficit in the soil, and a group of anaerobic denitrifying microorganisms will transform NO_3^- of the nitrogen fertilizer into gaseous N_2O or N_2, thus causing the loss of gaseous nitrogen. Generally, the soil might lose 9 ~ 15% nitrogen fertilizer by denitrification, and some might even lose up to 30% in raining seasons.

- Aerobic Nitrification:

When the soil is under the condition of good aeration, ammonium (NH_4^+) would be transformed into nitrite (NO_2^-) by a group of nitrifiers, and then into nitrate (NO_3^-) by microorganisms, which is called nitrification. In the process of nitrification, the microorganism will also give out N_2O, thus causing the loss of nitrogen fertilizer.

Guidelines for Improvement:
The nitrate fertilizer should not be applied during raining seasons and on paddy fields, and ammonium or urea fertilizer should be used for the application. And, it is important to apply the granulated fertilizer to the subsoil than to the topsoil (which would transform ammonium into nitrate and result in the denitrification). Nitrification inhibitors could also reduce the loss of nitrogen fertilizer from the gaseous nitrogen loss.

5.2.5 Guidelines of Improving the Availability of Nitrogen

The availability of nitrogen fertilizer is the highest among the nutrients, but nitrogen fertilizer is also the easiest lost. As a result, the availability of nitrogen is hugely influenced by the environment, and the ways of improving the availability of nitrogen are listed below:

Proper Time of Applying Fertilizers:
The effects of nitrogen fertilizer on the plant growth are obvious, but different timing and seasons may lead to different effects. Raining and drought, heat and

coldness, and sunny and cloudy weathers will all affect the availability of nitrogen fertilizer. In raining seasons, it is not recommended to apply nitrogen fertilizer excessively; in dry seasons, with soil moisture deficiency, it is not good to apply nitrogen fertilizer, either. The loss of nitrogen will be great if the temperature is high. Over-amount of nitrogen fertilizers should not be applied in the cloudy days.

Top dressing (applying additional fertilizer) is a way of applying fertilizer by fractions, and the availability will be the highest when top dress at the fast growth period.

Quantity of Fertilizer:
The amount of applied nitrogen fertilizer differs according to plant's types, growth period, and seasons. The amount of nitrogen applied to the legumes could be reduced greatly; for example, the soybeans only needs ca.40 kg/ha in every cropping, which is 1/2 ~ 1/5 less than other plants of the legumes. During the growth period, fractional application could be used. Don't use up all the amount of fertilizer at one time. Before the wet season, the fertilization of nitrogen fertilizer could be reduced accordingly. Especially, after mid-fruiting period of fruit trees, nitrogen fertilizer cannot be over-applied in order to avoid the excessive growth of plants and to reduce the quality of fruits. In a word, the amount of fertilizer application should be proper and under control.

Proper Fertilizer:
The availability of different nitrogen fertilizers in different environment is different. The nitrate fertilizer should not be applied in paddy fields or in rainy seasons, because nitrate could be easily lost by denitrification in the wet soil. Ammonium cannot be applied to the alkaline soil; otherwise, the ammonia evaporation would occur. Generally, the slow-release nitrogen fertilizer could enhance the availability of nitrogen.

Proper Application Method:
The method of applying nitrogen fertilizer will also affect the availability of it. The subsoil application, especially the subsoil application with the granulated fertilizer, is the most efficient. And, it is mainly ammonium fertilizer and urea. During the rainy seasons, topsoil application is the easiest to be washed. The concentration of nitrogen in the foliar dressing is low, so it can enhance the benefit of fertilization.

An Integrated Rational Fertilization:
To enhance the availability of nitrogen fertilizer, apart from proper timing, proper amount, proper position, and proper application methods, other fertilizers

should be applied properly, because there are mutually promoting and inhibiting potentials between different nutrients. For example, phosphorus deficiency affects plant growth and decreases the plant's ability to take up nitrogen. Thus, different kinds of nutrient application should be taken into consideration comprehensively, in order to apply fertilizer properly without overdose or deficiency of the other nutrients.

The Improvement of the Soil:

The soil's physical, chemical, and biological improvement will also affect the fertilizer availability. For example, the soil compaction, too high or too low pH value, and pest and disease are all the factors that could do harm to the uptake of nutrients.

The Usage of Material Control Method:

There are different ways to reduce the loss of nitrogen fertilizer. Adsorption materials such as porous minerals, inhibitors of nitrification and urease, etc. can be used to prevent the transformation of nitrogen. The inhibitors could not be over-applied; we need to pay attention to its side effects and apply it properly.

5.3 Nitrogen Transformation and Soil Microorganisms

In nature, organic nitrogen in the soil occupies 40% of the total biotic nitrogen, the benthic organic nitrogen in the ocean occupies 47%, and the rest about 13% belong to the living body and inorganic nitrogen. In the air, there is about 80% N_2. The organic matters in the soil occupy 98% of the nitrogen in the soil, and the proportion of inorganic nitrogen is small. The relationship between the nitrogen cycle (Figure 5-2) and soil microorganism is very close, and the nitrogen cycle's relationship with the nutrients is more vital.

The nitrogen transformations are illustrated as below:

Mineralization:

The residues of animals and plants will be decomposed into inorganic matters, and microorganisms play the main role in the decomposition. The outcome from the decomposition will be offered to the plant for nutrient uptake. The mineralization is the process of providing inorganic minerals and fertility of soil continuously. The more organic matters it contains, the more obvious mineralization will be. Therefore, the soil organic matter content is an important criterion of soil fertility. And, microorganisms for organic mineralization become an indispensable cleaner in nature.

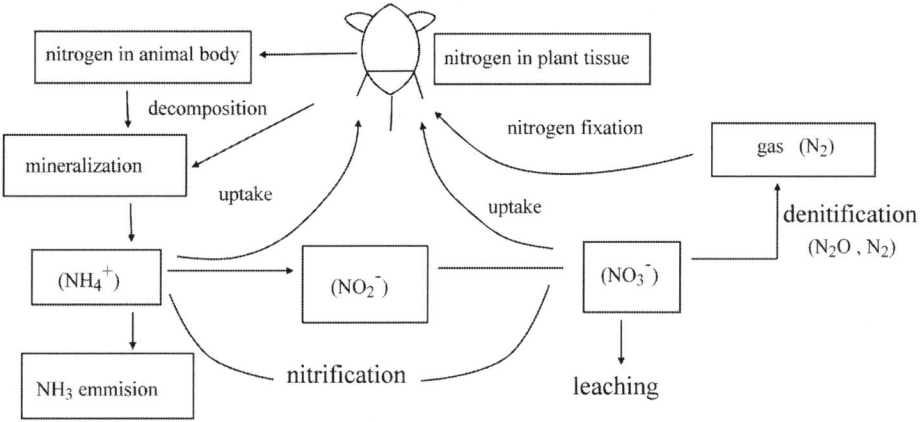

Figure 5-2. The Process of Nitrogen Cycle

N₂-Fixation:

There is a great deal of nitrogen gas (N₂) in the air, which cannot be utilized by the plants. Only some microorganisms have the function of N₂-fixation, which could be the source of soil nitrogen. The nitrogen in parent minerals of the soil is little. Among the biotic nitrogen in nature or the nitrogen fertilizer in the soil, N₂-fixing microorganisms play an important role. N₂-fixing microorganisms include the non-symbiosis, association, and symbiosis, which are all prokaryote (i.e., microorganism without cell nucleus). For example, bacteria, actinomycetes, and the blue-green algae, since a long time ago, have been arduously working in the soil. Calculating the amount of nitrogen fixation by unit area, the nitrogen fixation of symbiosis ranks the highest. Therefore, the utilization and development of microorganism with nitrogen fixation is very significant (Illustrated in details in Chapter 9).

Nitrification:

Sources of ammonium are transformed from organic nitrogen by mineralization, urea hydrolysis by urease, or ammonium fertilizers. The ammonium will be utilized by nitrifiers, and transformed into nitrite, and finally nitrate. This transformation is called nitrification. The plants use various nitrogen forms. Generally, upland crops are more prone to uptake nitrate. If ammonium or nitrite concentration accumulates too much, the plants might be poisoned; thus, microorganisms of nitrification could not be ignored. However, the nitrate is easily leached and lost, so, during the rainy seasons, the loss of nitrate is serious. The development of nitrification inhibitor is a compensation for this disadvantage of the nitrogen lost. Good aeration and sufficient oxygen are favorable conditions for the nitrifier to perform nitrification. When the oxygen is in shortage, the nitrification cannot take place efficiently.

Denitrification:

When the soil has poor drainage, denitrified microorganisms will transform nitrate (NO_3^-) into gaseous nitrogen (N_2O or N_2) and emits. This could be seen as a kind of waste from the utilization of nitrogen fertilizer. Thus, when urea is applied to paddy fields, some of the urea will be transformed to nitrate through the surface oxidizing zone. And, nitrate could go deep into the reduction zone, which would cause a waste of denitrification. Therefore, both urea and ammonium needs to be applied with deep fertilization; it is beneficial to the protection of nitrogen fertilizer, and can reduce the occurrence of nitrification. The big granulated fertilizer is also aimed at deep application. It is more apt to go deep into the reducing zone on the paddy field. Thus, decreasing nitrification and developing big granulated fertilizers with slow effect would be a win-win.

N Immobilization:

The N immobilization includes N fixation and the absorption of microorganisms. Ammonia and nitrate are taken up by microorganisms and are largely immobilized or made unavailable to plants. Some soils could fix ammonium to prevent the loss of nitrogen. When fresh organic fertilizer is applied to the soil, some microorganisms reproduce excessively; especially at the primary period, nitrogen is absorbed immensely and immobilized by microorganisms. And, if plant needs nitrogen from the soil, this microbial competition for nitrogen will trigger a deficiency of nitrogen, and the yellowing of leaf would occur in plant leaves. Therefore, organic fertilizers should be applied before the plant demands for nitrogen nutrients. For instance, in the orchard, organic fertilizer is always applied before spring, which is to avoid the nitrogen competition between plant and microorganism.

From the nitrogen transformation above, it is obvious that the relationship between soil microorganisms and nitrogen fertilizer is very close and cannot be separated. Managing well the soil, utilizing the ability of the soil microorganisms, reducing the waste of nitrogen fertilizer, and gaining the economic benefits all are the goals of the fine agriculture.

5.4 Features and Guidelines of Application for Phosphorus

5.4.1 The Relationship between the Phosphorus in Soil and Plants

Among the main nutrients in soil, the characteristics of phosphorus and nitrogen differ very much -- the transformation or loss of nitrogen is apt to occur, while phosphorus is not. Therefore, it is necessary to know the performance of phosphorus in soil in order to apply the phosphorus fertilizer properly and make the most of it.

Phosphate plays a very important role in plant nutrients; it is related to the biochemical reaction of plant energy. The phosphorylation is needed in many metabolisms of enzymes, and phosphate is the component of nucleic acids (biotic heredity matter), which is very important to the cell division and the meristem. Therefore, it is one of the indispensable elements for the plant growth and reproduction.

When the soil holds not enough phosphorus, symptoms of phosphorous deficiency in the plant become apparent, and the deficiency becomes more severe in the plant growth stage. The symptoms of the phosphorus deficiency are: some plants are apt to appear purplish red on the stem or leaves; some cannot be judged from the appearance, and plants analysis diagnosis and soil diagnosis should be used. The symptoms of phosphorus deficiency are that stems, roots, and leaves of the plant appear short and small, and there is red color on the leaves or dried leaves of the grass family. When the annual short-term plants are deficient in phosphorus, it will be too late for modification. If the fruit trees are deficient in phosphorus, the blossom and fruit setting will be seriously affected. Besides, phosphorus is also important for the tuber of potato and other tuber plants.

There will always be a large deal of phosphorus in the soil after the field is applied with excessive phosphorus fertilizers. The accumulated phosphorus is not easy to be dissolved. It is not easy for the plants to take up, either. If there is too much phosphorus in the soil, the plant growth will also be restrained, which often occurs during dry seasons. In the light soil, the growth period and mature period of the plants will be shortened.

5.4.2 The Forms and Transformation of Soil Phosphorus

The Forms of Soil Phosphorus:

The forms of soil phosphates can be divided into three main types:

- The organic phosphates in soil organic matters.
- The inorganic phosphates and phosphates fixed with calcium, magnesium, iron, aluminum, and clay.
- The organic and inorganic phosphates in the living organisms.

The organic phosphates in the organic matters will be decomposed by the soil microorganisms, and then it will be transformed into inorganic phosphates, which is called the mineralization of organic phosphates. The forms of the phosphates absorbed from soil by the plant are dihydrogen phosphate ion ($H_2PO_4^-$) and monohydrogen phosphate ion (HPO_4^{2-}), and it is easier for $H_2PO_4^-$ to be taken up than HPO_4^{2-}, and some small amounts of organic phosphates can also be taken up by plants. The proportion between $H_2PO_4^-$ and HPO_4^{2-} is affected by the value of soil pH. In the acidity, $H_2PO_4^-$ takes up more the proportion; if otherwise, HPO_4^{2-} takes up more.

The Movement of Phosphate in the Soil:

The movement of plant nutrients in soil is an important criterion for deciding the right way of application. The nutrients are taken up by plants in the rhizosphere; as a result, nutrients on the rhizosphere decrease gradually. And, nutrients will move from the rhizosphere to roots. The fastest is the mass flow with water; for example, the movement of nitrate belongs to mass flow. Another movement belongs to the diffusion from high concentration to low concentration. This diffusion of movement is very slow, and the movement of phosphate in soil is dependent on the diffusion. The ability to provide the root nutrient uptaken by the rhizosphere is weak. As a result, root interception is the main way to make phosphate absorbed, grow through the soil, and incidentally come into contact with nutrients. The coefficient of diffusion of phosphate in the wet soil is 1,000 ~ 2,000 times less than that of nitrogen. The phosphates are not apt to move from the topsoil to the subsoil, especially in the soil with much clay. The organic phosphates move more easily than the inorganic. And, the organic matters are beneficial to the movement of phosphates in the soil. The amount of organic phosphates varies according to the structure and features of soil. It holds about 15 ~ 18% of the total phosphates in the soil. The organic phosphates are especially high in peat soil; it holds 5 ~ 10% of the soil microorganisms.

5.4.3 The Types of Phosphorus Fertilizer

The sources and producing ways of phosphorus fertilizers are different, so the types of phosphorus fertilizer are varied. The solubility and availability of them also differ. The main source of phosphorus fertilizer is rock phosphates. It is a kind of mineral banked up from phosphates in the livings of the ancient geological age, and up to 80% of which contain phosphorite. The exploitation of rock phosphates is the same as that of petroleum; depletion may occur one day. In rock phosphates, the content of water-soluble phosphate is little. After adding different acids, different inorganic phosphorus fertilizers are produced. For example, the single superphosphate is the product after rock phosphate is treated with sulfuric acid. The rock phosphate after acid treatment will become water soluble phosphorus fertilizer, and it is easy to be fixed by the soil, particularly if applied in powder forms. And, it is better in granular form than in powder. Apart from a great deal of phosphoric acid produced from the rock phosphates, there still exists much calcium, which cannot be ignored either. Another source of phosphates is the byproducts from the iron industry -- the phosphorus-bearing slag. The slag contains matters such as phosphates, calcium, and silicate. It is an alkaline material, thus its effect would be like using lime, which can increase the pH of the soil. It is not recommended to be applied to alkaline soils; in contrast, it will be beneficial to acidic soils. Therefore, phosphorus-bearing slag can be the source of lime, phosphorus, calcium, and silicon fertilizers.

The water-soluble phosphorus fertilizer is applicable to all kinds of soil. The phosphates included in rock phosphate powder are mostly not water-soluble and could only be applied in acidic soils. Acidic soils are distributed widely; especially, the acidity is pretty high in red soils. If applying the rock phosphate powder directly, the costs for processing the phosphorus fertilizer could be saved. Moreover, advantages and disadvantages still need to be assessed.

Recently, the content of heavy metals (such as cadmium) in the rock phosphates is a bit too high. The quality of acids (sulfuric acid, phosphoric acid) added in the phosphorus fertilizer while producing could have an impact on the quality of phosphorus fertilizers. It is a problem that should be taken into account.

5.4.4 The Availability and the Fixation of Phosphorus Fertilizers

The availability of applying the phosphorus fertilizer to the soil is closely related to the solubility and exchange of phosphates. Generally, phosphates that can be taken up and utilized by plants is called "available phosphates," while phosphates adsorbed or fixed with other matters or elements is called "unavailable phosphates." In fact, the available and unavailable phosphates are not absolute phenomena. They are exchanged occasionally, which means, a part of available phosphates would be exchanged with some other parts of unavailable phosphates. Or, when the available phosphates are absorbed by the plants, some unavailable phosphates would be exchanged. Therefore, the phosphates in soil can be divided into three types according to the availability:

- Available phosphates in the aqueous solution.
- Able-to-be-exchanged or unstable phosphates.
- Unable-to-be-exchanged or stable phosphates.

The phenomena of exchange and liberation among these three types are listed as follows:

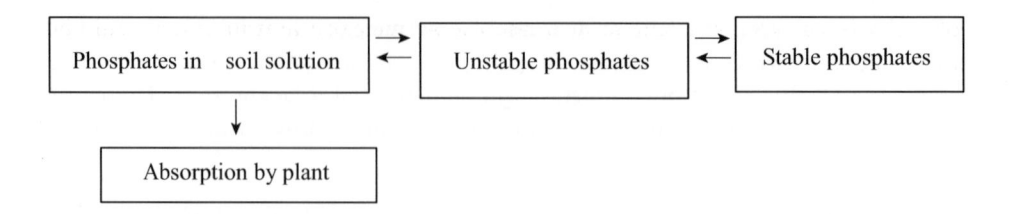

Phosphates in soil solution ⇄ Unstable phosphates ⇄ Stable phosphates

↓

Absorption by plant

When the phosphorus fertilizer is applied to the soil, some of the soluble phosphates would go into soil solution; some will be fixed as unstable or stable exchangeable phosphates. Generally, after the phosphorus fertilizer is applied to the soil for annual plants, less than 20% phosphates could be recovered by plant, commonly about 10%, which means that the annual plants could only take up 10 dollars' phosphorus fertilizer if 100 dollars' phosphorus fertilizer is applied. The phosphates that are not taken up will be fixed by the soil, including soil clay and mineral surface, ions and oxide hydroxide of the calcium, iron, aluminum, some non-crystalline compounds, etc. The fixation includes sorption and precipitation, it makes the soluble form turn into insoluble, valid into unavailable, and weaken the effects of the applied phosphorus fertilizer. From the above description, we can infer that the soil's ability to fix phosphates is closely related to the types of the soil minerals, the amount of clay, the activity and content of alkalinity and acidity, calcium, iron, and aluminum, the content of water in the soil, and the amount of the organic matters. If the soil acidity is high, it will definitely enhance the fixation of phosphates. Therefore, how to apply phosphorus fertilizer effectively and enhance the availability of phosphorus fertilizer is the important topic of plants production and producing quality improvement.

5.4.5 Guidelines of Enhancing the Availability and Function of Phosphorus Fertilizers

Knowing the actions of phosphates in soil would be beneficial to enhance the availability and fertilizer function of phosphates. To enhance the fertilizer efficiency, the phosphates should be reduced to be fixed, and the solubility and nutrient uptake of phosphates should be improved. The ways of agricultural management are listed below:

Adjust the Alkalinity or Acidity of the Soil:

The soil of too acidic (pH less than 5.0) or too alkaline (pH over 7.5) contains too much activity of iron and aluminum (most in acidic soils) or of calcium (most in alkaline soils), which has a strong ability to fix phosphates. Thus acidic soil should be applied with lime materials, or alkaline fertilizer to improve the acidity. And the alkaline soil should be applied with the acid-forming fertilizer and sulfuric materials to improve the soil, in order to decrease the activity of the iron, aluminum and calcium that fixes the phosphate.

Increase Mycorrhizal Fungi and Other Groups which are Beneficial to Soil Microorganisms:

Recently, with the development of soil microbial research, finding mycorrhizal fungi, beneficial microorganisms, and phosphate-solubilizing microorganisms of the rhizosphere is very important for the plants to take up the phosphates. As for the protection of soil microorganisms, it benefits not only the plant's resistance to disease, but also the plant's uptake of nutrients. When the soil microorganism is applied on poor soil, the beneficial microorganisms could enhance the uptake of phosphates and even transform insoluble phosphates into soluble phosphates, thus enhancing the availability of phosphates and the function of increasing production.

Apply Organic Fertilizers:

There are five advantages of applying organic fertilizers:

- The direct effect is that it could provide phosphates after the organic matters are decomposed.

- The organic acid produced by the decomposition of organic matters could be fixed with the phosphates which is combined with calcium, iron, and aluminum; and, the liberation of the soluble phosphates can therefore be easily absorbed by plant.

- Carbon dioxide will be produced after the decomposition of organic matters, and some form carbonic acids, which could also enhance the solubilization of phosphates fixed with calcium, iron, and magnesium.

- Organic matters will provide nutrients for the propagation of the beneficial microorganism, and enhance the growth of the roots, which will all improve the root uptake of phosphates.

- The humus of organic matters could wrap up soil colloid oxides and form protection on the surface, thus reducing fixed phosphates. Therefore, applying the phosphorus fertilizer with organic fertilizer has so many advantages.

Apply Proper Fertilizer:

After the fertilizer is taken up by plants, there could be mutual conflicts and supplements between all the elements. How to take the advantage of mutual supplements and avoid mutual conflicts is also one of the guidelines for fertilizer application.

According to records as early in 1929, silicon fertilizer (such as soluble silicate and slag) has been proved to be beneficial to enhance the production of plants. The slag used in the paddy field can help improve the pH of soil, and is beneficial to acidic soils. It contains calcium silicate, phosphates, and some other minerals, which could increase the amount of silicon in the grass family, strengthen the hardness of plants, foster the resistance to pests and diseases, and enhance the availability of phosphates in soil, which had all been reported in the past. Other ion such as NH_4^+ is beneficial to maize for its phosphates uptake.

Proper Ways of Application:

The phosphorus fertilizer is not easily moved in soil, so plant's absorption of the phosphates mainly depends on the roots touching or interception. The amount of phosphates that are diffused to roots in the soil is little; therefore, the phosphorus fertilizer should be applied deep root fertilization. Top-dressing is effective to the short-term plants or plants with shallow roots. However, the plants with deep roots, such as fruit trees, should be applied with phosphorus fertilizers to the subsoil, because the phosphate's ability to move from the surface into deep soil is weaker than that of nitrogen.

The Rotation with Paddy Field or Flooding:

The availability of phosphate in the paddy field is better than in the dry field. The reason is that, under the condition of paddy field or flooding, the solubility of fixed phosphates increases, and organic matters and organic acids are not easy to be decomposed in the paddy field, which could enhance the availability of phosphates.

Others:

Much management to the soil can strengthen the plant's root growth. And, with the increase of the uptake area, the phosphates will also be taken up more easily. The plant's ability to utilize and take up the phosphorus fertilizer varies according to the varieties of crops. Therefore, picking a good variety is also vital to enhance the availability of phosphorus fertilizer. A lot of weeds can survive the soil of low phosphates content; this ability should be valued.

5.5 Features and Guidelines of Application for Monopotassium and Monocalcium Phosphates

5.5.1 The Features of Monopotassium and Monocalcium Phosphates

Phosphate has many salt types, such as potassium phosphate, calcium phosphate, ammonium phosphate, etc., because phosphoric acid exists in forms of ions of negative monovalence (dihydrogen phosphate anion), negative bivalence (hydrogen phosphate anion), and negative trivalence (orthophosphate anion). Therefore, potassium salts include monopotassium, dipotassium, and tripotassium phosphates, and calcium salts include monocalcium, dicalcium, and tricalcium phosphates.

The monopotassium phosphate and monocalcium phosphate are usually applied in agriculture because the dihydrogen phosphate anion ($H_2PO_4^-$) is soluble in water and easily absorbed by plant. The ability of plant absorption of the negative divalence and trivalence of phosphoric acid is less than that of monovalence ones. Besides, the solubility of monopotassium phosphate and monocalcium phosphate are higher than that of dibasic and tribasic potassium and calcium.

Monopotassium phosphate is colorless crystal and acidic. Monocalcium phosphate is colorless crystal powder. Both of them dissolve in water. Single superphosphate is made by the rock phosphate mixed with sulfuric acid. The major components of single superphosphate are monocalcium phosphate and calcium sulfate. Triple superphosphate is rock phosphate treated by phosphoric acid rather than by sulfuric acid. The major components of triple superphosphate also are monocalcium phosphate; therefore, monocalcium phosphate is a common fertilizer.

5.5.2 The Function of Monopotassium and Monocalcium Phosphates

The Use of Fertilizer:

Monopotassium phosphate and monocalcium phosphate are fertilizers; both of then contain two types of nutrient elements. They are phosphorus fertilizer as well as potassium or calcium fertilizer. They can be applied to soil or on foliar dressing as compensation. The phosphorus fertilizer could enhance the plants maturity, and potassium fertilizer could restrain the overdose of nitrogen fertilizer to achieve a balance. Calcium fertilizer could adjust the osmosis of cells.

The amount of monopotassium phosphate or monocalcium phosphate applied varies according to different soils and species of crop and seasons. The dilution of monopotassium phosphate and monocalcium phosphate also can be applied on foliar dressing.

Special Usage:

Control the Growth of New Branches and Enhance Maturity:
When fruit tree grows, it consumes energy. After leaves are mature, it is then favorable for the storage of assimilated photosynthates. When more photosynthates are stored, more flowers and fruits are produced. If the branches and new shoots are excessively grown, it may lead to flower and fruit drops, poor flowering, and poor quality. Monopotassium phosphate and monocalcium phosphate could restrain the growth vigor of branches and

promote the maturity; it could also balance the overdose of nitrogen, which is reported to be applied to grapes, pears, lemons, etc.

Enhance the Leaf Drop and the Emerge of New Branches to Make Earlier Blossom:
Some crops with flowers bloom after the leaf drop; thus, making leaves drop early is a method to promote the growth of new branches. There are many deciduous plants that could be used. According to the reports, high concentration of monopotassium phosphate and monocalcium phosphate could enhance the leaf drop and emerge of new branches, thus make blossom take place earlier. For example, applying monopotassium phosphate and monocalcium phosphate to guava is able to make early blossom. Many fertilizers with over-application applied by foliar dressing would also cause leaf abscission. For example, using high concentration of nitrogen fertilizer as defoliant agent for the goal of promoting maturity is not appropriate, because nitrogen fertilizer get into the soil after leaves drop. Plant would absorb too much nitrogen fertilizer and then inhibit flowering or cause fruit drop. Therefore, it is no problem to apply monopotassium phosphate and monocalcium phosphate as defoliant agent.

5.5.3 Guidelines of the Application of Monopotassium and Monocalcium Phosphates

Guidelines of Application to the Soil:

Monopotassium phosphate and monocalcium phosphate are applied to the soil as fertilizers. Furrow application and broadcast application could be adopted. When it is aimed at phosphorus fertilizer, the furrow application might be the best choice. Deep placement might be better, when applied to the soil of fruit trees as liquid fertilizer. Overdose of the concentration of liquid application should be avoided, and the cost of application should also be considered. The fertilizers with high quality and high cost are applied at the time of high economical benefit. Sometimes, applying fertilizer with high quality relates to the crop quality and production. If the cost of soil fertilization is too high, foliar dressing or other fertilizers can be used as alternatives.

Foliar Dressing:

When the certain stage of plant needs more phosphorus, potassium, and calcium, foliar dressing is an urgent remedy for the deficiency. It could also balance the overdose of nitrogen immediately. The effect of phosphorus fertilizer applied in the acidic or alkaline soil is more obvious.

Control the Growth Vigor of New Buds:

Methods for controlling the growth vigor are used to the excessively growing buds of fruit tree for reducing the speed of growth and enhancing the maturity of leaves. In order to increase the growth of fruits and reduce photosynthates partitioning to branches and leaves. Please attention to the applying time. For example, if new buds of pear appear before flower blossom or excessive growth after new budding, monopotassium phosphate and monocalcium phosphate can be applied to inhibit new budding and prevent the competition between fruits and leaves. The number of hairs, the length of internode, or the big seize leaf of new buds of fruit tree is an index of growth which closely relates to nitrogen fertilizer and excessive growth. If fruit tree shows excessive growth vigor and too much nitrogen after diagnosis, the monopotassium phosphate and monocalcium phosphate are able to solve this unbalance problem.

Enhance the Leaf Drop and Adjust the Fruiting Period:

Monopotassium phosphate with high concentration can enhance leaf drop of some fruit trees. Appropriate leaf drop to reduce nutrient consumption was beneficial to the growth of new budding, and blossom in earlier was achieved in guava tests.

5.5.4 Guidelines for Application of Monopotassium and Monocalcium Phosphates

The Concentration and pH should be Noted in the Application of Foliar Dressing:

Different plants might demand for different concentration and pH, when applying the foliar dressing. The new leaves differ from the old leaves. Generally, the tolerance of plants of the grass family is higher than that of the fruit trees, grapes, vegetables, and flowers. Each concentration may fit only some kind of crops, not all crops. The solution of foliar dressing which is too acidic (pH lower then 5) or too alkaline (pH higher than 8) will both cause injury on the foliage. It is necessary to test your own crops and soils in practice, and find out adequate methods in different climates.

It is not Recommended to be Mixed with Pesticides:

Consult the reference book before applying the fertilizer is necessary, because many injuries by carelessness happen while spraying. When mixing pesticides with fertilizers, we should pay attention to the pH and osmotic pressure. Too acidic, too alkaline, or too high of the osmotic pressure (too much solutes) might cause chemical injury or fertilizer injury. We recommend concerning the negative effects of mixing fertilizers and pesticides.

The Amounts of Phosphorus, Potassium, and Calcium in the Soil should be Considered on the Foliar Dressing:

When the foliage is sprayed with monopotassium phosphate and monocalcium phosphate, after the rain and drip washing, the fertilizer that has not been taken up will go into the soil, and increase the amount of phosphates, potassium, and calcium. Therefore, the overdose of these elements could cause the unbalance of the uptake of nutrients. Too much phosphate inhibits the absorption of trace elements; too much potassium or calcium affects the absorption of magnesium by the plant.

Match up the Demands of Plants and Adjust the Fruiting Period:

If it is applied to enhance the leaf drop and promote the growth of new shoots, then the seasons and the tendency of growth vigor should be matched up. If it is applied to restrain the growth of new shoots, then the demands of crop should be matched up. And, avoid applying during the week of blossom for fear the results with flower drop or small fruits.

Monopotassium and Monocalcium Phosphates and Microelements:

The nutrients in monopotassium and monocalcium phosphates are not microelements. Microelements include boron, copper, iron, zinc, manganese, and chlorine. If necessary, they can be mixed mutually. For example, when adjusting the new shoots of fruit tree, the boron could be applied to plants which are deficient in boron, and the effect of promoting fruits could be achieved.

Cooperate with the Improvement of the Soil:

The phosphate in monopotassium and monocalcium phosphates is negative monobasic, which is the most effective form of phosphorus for plant absorption. But, when it is applied to the alkaline soil, it will become dibasic or tribasic, thus decreasing the function. Therefore, when applying even the best fertilizer, the alkaline soil (> pH 8.0) or the acidic soil (< pH 5.0) should be neutralized. Otherwise, the phosphates might be fixed or transformed to low available phosphates, which can be seen as a loss of the fertilizer. In a word, the concept of macroscopic function and mutual supplement should be set up.

5.6 Features and Guidelines of Application for Rock Phosphorus Fertilizers

Rock phosphate powder is the natural source of phosphorus fertilizer, and it is the raw materials for producing single superphosphate and other phosphorous fertilizers, generally, compounded by adding acidic materials to the rock phosphate powder. And, single superphosphate is compounded by adding sulfuric acid. Since a long time ago rock phosphate powder has been applied to the soil in many countries. Because there are many acidic soils in tropical, subtropical, and temperate area, it is beneficial for the application of rock phosphate powder. Therefore, this chapter will introduce rock phosphate and guidelines of application for reference.

5.6.1 The Types of Rock Phosphates

It could be divided according to the formation and composition:

According to the Occurrence of Minerals:

It could be divided into sedimentary, metamorphic, and igneous rock.

Sedimentary Rock:
It is the most important of these three types, and the most ecological for exploitation. It appears small crystallized and loose-structured, with large surface area. Crystalline phosphorite is one example, which also contains some other matters (silicate and carbonate soil). It could be applied to the soil directly.

Metamorphic Rock:
It is not easy to be dissolved physically and chemically, and appears crystalline.

Igneous Rock:
Also not easy to be dissolved as metamorphic rock, and appears crystalline. But, its composition of phosphates is high, could be used in strong acids treatment, and produce single superphosphate and triple phosphate, etc.

According to the General Formula of Minerals:

Iron and Aluminum Phosphate:
Example: Wavellite [$Al3(PO4)2(OH3) \cdot 5H2O$], amblygonite ($ALPO4) \cdot 2H2O$, and barrandite ($FePO4 \cdot 2H2O$) are more important in economic.

Calcium-Iron-Aluminum Phosphate:
Example: Crandallite rock [$CaAl_3(PO_4)_2 \cdot H_2O$] and millisite [$(Na,K)CaAl_6[PO_4]_4(OH)_9 \cdot 3H_2O$].

Calcium Phosphate:
Example: Apatite [$Ca_{10-a-b}Na_aMg_b(PO_4)_{6-x}XF_{2+y}$]. The above two rock phosphates are not easy to form in weathering condition. The most economical value is deposited apatite.

From the classification above we can see that there are many types of rock phosphates, but it will be affected by different environments in the nature. And, the quality or the amount of soluble phosphates varies greatly. For example, the guano (island rock phosphate) is the excrement (feces and urine) of sea birds, seals, and cave dwelling bats for a hundred year's inhabitation. This accumulation of the guano with high availability of phosphorus appears. Because the solubility of phosphates in different kinds of rock phosphates is different, not all rock phosphates are applicable directly.

5.6.2 The Function of Rock Phosphates in the Soil

The rock phosphate could be applied directly to the soil; it is dissolved by the conditions of soil and then release the phosphorus, and the factors that would affect the effect of its application are listed as follows:

Soil pH Value:

The lower the pH value is, the easier rock phosphates would be dissolved. Therefore, it is better to apply the rock phosphate in acidic soils.

The Concentration of Calcium Activity:

The calcium activity of soil solution is low, which will increase the solubility of rock phosphates.

The Concentration of Phosphate Negative Ions in Soil Solution:

The lower the concentration of phosphate negative ions in soil solution is, the more beneficial the solubility of rock phosphates would be.

The Amount of Organic Matters and Phosphate-Solubilizing Microorganism in the Soil:

The more organic matters content are in the soil, the better it is for the solubility of rock phosphate; and the organic acids produced after the organic matters are decomposed is beneficial to the solubility of rock phosphates. There are many types of phosphate-solubilizing microorganisms, including bacteria, actinomyces, and fungi, and their ability to dissolve the rock phosphates also varies.

The Carbonic Ions in Rock Phosphates:

Under certain pH the solubility of rock phosphate could be enhanced by increasing carbonic ions.

5.6.3 The Features of Rock Phosphates

The major constituent of rock phosphates is phosphorus nutrient, and some other rock phosphates contain high calcium, magnesium, and iron as accessory ingredients. The function of it is the same as phosphorus fertilizers, and some has the function of calcium and magnesium, which is the most suitable to be applied to strong acidic soils. Other nitrogen and potassium fertilizers should also be incorporated.

Rock phosphate belongs to the natural phosphorus fertilizer with longer and slower release effect, lasting for 1 ~ 2 years. When it is applied to the acidic soil in hillside orchards or grass fields, this phosphorus fertilizer will be released slowly. If applied to the short term plants, it is better to apply in advance. But the effect varies according to different soils.

Rock phosphates are natural minerals, if mixed with organic fertilizer, phosphate-solubilizing microorganism, or mycorrhiza, the effect would be better.

Generally, commercial rock phosphates are calcium phosphates salts with low alkalinity. And, it is also beneficial to improve acidic soils.

5.6.4 The Quality of Agricultural Rock Phosphates

The rock phosphate should be grinded into powders, so that it could be easily applied into the soil, and the quality should take the following factors into consideration:

The Amount of Phosphates and its Availability:

The amount of phosphates in rock phosphate powder varies according to the different types of rocks. Generally, the levels of phosphorus fertilizer can be divided by water solubility, ammonium citrate solubility, and ammonium citrate insolubility. The more the phosphates of water solubility and ammonium citrate solubility, the better the effect will be.

The Size of Particles:

When the rock phosphate is grinded into powders, the size of particles will affect the surface area that interacts with the soil, and releases phosphates. The smaller particle it is, the better the effect, and the application rate could be reduced. But, the size of particles is not the only criterion of quality; the availability of phosphates should also be taken into account; if particle is small but the availability is low, then it is not the best rock phosphate.

The Types and Amount of Impurities:

The rock phosphate is natural rock, and there are always some impurities. Some of the impurities are agricultural beneficial ones, such as calcium, carbonate, magnesium, etc. However, some are detrimental, such as fluorine and poisonous heavy metals; the more it contains, the worse the quality is.

5.6.5 The Guidelines of Application for Rock Phosphates

The Plants to be Applied to:

Rock phosphate is applicable to all upland and lowland plants, including all kinds of agronomy, horticulture, grass, forests, etc. Especially, the long-term effect of rock phosphate could be achieved if mixed with sands before the planting of grass for the maintenance of the golf course.

The Period of Application:

Rock phosphate could be applied anytime. The best time is applied on basal fertilizer by plowing into the soil.

The Soil to be Applied to:

The acidic soil is preferable, and the alkaline soil is not recommended unless the availability of phosphates is high or it has been acidic treated. And, it could perform well if the alkaline soil is mixed with the organic fertilizer. Because the effect of rock phosphate differs according to different soils, the observation of the effect should not only be one time. Generally, there will be good effect, but also some residuals.

The Amount of Application:

The amount of application of rock phosphate for the fertility of soil, the plant demand, the source and the size of particles rock phosphates are different. In principle, the soil that demands phosphorus fertilizer and the long-term perennial plants can be applied with a little more, generally 50 kg per 0.1 ha, or the same amount as single superphosphate, or one time more than that, which depends on the plants and soils.

The Method of Application:

- For the farming field, it could be evenly scattered on the soil and even plowed into soil, and the effect might be better if there is irrigation or rain after application.
- The effect could last for longer time, if the materials for growing seedlings are plowed and mixed well with soil.
- The effect of rock phosphate could be improved by mixing up some organic fertilizer with any proportion.

5.6.6 The Additives Affect the Solubility of Rock Phosphates

If rock phosphate is incorporated with some proper additives, the effect will be enhanced; otherwise, the effect might be reduced. The influences of the additives are listed as the following:

Influenced by the Forms of Nitrogen Fertilizer:

Acid-producing nitrogen fertilizers could increase the solubility of rock phosphate and enhance the availability of phosphorus fertilizers. The effect is obvious especially when it is applied with the formulation of phosphorus minerals with nitrogen.

Influenced by Sulfur Powder:

The sulfur powders mixing with the rock phosphate powders is beneficial to the solubility of rock phosphate, because sulfur powder will be oxidized by soil microorganisms and transformed into sulfuric acid, which thus speed up the solubility of rock phosphate powder. But, pay attention that this is used only in the alkaline soil, not in the strong acidic soil.

Influenced by Organic Fertilizers and Green Manures:

The organic fertilizer and green manure produce organic acids during the decomposition, which could enhance the solubility of rock phosphate. This is used to all kinds of soil.

Influenced by Phosphate-Solubilizing Microorganism and Mycorrhizal Fungi:

In recent years, there are many researches indicating that adding microorganisms could enhance the utilization of rock phosphate powder. This is used to all kinds of soil.

5.7 Features and Guidelines of Application for Potassium and Magnesium

5.7.1 The Forms and Uptake of Potassium and Magnesium in Soil

In the nature, potassium and magnesium cannot exist as a single element, but only in compounds with other elements. The potassium and magnesium in soil are from the parent rock or the alluviation, and exist in all kinds of salts and minerals. The solubility is different accordingly. And, it can be divided into soluble and insoluble types. The soluble salts are such as carbonate, sulfate, chloride, etc. And, the insoluble type always exists in all kinds of minerals. Potassium and magnesium will be dissolved into cations (K^+, Mg^{2+}) in the soil water solution, or chelated by organic acids. The organic matters chelate with these cations which can be taken up by plants.

The soluble part and the part adsorbed on the exchanging compounds of the potassium and magnesium in the soil can be utilized by plants. Apart from the manual application, the amount of potassium and magnesium in soil differ according to the source of its parent matters, the amount of leaching, and the uptake by plants.

Generally, the content of potassium and magnesium in the sandy soil is very low, unless the sands contain the minerals of potassium and magnesium. The potassium and magnesium in the sandy soil is apt to move with water, and are easy to be washed away. Generally, the potassium in soil moves more slowly than nitrogen, but faster than phosphates.

When the potassium and magnesium fertilizers are applied to the soil, they will be dissolved and go into the soil water solution. And then, most of the potassium and magnesium will be absorbed directly on the exchange positions of soil. This exchange form is easy to be utilized and taken up by plants. The soil capacity of holding this fertility is related to the amount of the materials for the exchange. So, the cation exchange capacity (CEC) of soil is always seen as a criterion of the capacity of holding these cation fertilizers. Some potassium can be fixed on the clay mineral, and the availability will be reduced.

Plant absorbing a great amount of potassium and magnesium will exude H^+ and organic acids. The exchange materials of soil will be occupied by the H^+, which could make the soil acidity. The application of potassium and magnesium could take place of H^+ again.

There are competitions between the cations of the plant nutrients. For example, the excessive uptake of potassium will reduce the uptake of magnesium and calcium. As a result, the deficiency of magnesium is the most obvious when the potassium is over-applied. Therefore, the balance of potassium and magnesium in the soil is very important.

5.7.2 The Influence of Potassium and Magnesium on Plants

The requirement of plant for potassium is comparatively high, and the accumulated amount of the potassium and magnesium in the plants are different according to different plants: the potassium is consumed fast after the soil is planted with crops, and the perennial plant is the most obvious. Besides, the spinach and celery also have the tendency. Many people regard the content of potassium or the potassium/sodium proportion as a criterion of wholesome food.

Potassium is an important nutrient for the plant growth and disease resistance. The potassium is related to photosynthesis, protein synthesis, conduction, and transpiration of plants. And, the ability to strengthen plant disease resistance has been regarded as a very important function. Although the plant's demand for magnesium is lower than potassium, the soil will always has magnesium deficiency under the wet and raining environment. Magnesium is a component of the plant chlorophyll, and is closely related to the metabolism of phosphates, respiration and nitrogen fixation, etc.

Potassium and magnesium is movable in plants; therefore, the symptom of deficiency will appear on the old leave at the bottom of the plants. The symptoms of potassium deficiency are poor root development, weak stem, lodging, poor fruit production, poor disease resistance, etc. The symptoms of magnesium deficiency are that old leaves turn yellowish-brown but the veins remain green. The unbalance between calcium and magnesium will aggravate the symptoms for the deficiency of magnesium. And, the overdose of potassium and ammonium fertilizer will also cause such deficiency of magnesium.

5.7.3 Types of Potassium and Magnesium Fertilizers

The main sources of potassium and magnesium fertilizers are minerals or salts. According to different water solubility of fertilizers, components and the availability of fertilizers are different. Based on the difference of solubility, the fertilizer could be divided into fast release and slow release fertilizers. The fast release fertilizer could be dissolved into the soil water solution at a fast speed; the slow release fertilizer is the opposite. The slow release fertilizer is generally the powder of mineral rocks or some artificial fertilizers.

The Potassium Fertilizer:

Potassium fertilizers are made from inorganic and organic matters. And the inorganic potassium fertilizer is more common. Even the plant ashes could provide potassium. There are two types of potassium fertilizer: the refined potassium fertilizer and the mineral powder fertilizer. The refined potassium fertilizer includes potassium chloride, potassium sulfate, potassium magnesium sulfate, potassium nitrate, potassium carbonate, etc. The potassium minerals include langbeinite, sylvite, brines, etc.

The Magnesium Fertilizer:

Magnesium fertilizer is such as the potassium fertilizer, mainly from minerals and refined manufacturing. The refined ones include: magnesium sulfate, potassium magnesium sulfate, etc. The mineral powders include dolomite, magnesite, talcum, serpentine, and some limestones. There are some calcium magnesium fertilizers in the industrial byproducts.

5.7.4 Guidelines of Application of Potassium and Magnesium Fertilizers

Method and Amount of Application:

The method for applying potassium and magnesium fertilizers varies according to the different soil, plants, and cultivation. Furrow application and broadcast application could be used. Furrow application could provide the nutrient of higher concentration, which will speed up the growth of seedlings. But pay attention, the fertilizer could not be too close to seeds or roots in order to avoid fertilizer injury. The broadcast application is a convenient method. Pay attention, some soils might fix a great deal of potassium, thus broadcast application could reduce the availability. The effect would be better if applied by the combination of row and broadcast application. The amount of application could depend on the experience in the past years or the amount recommended, but soil and plant analysis should be conducted every 3 ~ 4 years to test, whether there is overdose or unbalance of nutrients.

Choosing the Fertilizer:

According to reports, choosing potassium fertilizer is mainly based on the plant tolerance and sensitivity to the chloride. A recent research points out that chloride ion is harmful to the woody plants (such as citrus and grapes). Choosing magnesium fertilizer depends on whether the soil is applied with limes. As for the limes-demanding soil (acidic soil), choose the calcic fertilizer with magnesium; as for the non-lime demanding soil (neutral or alkaline soil), choose magnesium fertilizer with acidic materials (magnesium sulfate) or magnesium fertilizer without limes. Besides, the acid-forming fertilizer could not be applied in the acidic soil for a long period for fear that the soil becomes acidification.

Overdose of Potassium and Ammonium Fertilizer is not Recommended:

The competition of cations discussed above could lead to deficiency of magnesium if the potassium or ammonium is applied excessively to soils of magnesium deficiency. Thus, the concept of balanced application should be adopted.

Application on the Foliage:

When the plants cannot take up sufficient fertilizer from the soil, the fertilizer on the foliage could reach quickly and make compensation. The foliar dressing could be used, if the fertilizer could be soluble or dispersed in the water, and taken up by the foliage. The magnesium is often used for foliage fertilizer. The foliar dressing is not only a method to make up the nutrients, but also could force the foliage to take up a great amount of fertilizer, so that the production and quality could be improved.

Incorporation with Organic Fertilizer:

The solubility of mineral fertilizer is limited. When organic matters are decomposed, they will release a large amount of carbonates, nitrate, and some organic acids, which is beneficial to the effect of dissolving the potassium and magnesium by soil minerals. And it is helpful to the solubility of minerals and magnesium.

5.8 Features and Guidelines of Application for Calcium and Sulfur

Calcium and sulfur are the microelements of plants nutrients; it does not mean they play the less important role in the plant growth, but that the plant demand for it is less than nitrogen, phosphorus, and potassium. As for the nutrient needed by plants, all kinds of elements are indispensable.

The content of calcium in plants varies according to the types of plants. Citrus and legumes contain the most calcium, the grass family and tea lower. The plant's demand for calcium also differs: Legumes and *Brassica* needs more, but the rhododendron needs less, which could be planted in the acidic soil which the availability of calcium is low.

The content of sulfur in residues, straw, and stems of plants is similar or two times as much as the content of phosphorus. Some plants such as onion, cabbage, and clover contain high amount of sulfur. Most seeds contain sulfur with half of the amount of phosphorus.

The calcium is taken up in cations (Ca^{2+}), and sulfur in negative (SO_4^{2-}); generally, the agricultural calcium is calcic alkaline matters, but matters with sulfur are always acidic. Both of them play an important role in adjusting the pH of soil. And the availability of calcium and sulfur is related to the pH of soil. But, the content of calcium and sulfur in the soil is also related with the amount of plant uptake. The higher the content is, the more the plants absorb.

5.8.1 The Function of Calcium and Sulfur on Plants

The Function of Calcium:
- Calcium is one of components of the plant cell wall, which could strengthen the plant for enhancing the growth of roots and leaves.
- Calcium could help reduce the content of nitrate in the plants, activate some enzyme systems, and neutralize organic acids of plants.
- Calcium is the indispensable element for seed development of peanuts and the growth of root nodule of legumes.
- Many matters of calcium could reduce the acidity of soil, and affect the production indirectly; they can also reduce the solubility and poison of manganese, copper, and aluminum.
- Calcium could enhance the environment of root growth, the activity of microorganism, the availability of molybdenum, and the uptake of other nutrients.

The Function of Sulfur:
- Sulfur is the ingredient of some amino acids, which is indispensable for the synthesis of protein.

- Sulfur could help with the synthesis of enzymes and vitamins.
- Sulfur is a necessity for chlorophyll, even though it is not a component of it.
- Sulfur could enhance the formation of the root nodule of the legumes and the production.
- The materials with sulfur often appear acidic, which could neutralize alkaline soil and indirectly enhance the availability of other elements such as iron, copper, zinc, etc.

5.8.2 The Symptoms of Calcium and Sulfur Deficiency

The Symptoms of Calcium Deficiency (Occurred More Frequently in the Acidic Soil):

The most common symptom of calcium deficiency is poor plant development. More seriously, the growing point withers, roots turn deficiency and rotten. Because calcium has difficulty moving in the plants, when then symptoms of calcium deficiency appear, the new tissues would present at first, and the new tissues would need pectinate to synthesize the cell wall. Therefore, the deficiency of calcium would lead to the withering of growing point and new leaves. The deficiency of calcium often occurs on peanuts. The other plants in the field are not easy to be observed; therefore, in the soil with strong acidic, the plants will be harmed by acid at first.

The Symptoms of Sulfur Deficiency (Occurred More Frequently in the Alkaline Soil):

The symptom of sulfur deficiency is that new leaves turn light green. Generally, the whole plant can bear the light greening and will not die of it, but a large amount of carbohydrate and nitrate will be accumulated in plants. More seriously, the leaves will be shrinking and wither to death. The deficiency of sulfur always happens in poor organic soils or wet fields.

5.8.3 The Activity of Calcium and Sulfur in the Soil

Calcium in the Soil:

The content of calcium in the soil differs a lot. The lowest is less than 0.1%, while the highest could be 25%. Generally, the content of calcium in the clay soil is higher than in the sandy soil. Calcium is the main cation in soil, and exists as positive bivalence. The same as other cations, calcium will be affected by the exchange of cations. The calcium could be adsorbed on the clay minerals and the negative valence of the organic matter surface. Its ability to move in soil is slower than nitrogen and potassium, but faster than phosphates. Therefore, the content of calcium fertilizer leached by the rain and irrigation is lower than that of nitrogen and potassium fertilizer. The movement between elements is comparative, but not absolute. In fact, the movement of nutrients in the soil is related to the moisture, the texture, the organic matter content, and the parent matters of soil.

Sulfur in the Soil:

The inorganic sulfur exists as the form of SO_4^{2-}. There is considerable content of sulfur in the organic matters in the soil, which is the source for the plants to take up sulfur. Therefore, the organic matter content and the solubility of sulfur affect the availability of sulfur.

The SO_4^{2-} presents negative valence, which exists in the soil solution. Therefore, the most of SO_4^{2-} is not easy to be adsorbed by the clay and the organic matters, thus it will move with the water in soil. So, it is easily lost by leaching. This is the reason why the content of sulfur in topsoil is less than that in the subsoil.

The sulfur could also derive from the sulfur dioxide in the air, which falls with the rain, or derive from the fertilizer with sulfate and pesticides. The content of sulfur in different districts is different. Generally, there is much sulfur dioxide in the downtown or industrial areas. Besides, single superphosphate also contains sulfur about 1.9%. Therefore, the deficiency of sulfur is rare.

5.8.4 The Sources and Types of Calcium and Sulfur Fertilizers

The Calcium Dertilizers:

Most of the calcium fertilizer comes from lime materials, such as limestone, dolomite, gypsum, and phosphorite, etc., most of which could only be utilized after grinding and heating. Because the ways of manufacturing and the parent matters are different, the availability of calcium is different. The lime materials appear alkaline, which are great materials for neutralizing the acidic soil. The ability of neutralizing differs according to different lime materials.

The Sulfur Fertilizers:

The sources of sulfuric fertilizer could be divided into soluble and insoluble sulfur. The forms of the soluble ones are the sulfate with calcium, potassium, magnesium, ammonium, zinc, copper and manganese, which has high availability to plants, but easy to be leaching, especially in the sandy soil and the rainy places. The insoluble sulfur is sulfur element, which cannot be utilized directly by the plants. It should be oxidized by the soil and becomes sulfuric acid, which could be used to improve the strong alkaline soil. Most of the oxidation of sulfur is transformed through the microorganisms. Sufficient water and aeration, higher soil temperature and fine mineral powders can help transform the sulfur element into soluble SO_4^{2-}.

5.8.5 Guidelines for Applying the Calcium and Sulfur Fertilizers

Do not Excessively Apply the Lime Materials or Acid Materials with Sulfur:

If the acidic soil is applied with lime materials excessively, a deficiency of elements (iron, zinc, copper) in plant would occur. The neutralizing materials cannot be excessively applied to the alkaline soil either. While applying neutralizer such as limes and sulfur materials, the pH that suits the plants should not be ignored. Not all plants are adapted to the neutralized or acidic soil.

Incorporate Organic Fertilizer with Lime Materials:

In subtropical and tropical zones and the dry areas, the content of organic matters in soil is low. So the acidic soil cannot be only applied with lime materials to neutralize the acid. If the acidic soil is applied only with lime materials to correct the pH of soil, the production and quality of plants or fruit trees improved only a short term. The reason is that after being applied with lime materials for a longer period, the organic matters in the soil are fast decomposed. With the deficiency of organic matters, the physical properties of soil will be destroyed seriously, the capacity of holding water and fertility weaken, which causes poor growth of plant. This cannot be ignored when applying the lime. Don't forget to add organic matters when applying lime materials, in order to increase the production and quality for the sustainable agriculture. This is also beneficial to the uptake of calcium, and thus prevents the loss of it. Otherwise, the pH of soil will be increased and the physical property of soil destroyed.

The Position and Time of Applying for the Neutralizer:

The factors that determine the availability of the neutralizer (lime or sulfur powder) are the method and position of application. The neutralizer should be fully mixed and touched with the soil. Before planting, if the field is applied with a great amount of neutralizing matters, then plowing is needed to make it well mixed with soil. But as for fruit trees, broadcast application could be used, and the deeper application of limes or sulfur powders are, the better the effect will be. Water is the precondition of the soil neutralization, because the neutralizers such as limes or the sulfur powders do not easily dissolve in water. If the soil is dry, then the neutralization will be weak, which is harmful to the roots of plant. The neutralizer should not be applied during the bloom season or dry seasons. It is better to be applied in the period of dormancy of plant or after harvest.

Applying Calcium Fertilizer under Special Conditions:

Generally, the alkaline soil is not deficient in calcium, but when the alkaline soil of whose parent rock is deficient in calcium, or when the soil which doesn't suit liming needs a great deal of calcium, the lime materials could be replaced by the calcium sulfate or calcium carbonate.

Reduce the Overdose of the Elements that Prevent the Uptake of Calcium:

Calcium is antagonistic with potassium and magnesium; if the potassium and magnesium fertilizer is applied excessively, then it is not good for the uptake of calcium by the plant.

5.9 The Microelements

5.9.1 The Microelements are Indispensable

Microelements refer to the elements needed little by the plants, but it doesn't mean that they are not important. There are at least 7 confirmed microelements needed by higher plants. And there could be more in the future research and development as expected. Some elements are less needed by the plants, but the importance cannot be ignored. The microelements could affect the plants growth, production, and quality. Of course the amount of microelements could not be excessive in the soil; otherwise, they will be toxic to plant.

The confirmed microelements include: boron, molybdenum, chlorine, copper, zinc, manganese, iron, etc. The others are such as silicon, sodium, cobalt, selenium, vanadium, and other beneficial elements. When the fields are cultivated with a large amount of plants for a long term, the microelements will thus be absorbed excessively and taken out from the land. But some soils could not release or supply enough elements for plants growth, and farmers typically do not concern the importance of microelements in the soil. Organic fertilizers are not widely adopted because chemical fertilizers are more convenient. As a result, the sources of microelements decrease and cause commonly the deficiency of microelements in the soil.

The principal elements that constitute the human body cell are: carbon, hydrogen, oxygen, nitrogen, phosphates, sulfur, potassium, sodium, calcium, magnesium, iron, etc. There has been over 20 elements (more to be found). When people are deficient in some of these elements, disease might occur. The former 4 elements take up to 99% of the whole, weigh 96%, and the rest 4% are the calcium and phosphorus to constitute bones. And, as for the microelements (such as selenium, molybdenum, iodine, copper, manganese, zinc, boron, vanadium, nickel, chromium, cobalt, silicon, fluorine, chlorine, and arsenic), they are needed a little by the humans. The overdose of heavy metals would cause harm to human body. Death even might occur if serious. For example, the lead poisoning and the event of cadmium poisoned rice. Therefore, the contamination of heavy metals should never be ignored in agricultural production.

94 Soil and Fertilizer

The following two tables (Tables 5-1 and 5-2) illustrate diseases caused by the deficiency of elements and the recommendation content of elements needed by humans. The content is only for reference and will be adjusted according to the research of medicine.

The diagnosis and treatment of the deficiency disease will be introduced one by one. So, the farmers could cultivate according to the demands of plants.

5.10 Features and Guidelines of Application for Boron and Molybdenum

5.10.1 Features and Guidelines of Application of Boron

The Boron in Soil and Plants:

The content of boron in soil remains ca. 20 ~ 200 mg/kg. Most of boron is unavailable to plants, and only 0.4 ~ 5 mg/kg is available. The boron exists in the minerals, and some exist as soluble boric acid, which could be easily washed out from

Table 5-1. The Human Disease Caused by the Deficiency of Elements

Deficient	Effects
Ca	sluggish muscle growth
Mg	jerkiness
Fe	anemia and abnormality of immune system
Cu	languishment of artery channel border, diauxie anemia, abnormality of liver
Zn	injury of skin, hindrance of genitals maturity, mal development
Mn	sluggish growth of bones, sterile
Mo	problems of cell growth, tendency of decayed tooth
Co	malignant anemia
Ni	dermatitis, rustiness of growth
Cr	diabetes
I	abnormality of thyroid, sluggish metabolism
Si	abnormality of bones growth
Se	muscle languishment
As	languishment of growth for the animals, for human not confirmed
F	decayed tooth

Table 5-2. Essential Elements Needed by the Human Body and the Needed per Day

Elements		The Need per Day	
		Adult	Infants
K	Potassium	2,000 ~ 5,500	530
Na	Sodium	1,100 ~ 3,300	260
Cl	Chloride	3,200	470
PO_4^{-3}	Phosphate	800 ~ 1,200	210
Ca	Calcium	800 ~ 1,200	420
Mg	Magnesium	300 ~ 400	60
Zn	Zinc	15	5
Fe	Iron	10 ~ 12	7
SO_4^{-2}	Sulfate	10	
Mn	Manganese	2 ~ 5	1.3
F	Fluoride	105 ~ 4	0.6
Cu	Copper	105 ~ 3	1
Mo	Molybdenum	0.075 ~ 0.25	0.06
Cr	Chromium	0.05 ~ 0.2	0.04
Co	Cobalt	ca. 0.2(vitamin B_{12})	0.001
I	Iodine	0.15	0.07
Se	Selenium	0.05 ~ 0.07	

soil by rain and irrigation. Besides, some boron is gathered in organic matters, and will be released during decomposition.

Boron is quite important to the function of plants; it is closely related to the synthesis of nucleic acids, proteins, auxins, and transportation of plant. It is essential for the germination of pollen and the growth of pollen tube. It is related to the formation of seeds, and essential for the formation of cell wall.

The Symptoms of Boron Deficiency:

Boron is not easily mobile in the plants, so the symptoms of deficiency always appear in the new tissues and fruits; for example, the curving of growing point and new leaves, stop growing to death, curving of leafstalk, weakness and withering of roots, and lignification will happen in the root vegetables. With boron deficiency, the fruit shape of papaya will be rough and uneven in surface, the citrus will become small, dry, and hard, the grapes will become small and poor quality, and a crack will appear inside the leafstalk of cabbage. The overdose of boron will also cause poisoning. The symptoms are that the apex of the old leaves turns yellow, spots appear on the leaves, and the plant even withers to death.

Factors for the Cause of Boron Deficiency:

Soils:

The factors for the deficiency of boron in soil are various, and it is related to the physical and chemical environment of the soil.

- The soil is too alkaline (pH > 7.0) or too acidic (pH > 4.5), or it contains too much calcium.
- The content of boron could be reduced by leaching with rain and irrigation in rough sandy soils or soils with low humus.
- During the dry seasons, the growth and uptake of roots is reduced. And the organic matters are rarely decomposed, so the deficiency of boron occurs.
- Overdose of potassium and limes.

Plants:

When plants take up nutrients in an unbalanced way, the deficiency of boron is apt to happen. For example, the overdose of calcium and potassium will cause the reduction of boron.

The Treatments of Boron:

The Application of Boron:

Different plants demand different amount of boron. Plants that demand more boron are: cauliflowers, celery, turnip, rapes, and peas. Besides, the plants of grass family demand 6 ~ 8 times less than formers. According to the report, the plants that demand little boron will take up 50 grams of boron per year/ha., while the plants that demand much boron will take up 500 grams/year/ha. Plants that demand high boron should be fertilized with different amount of boron according to different soil. At most, 2 kg of boron could be applied per year/ha. In the acidic soil, poisoning could happen with over 4 kg boron/year/ha.

Apply Boron by the Foliar Dressing:

The tolerance of foliage should be taken into consideration while applying boron on the foliage; otherwise, the foliage might be damaged by fertilizer. Generally, 0.5% borax solution is sprayed and 3 kg/ha according to the demand of the plants. Commercial fertilizer of foliar dressing is available for fast supply of boron.

Supply the Organic Matters:
In the humus of soil, a deficiency of boron always happens, and when applying organic fertilizer, the decomposition of it will provide the boron needed, which is one of the fundamental tactics.

5.10.2 Features and Guidelines of Application of Molybdenum

Molybdenum in Soils and Plants:

The chemical change of molybdenum in the soil is very complicated. Generally, the content of molybdenum in soil is between 0.6 ~ 3.5 mg/kg, the average is 2.0 mg/kg, but the valid content is only 0.2 mg/kg. The forms of molybdenum depend on the pH of soil, MoO_{42}- (pH over 5.0), a compound of H_2MoO_4, $HMoO_4$-, MoO_{42}- (pH 2.5 ~ 4.5), and H_2MoO_4 (pH less than 2.5). The fixation activity of molybdenum acid in the soil will affect the availability of molybdenum.

The content of molybdenum in plants is about 0.1 mg/kg, and is taken by the plants in negative ions. Molybdenum is closely related to the metabolism of nitrogen. Molybdenum is the essential component of the nitrate reductase and nitrogenase.

The Symptoms of Molybdenum Deficiency:

The molybdenum is related to the metabolism of nitrogen; therefore, the symptom of deficiency is the same as when deficient in nitrogen. The growth and blossom will be inhibited, the leaves will curve, become small, spots will appear, and so on. The deficiency of molybdenum will often appear on crucifer, peas, and citrus, but seldom on the grass family.

Factors Affecting Molybdenum Deficiency:

Soils:
The deficiency of molybdenum will happen more frequently in the acidic soil (pH < 5.5). The more acidic, the lower the availability of molybdenum would be. Adjusting the pH of soil could enhance the availability of molybdenum. The molybdenum is apt to be fixed by the oxide and clay minerals, and the negative ions could replace the absorption of molybdenum, thus enhancing the availability of molybdenum. Do not apply limes excessively; otherwise, the overdose of molybdenum could cause harm to human as well as other animals.

Plants:
The molybdenum is prone to move in plants and be taken by plants in negative ions. The uptake will be inhibited by phosphates and sulfates, and it is antagonistic with the uptake of copper and iron. The deficiency of potassium is always with and overdoses of molybdenum, which is caused by the unbalance of the nutrient uptake.

The Application of the Molybdenum:

The Application of Molybdenum:
While applying molybdenum, the content should not have over application; otherwise, it would be harmful to plants, human, and animals. At most, the sodium molybdate is needed 500 ~ 2,000 grams per ha annually, and at least 150 ~ 200 grams. Because the amount of application is little, it can be applied in the soil after solubilization and dilution, or spurred with fine soil. The plants could be healthy after applied with it.

Applying Molybdenum on the Foliage:
0.5% ammonium molybdate could be sprayed on the foliage, especially for the short-term plants.

Applying Lime Materials:
It will cause a deficiency of molybdenum in the acidic soil and reduce the availability of molybdenum. Lime materials could be used to increase the pH of soil, which could enhance the availability of molybdenum. It is also an indirect effective way. But over application should be avoided.

5.11 Features and Guidelines of Application for Microelements of Iron, Copper, Zinc and Manganese

Among the microelements, iron, copper, zinc, and manganese are heavy metals, which are the indispensable elements for the plants growth. The deficiency of anyone of them will affect the growth of plants. The content of these microelements needed by the plants is very low, and the overdose could cause poisoning.

The plants uptake iron, copper, zinc, and manganese in positive valence (or positive bivalence), sometimes will also be taken up from the chelated forms. The activities of these microelements in soil are quite similar. The content and availability of them are affected by parent rocks and weathering of soil. Generally, the soils in areas of higher leaching, temperature and moisture contain little microelements. For example, the deficiency of microelements is likely to happen in sandy soils. It should be noted that the high content of microelements in soil does not mean high availability of the elements; and also affected by soil pH and some other factors. Therefore, the balance of different elements is an important part of plant growth.

5.11.1 The Function on Plants and Symptoms of Microelement Deficiency

The Functions and Deficiency Symptoms of Iron:

Iron is one of the important elements in enzymes of photosynthesis and respiration of plant cells. It is related to metabolic systems of utilizing light energy and energy synthesis of plants. And, it will improve the oxidation-reduction and chlorophyll synthesis.

Iron cannot be moved easily in soils. Therefore, the symptoms of deficiency will appear on the new leaves and tissues. The most obvious symptom is that the new leaves turn yellow and appear light green. More seriously, the whole plant might be yellowish-white and wither to death.

The Functions and Deficiency Symptoms of Copper:

Copper is one of the important ingredients of enzymes for amino acid synthesis and oxidation-reduction. It is related to the metabolisms of amino acid and protein, which is an essential element for chlorophyll synthesis, and directly takes part in the respiration system. The copper is the most available in the slightly acidic and neutral pH soils. It will be fixed by the ferric oxide, aluminum oxide, and silicate in strong acidic soil. Copper will also be chelated with organic matters and form complexes. The unbalance of iron, manganese, and aluminum will reduce the copper uptake by roots.

The deficiency of copper happens rarely, because many of the agricultural materials contain copper. The symptoms of deficiency appear yellowish on the new shoots and new leaves; and, the growth will be small and weak.

The Functions and Deficiency Symptoms of Zinc:

Zinc is one of the important elements in enzymes of the metabolism of protein, which is related to the synthesis of tryptophan (the main precursor of auxins). The zinc element could neither mobile easily in plants; therefore, the symptoms of deficiency will appear on the new leaves or new tissues.

The Functions and Deficiency Symptoms of Manganese:

Manganese is related to many systems of enzyme, which is a component of phospholipid enzyme. And, it is related to the photosynthesis and chlorophyll synthesis. It is not mobile easily in plants, and the symptoms of deficiency will appear on the new leaves or new tissues. For example, the vein turns yellowish, and brown spots might occur.

5.11.2 The Factors Affecting Microelement Deficiency

There are many factors that affect the deficiency of the microelements. Apart from the deficiency of parent rocks, it could be mainly caused by the physical, chemical, and biological factors of the soil. Any of those factors that affect the roots development and health will lead to deficiency in microelements. The sickness of plants might cause the deficiency of microelements, and the way around, the deficiency will cause sickness of plants in turn. It is beneficial to know the reasons for deficiency.

The abnormality of the pH of soil is the main factor affecting the deficiency of microelements. The followings are to be checked on the field:

- High content of organic matters (more than 6%).
- High content of sand.
- High content of phosphates, but low of microelements.
- Drought.
- High density of soil.
- Turnover of subsoil during the soil cultivation.

The conditions of soil which is likely to cause deficiency of microelements and the plants in accordance are listed below in Table 5-3:

Table 5-3. The Main Factors and Common Apparent Crops that Cause the Deficiency of Microelements

Microelements	Conditions of Soil	Common Apparent Crops
Deficiency of Iron	High pH, high phosphate.	Sweets corn, sorghum, soybean.
Deficiency of Copper	Sandy soil, high organic soil, high PH, clay.	Cereals, vegetables, fruit trees.
Deficiency of Zinc	Sandy soil, overdose of limes, high phosphate.	Citrus, sweet corns, soybean, fruit trees, and vegetables.
Deficiency of Manganese	Sandy soil, overdose of limes, high phosphate.	Soybean, cereals, fruit trees, sweet potato, leaf vegetables.
Deficiency of Boron	Sandy soil, overdose of limes, strong acidic soil, high organic soil.	Peas, tomatoes, citrus, fruit trees, sweet potato, leaf vegetables, grapes.
Deficiency of Molybdenum	Highly weathering acidic soil.	Peas, citrus, cauliflower.

5.11.3 The pH Affects the Availability of the Microelements

The deficiency of microelements is closely related to parent rocks and property of plants and soils. Among which, the pH is an important criterion determining the availability of microelements. When the soil pH is changed, the availability of microelements in the soil will be affected. As is illustrated in Figure 5-3, the thinner the deficiency column is, the lower the availability:

Iron:

The iron is easier to be soluble when the pH is low. As the pH increases, the iron will be oxidized by chemicals and microorganisms, and thus the activity will reduce. When the pH is higher than 6.5, it will appear as ferric oxide, which is not soluble and thus leads to poor uptake. The other elements and compounds will be combined with iron and become insoluble complex, which will also reduce the uptake of iron.

Copper:

The copper will be the most available in slight acidic and neutral soils. In strong acidic, it will be absorbed by the ferric iron, aluminum oxide, and silicate; the organic

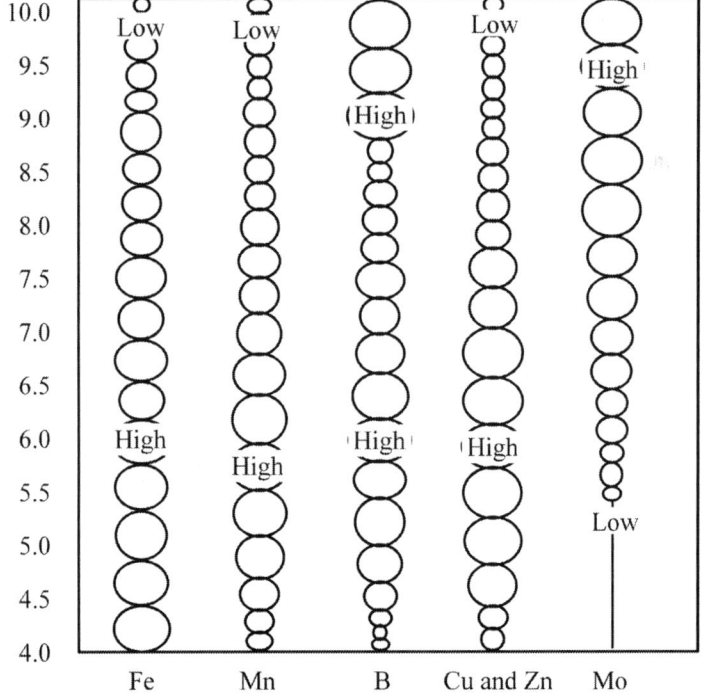

Figure 5-3. The Soil pH Affects the Availability of Trace Elements

matters will also be combined with copper and forms strongly bounded complex, which is not easy to be taken up. The root uptake of copper will also be reduced with the unbalance of copper and other metal elements.

Zinc:

The zinc will also be the most available in slight acidic and neutral soils. Its activity is similar to that of copper. In the alkaline soil, the deficiency of zinc should be taken seriously, and the overdose of phosphorus fertilizer should be avoided.

Manganese:

Most available in the pH of 5.0 ~ 6.5; below 5.0, manganese will become insoluble complex by ferric oxide, aluminum, and silicate; when pH increases, it will be oxidized by chemical and microorganism, and the availability of manganese might be reduced. So, there might be manganese deficiency in alkaline soils. The strong acidic soil will also cause poisoning of manganese, which is caused by the high content and activity. The unbalance of manganese with other elements (calcium, magnesium, iron, etc.) will also cause a deficiency of manganese. The deficiency will also happen when the high organic soil is in cold and flood environment. When the temperature goes up again and it becomes drought, it will go back to normal.

5.11.4 The Application of Iron, Copper, Zinc and Manganese, and the Tactics for the Deficiency

Overdose of microelements should be avoided, because the microelements such as iron, copper, zinc, and manganese are heavy metals that would cause contamination if applied excessively. Now the tactics for application are illustrated as follows:

Iron:

Soluble iron salts (such as ferrous sulfate, ferric (2+) sulfate) could be applied to iron-deficient soils. And it would be better if incorporated with acidic nitrogen fertilizer. Iron will immediately become unavailable in the soil. Therefore, foliar dressing should be applied, and the soluble iron salt or chelated iron could be used. There are many types of chelated iron, which could keep the water solubility and availability, thus enhancing the effect of uptake. The concentration of spray should vary according to the types and the levels of deficiency. It should be 1,000 ~ 4,000 times dilution. When the ferrous sulfate is sprayed in the 1,000 times dilution, the new leaves might be hurt, but the newly grown leaves could be getting the normal growth. Some plant leaves cannot be hurt, so the dilution should be applied more times, or adopt the safe chelated ferrous salts. Besides, as for the deficiency of iron caused by the alkalinity of soil, the best way to solve this problem is to adjust the pH of soil: apply the acidic materials and mix them with the soil to enhance the availability of iron.

Copper:

When plants are diagnosed as having a deficiency of copper, cupric sulfate or chelated copper could be used according to the level of deficiency and the demand of the plants. 0.5 ~ 1 kg per ha of Cupric sulfate is commonly recommended, and frequent application should be avoided. The pollution of copper happen because of the improper irrigation of polluted water. The cupric sulfate or chelated copper diluting more than 2,000 times could be used for spray application for plants. Adjusting the pH of soil could also enhance the supply and uptake of copper. 0.02 mg/kg of cupric sulfate could be applied in the water or hypotonic cultivation.

Zinc:

When plant is deficient in zinc, zinc compounds could be applied. For serious deficiency, 10 kg/ha of zinc sulfate could be applied. Both furrow application and broadcast application are all right. The foliage could be applied with zinc salts or chelated zinc, and 1,000 ~ 2,000 times dilution could be used. Overdose of it should be avoided. 0.05 mg/kg of zinc sulfate could be applied to the solution in the water or hypotonic cultivation. Adjusting the pH of soil could also improve the deficiency.

Manganese:

The deficiency of manganese always happens in the alkaline soil or the soil applied with too much limes. Sulfate could be used to adjust the pH of soil. And the collocation with acidic nitrogen fertilizer could enhance the availability of manganese. Furrow application or broadcast application with manganese sulfate monohydrate or manganese oxide could be adopted. It could also be mixed with basal fertilizers. When the deficient situation is serious, 10 kg of manganese sulfate monohydrate should be applied per ha. Generally, the foliar dressing could be adopted for urgent remedy. The dilution of more than 500 ~ 1,000 times manganese sulfate monohydrate or the chelated manganese salts could be used for spray application. In the water cultivation, the concentration of manganese could be 0.2 ~ 0.5 mg/kg.

The amount of application recommended above is not suitable for different kinds of plants, because the tolerance of all kinds of plants is very different. The tolerance of new leaves is weaker than the old; and, at noon or under strong sunshine, the spray application will have different effects. So, before applying widely, small-scale experiment should be performed in advance, or fertilizer damage might occur.

The application of the above microelements is mainly provided through the soil or foliage. And, the deficiency occurs in fields is often caused by the deficiency of several certain elements. When it is not easy to observe, the diagnosis of soil and plants should be immediately adopted, and foliage remedy should be adopted in order to improve the soil and solve the problem.

5.12 Features and Guidelines of Application for Silicon Fertilizer

5.12.1 Preface

Silicon (Si) is one of the elements distributed widely in the nature, and includes the most in the crust, which holds up to about 27% silicon. But, the solubility of silicon structure is extremely low, in the farmland of long-term intensive production, apart from the fertilizer of nitrogen, phosphates, and potassium, soluble silicon fertilizer is also needed to make the plants healthy. According to threshold limits value of deficiency of silicon in the soil, many farmlands could be deficient in silicon. It is reported that the soil contains silicon, but, because of the low solubility, silicon fertilizer is needed. It has been confirmed that the silicate slag have the capacity of improving soil properties, crop production, and crop quality. There are many reports for achievements at different soils. Even the plants of agriculture, horticulture, and forest have good effects.

The function of silicon on the plant production has always been ignored, because the soil contains large amount of silicon in the past. But, recently, the valid and soluble silicon in the farmland are becoming less and less. The recent reports of silicon lead to a heat spot of utilizing silicon again. Because silicon is beneficial to improve the soil property, decrease the uptake of heavy metals, and enhance the quality of plants, the development of silicon has attracted the attention from all aroud the world.

The silicon fertilizer is becoming better with the research and development of multi-functional humic acid silicon fertilizer.

5.12.2 The Function of Silicon on Plants

The function of silicon on plants include:

Silicon is an Important Element for Constituting Plants:

The content of silica in all kinds of plants holds up to 14 ~ 61%; plants of grass family (such as rice, sweet corns, sugar cane, sorghum, wheat, barley, bamboo, etc.) contain high content of silicon. Rice plant contains silicon in stem and leaf holds up to 15 ~ 20%, which exceeds the total content of nitrogen, phosphates, and potassium. Some other plants such as woody plants also contain high silicon content. But plants of legumes (such as soybean, clover, trefoil, common vetch, Chinese milk vetch, etc.) contain little content of silicon. The accumulation of silicon in plants is thought to have the function of reducing the uptake of nitrogen, decreasing the excessive growth, lodging, and enhancing the quality of fruits.

Silicon Fertilizer can Increase the Photosynthetic Efficiency and Quality of Plants:

The silicon fertilizer could fortify the strength of cell wall and form structure that is beneficial to ventilation and good transmission of canopy in the farmland, which helps the plants harvest the light and carbon dioxide, and enhances the fixation of carbon dioxide to produce a large amount of carbohydrate under the photosynthesis. Thus, the photosynthetic efficiency of plants could be improved to increase the production and sweetness. The enhancement of photosynthesis will increase the content of protein, fat, and starch. The improvement of rice quality and taste has also been reported.

The Silicon Fertilizer could Increase Pathogen Resistance of Plants and Reduce the Application of Pesticides:

Resist Lodging:
The lodging of crop is a main reason for the reduction of production. The silicon could fortify the strength of the roots of plants and improve the ventilation inside the plants, and thus prevent the lodging of crops.

Resist Root Rot:
The roots of plants are apt to be rotten after rainy seasons, thus leading to the withering of plants. Especially for the melons, silicon could help improve the ventilation and strength of roots and restrain the overdose of nitrogen, which could reduce the root rotten.

Resist Pathogen:
Silicon could increase the thickness of cell wall of plant epidermis, thus reducing the invasion of pathogens or insects. The prevention of rice blast, stem rot, powdery mildew, etc., has been reported. Therefore, the silicon can be applied to reduce the use of pesticides.

Resist coldness, drought and salt:
The silicon fertilizer could adjust the open or close of the stomata and the evaporation of water, making the plants resistant to coldness, drought, and hot blast, and reducing the premature drop of fruits.

Tolerance for Storage and Transportation:

The strength of epidermis and inhibit the emission of water could make the plants tolerant for storage and transportation.

Silicon Fertilizer could Enhance the Availability of Phosphorus Fertilizer and Increase the Sugar Content and Quality of Plants:

When silicon is taken up as SiO_4^{4-}, the form of it resembles the negative valence of phosphate (PO_4^{3-}). As for the soil that has a good capacity of fixing phosphates, silicon fertilizers could help increase the availability of phosphates and increase the utilization of phosphorus fertilizer. Phosphorus is one of the three essential elements that dramatically affect the production and quality of crops, but the efficiency of utilization is the lowest one from soil. Especially in strong phosphate fixing soil, silicon fertilizer could increase the efficiency of phosphorus fertilizer.

Silicon could Improve the Soil Comprehensively:

Generally, the silicon fertilizer contains not only silicate, but also calcium, magnesium, iron, boron, zinc, manganese, copper, and some other nutrients, which could improve the supply of nutrients, and especially for the strong acidic soils.

Silicon Fertilizer could Increase the Rate of Fruiting and Reduce the Excessive Growth:

Silicon fertilizer could increase not only the uptake of phosphate, but also the activity of pollination, and thus enhancing the fruiting of melons and fruits. Silicon fertilizer could also reduce the excessive growth and increase the weight and sugar content.

Silicon Fertilizer could Prevent the Contamination of Heavy Metals:

The alkaline materials in the silicon fertilizer could adjust soil pH and reduce the availability of heavy metals, and could form precipitated matters such as aluminum silicate, cadmium silicate, and mercury silicate, and reduce the risk of being taken up by plants.

Silicon Fertilizer could Promote the Decomposition of the Organic Fertilizer:

The alkaline materials in the silicon fertilizer could adjust the decomposition of organic fertilizers. But, overdose should be avoided, because of the risk of over-alkaline.

From the points above, silicon fertilizer is a kind of fertilizer that can increase the production and enhance plant's quality.

5.12.3 The Features and Ingredients of Silicon Fertilizer

Source:

The silicon fertilizer is mainly the silicate mines and slag, while all kinds of minerals are produced by the industry, which has been researched and produced.

Components:

Silicon fertilizer contains sufficient calcium and magnesium; the most common are calcium silicate, magnesium fertilizer. The main ingredients of it are $CaSiO_4$, $CaSiO_3$, $MgSiO_4$, and $Ca_3Mg(SiO_4)_2$. Because it is manufactured out of all kinds of slag, it contains many microelements (iron, boron, copper, zinc, and manganese) which could be comprehensively functional.

Features:

Silicon fertilizer is an alkaline product, and most of it is not soluble in water. Therefore, the loss of it is rare. It is soluble in the acid, so the availability in the acidic soil is high. The raw materials are different, so the silicon fertilizers appear white, gray, and black. The density of silicon is high; weigh is heavy, $3,000 \sim 5,000$ kg/m^3.

Commercial Products:

Most of the commercial products of silicon fertilizer are manufactured in the silicate mines and slags, and different countries of the world make a criterion of available ingredients of silicon fertilizer, according to the concept of the resource utilization. The SiO2 should exceed $15 \sim 20\%$, the CaO + MgO more than 35%. The grain size is also different. All of them have to be through the 10 meshes. Among which, 60% has to be through 30 meshes, the first grade product 85% through 100 meshes; the second grade product 85% through 65 meshes.

5.12.4 The Application and Guidelines of Silicon Fertilizer

Because the sources of silicon fertilizer are different, the chemical property should be suitable to the soil to be applied. For example, if the pH of the product is high, then it is not suitable for the alkaline soil. The time, the method, and the content of application are illustrated as follows:

Time of Application:

Silicon fertilizer could be used as basal fertilizer or additional fertilizer. The best period is on the basal fertilizer. Fully broadcast before tilling to the field in order to mix fully with the soil. As to the long-term perennial plants, fully broadcast after harvest or after leaf-drop of deciduous fruit trees.

Method of Application:

Uniform broadcast sowing application should be adopted. The side-dressing and furrow application are not recommended. In the orchard, during the rainy seasons, it could be uniform broadcast sowing on the surface at 50 cm away from the base of tree.

Amount of Application:

The amount of fertilizer application should take the property of soil, the demand of plants, the supply of irrigation, and the supplying speed or content of the fertilizer into consideration. The pH of soil should be the first factors to be considered. Because most of silicon fertilizer is alkaline, the acidic soil is the most applicable. Generally, 400 ~ 500 kg/ha is the most common, and the strong acidic soil or the soil with silicon deficiency should be applied about 800 ~ 1,000 kg/ha.

5.12.5 Notices for Application of Silicon Fertilizer

Incorporated with Organic Fertilizers:

There are mutual counteractions and neutralizations between fertilizers. The best fertilizers to be incorporated with the silicon fertilizer are organic fertilizers, farmyard manures, green manures, etc. The organic fertilizer could produce organic acids and carbon dioxide, which could increase the solubility of silicon.

Silicon Should Be Incorporated with Nitrogen, Phosphates and Potassium Fertilizers:

The cooperation with these three elements is the fundamental method to increase the quality and production.

The Dust Problem in Applying Powders of Silicon Fertilizer:

The wind direction should be concerned during broadcast application. The application should be conducted while the wind is moderate. To avoid the dust flying upwards, wetting soils and mixing with wet organic materials could be incorporated with the silicon fertilizer.

Avoid the Overdose of Silicon Fertilizer:

The silicon fertilizer could restrain the overdose of nitrogen and reduce the excessive growth of shoots. But the overdose of silicon could also harm the plants. And it is not recommended to apply at the basal location of stem.

5.13 The Siderophores and the Health of Plants

5.13.1 The Necessities for the Plant Health

The plants could only grow healthily under the proper growing environment, the proper environment needs soil that could sufficiently provide nutrients, water, and oxygen needed by plants, and good climate conditions. The health of plant is affected the most by the pest and disease control in root and shoot. And the pest and disease control is closely related to the conditions of the soil.

The availability of nutrients is important no matter how much the plants demand for nutrients. The fixed or insoluble nutrients are not easy to be taken up by plants, and the availability of nutrients is related to soil pH and the existing of organic chelated matters.

5.13.2 Iron is an Indispensable Element for Organisms

Iron is related to many important metabolisms of plants, which is an indispensable element. The fixed iron cannot be easily taken up by organisms. But fungi and bacteria in the soil could utilize the fixed iron to perform their metabolisms. These microorganisms secrete organic chelated matters and form stable compounds with iron. The secreted organic chelated matters are called siderophores. These compounds can be decomposed and taken up by the microorganism, and the released iron can be used in metabolisms. In a word, iron is also an indispensable element for microorganisms.

In agriculture and the ecosystem of nature, these microorganisms always exist in the environment of rhizosphere. Therefore, scholars considered that siderophores are related to the health and nutrient of plants. Over the past 15 years, soil scientists, plant physiologists, plant pathologists, and microbiologists have deeply researched the importance of siderophores in the ecosystem of soil. At present, their structures, the existing in the soil, the roles in the plant-microorganism-soil interaction, and the importance to agriculture have been drawn to study. Especially, for the soil in which the availability of iron is low, it has been regarded as very valuable.

5.13.3 The Features of Siderophores

Some soil microorganisms or roots have the ability to secret siderophores. The organic compounds of the secretion have the capacity of chelating the iron to form complexes. When the soil is deficient in the available iron, microorganisms could secrete more siderophores. This chelating function of siderophores is very important in the neutral and alkaline soils, because the availability of iron in such soils is low. The types of siderophores secreted by microorganisms vary a lot. And the size of molecules is also very different, so as the structure. But they all have the function to chelate with iron.

The stability of siderophores after chelating varies, too. Some of them could form very stable compounds. Some siderophores secreted by microorganisms become Fe-complexes. Some microorganisms might be not able to utilize it, and forms specificity. This kind of siderophores with specificity might be utilized by some microorganisms and become the tool to resist other microorganisms in soil for taking up iron.

5.13.4 The Function of Siderophores

The siderophores secreted by soil microorganisms and plants have the capacity of being chelated with iron, and the effect of this kind of chelated compounds is still being studied. The theoretical application of its function includes:

- The siderophores provide the iron element of plants, and enhance the nutrient of plants
- The siderophores take the iron source from plant pathogenic microorganisms, thus protecting plants.

The View of Providing Plant Nutrients:

According to reports, the content of soluble iron in soil far exceeds the inorganic balance. The increase of the availability of iron might be attributed to the chelated organic siderophores. In soil test, it shows that the existence of siderophores is able to affect the iron nutrition of plant. According to some reports, the chelated iron is easier to be taken up by the plants than the inorganic iron.

The View of Disease Resistance:

Iron is also essential for the growth of microorganism. The scavenger of iron could also be regarded as a mechanism of defense. According to some reports, this theory means that some microorganisms take the advantage of secreting the siderophores with specificity and scavenge the iron nutrient, prevent the other group of microorganisms (pathogens) from utilizing it, thus achieving the function of reducing diseases.

Some researches call one kind of special-structured siderophores released pseudobactin by *Pseudomonas putida* to control one kind of pathogens in soils, and to reduce the soft rot disease of radish, cucumber, tomatoes, and potatoes.

The siderophores could affect the iron nutrition of the microorganism closed to them. Plant pathologists think that the reason why *Pseudomonas putida* and microorganisms produced siderophores could promote the growth of plants is that the iron needed for pathogens is scavenged by those microorganisms. The plant pathologists suggest that the siderophores are important, because they can enhance the availability of iron for plants. In the relationship of plants-soil-microorganism, it might a play vital role in the health and the uptake of plants which is need to be researched and developed thoroughly.

5.13.5 The Application of the Siderophores

The application of siderophores still remains the first step. According to the research, *Pseudomonas putida* was used in the inoculant of seeds, which could enhance the growth and production of tomatoes, beet, and radish. Some increase amounts to 144%. The increase of production can be ascribed to the inhibition of pathogens. And such huge increase of production can help prove the value of *Pseudomonas putida*. In the aspect of plant nutrient, tackling the deficiency of iron as foliar dressing, by utilizing the compounds of iron and siderophores, has gained a patent in the USA. And it has been proved that the effect is better than a chelator, EDTA.

It is also reported that applying organic matters in the soil with iron deficiency could effectively increase the content of siderophores, and thus enhancing the availability of iron. In recent years, the study for commercial utilizing siderophores directly was under development and research. The microorganism that is applicable to the soil is also being researched, so that beneficial microorganisms could be applied to the farmland.

5.13.6 Conclusion

To lower the cost of crop production and increase the utilization of biological control, natural ways and methods for the plant nutrient and protection should be conducted. And the exploitation of siderophores and other natural chelated matters is one of the important applications, which will greatly improve the health of plants and soils.

5.14 Methods and Guidelines of Fertilization for Solid Fertilizers

Solid fertilizers include chemical fertilizer, organic fertilizer, and microbial fertilizer. These three kinds of fertilizer need different methods of fertilization, therefore I will introduce them respectively:

Methods and Guidelines of Fertilization for Chemical Fertilizer:

Chemical fertilizer is commonly fine-grained, and the method of fertilization for chemical fertilizer are mainly broadcast, hole, and furrow fertilizations:

Broadcast Fertilization:

Chemical fertilizer is uniformed broadcast on the farmland in order to grow crops during soil cultivation or to grow crops on a large scale. Broadcast fertilization needs to avoid leaching problems which are caused by rain. Therefore, it is better to broadcast before turning plow, or on the grass of orchards to prevent leaching and hurting root.

Hole Fertilization:

Hole fertilization is a fertilization method to put the fertilizer in one hole, which is applied among the crops of the largest row spacing. The hole fertilization must avoid the fertilizer being too close to the crops, or it may cause some harms to crops. It is better to cover the fertilizers in the soil to prevent from washing out the fertilizer and damaging root.

Furrow Fertilization

Fertilization in furrows is used to plant those crops in furrows or hill cultivation. Fertilization in furrows must avoid the fertilizer being so close to the crops that it may cause some harms to crops. It is better to dig a furrow, apply fertilizer, and then cover with soil.

Methods and Guidelines of Fertilization for Organic Fertilizer:

Organic fertilizer is commonly in loose powder or extruded forms. The fertilization method of organic fertilizer is the same as that of chemical fertilizer. However, the fertilization volume of organic fertilizer is much more than the volume of chemical fertilizer. Therefore, it is necessary to pay attention to fully mix organic fertilizer with the soil. When applying hole or furrow fertilization on long-term fruit trees, it is better to dig a different hole or furrow each year to do the fertilization.

Methods and Guidelines of Fertilization for Microbial Fertilizer:

Microbial fertilizer is commonly cultivation of living microorganisms for the inoculant. The fertilization method of microbial fertilizer is very different from chemical fertilizer and organic fertilizer. Microbial fertilizer or diluted microbial inoculant should be contacted directly with roots of plants to get the best effects. It would have significant effects, when the microbial fertilizer is contacted over 20% of the root system at least.

5.15 Methods and Guidelines of Fertilization for Liquid Fertilizers

5.15.1 The Development History of Liquid Fertilizers

In recent years of the fertilizer development for crops of agronomy and horticulture, there are liquid fertilizer products which have already been put into the fertilizer market besides the solid fertilizer. Just as the solid fertilizer, liquid fertilizer has its advantages and its methods of application. The correct fertilization method can help liquid fertilizer get the best effects.

According to their different states, liquid fertilizer or fluid fertilizer can be classified into liquid soluble fertilizer and suspension fertilizer. Liquid soluble fertilizer is solution without solid particles; while suspension fertilizer means that the particles of solid fertilizer are in suspension solution, indicating that there exist fertilizer particles which are not dissolved, among which some are the characteristics of the fertilizer itself but some are additions to increase the viscosity of the liquid and to prevent the crystallization and aggregation of oversaturated solution.

Along with the development of technology, the compositions of liquid soluble fertilizer can be nutrient elements, microbial fertilizer, and plant stimulator to enhance nutrient absorption and improve soil fertility. Therefore, the development of liquid fertilizer has a rather high potential.

5.15.2 The Types of Liquid Fertilizer

Based on the content of types of liquid fertilizer, it can be classified into single and compound liquid fertilizers. Single liquid fertilizer only contains one nutrient of fertilizer, but compound liquid fertilizer contains two or more kinds of nutrients of fertilizer. The liquid fertilizers are commonly used with compound liquid fertilizers. According to the different materials of fertilizer, there are three types of liquid fertilizers:

Chemical Liquid Fertilizers:

Chemical liquid fertilizer is the fertilizer which uses the soluble chemical fertilizers to form the single and compound liquid fertilizer such as nitrogen, phosphorus, potassium, calcium, magnesium, and trace elements.

Organic Liquid Fertilizers:

Organic liquid fertilizer is made on the basis of organic matter or by fermentation. This kind of liquid fertilizer uses the residues of animals, plants or microorganisms, such as fish, by-product meals, yeast powder, humus, sea-weed, milk powder, and eggs. The compositions of organic fertilizer contain pure organic liquid fertilizer and organic compound liquid fertilizer with the addition of chemical fertilizers.

Microbial Organic Liquid Fertilizers:

Microbial organic liquid fertilizer is a kind of fertilizer that combines microbial fertilizer and organic compound liquid fertilizer together. It uses the microbial fertilizer including nitrogen fixing, phosphate solublizing, potassium solublizing microorganisms, or other beneficial microorganisms with organic liquid fertilizer or organic fermented liquid fertilizer.

5.15.3 Equipments and Methods of Application for Liquid Fertilizer

Equipments of Application:

The equipments of application for liquid fertilizer are different from that of solid fertilizer, which vary according to crops or plants in different areas and different fields. The basic equipments are tank, pipes, irrigator, or sprayer. For example, if there are equipments for a larger area, it is necessary to use sprayer machines, pump, blender, etc. It is better to pay attention to the quality of all the equipments employed in the process in order to prevent the corrosion or blocking in the system of liquid fertilization.

Methods of Fertilization:

Fertilization to the Soil Directly:

It is common that diluted liquid fertilizer can be fertilized directly into the soil of the fields where crops grow. Usually, diluted fertilizer is employed or poured into deeper soil after planting, and undiluted fertilizer can be used before soil cultivation in order to plough and spade to attain the goal of even distribution.

Fertilization in the Irrigation Water System:

Irrigation uses the method of line pipe or without line pipe. Irrigation with the addition of liquid fertilizer can get to the purpose of fertilization, which can also be fulfilled by drip irrigation in order to the supply of liquid fertilizer.

Fertilization to the Whole Plant by Spraying Method:

The method of spraying the dilution liquid fertilizer to the whole plant or the surface of the soil, just such as the fertilization on leaf, can be applied in emergency replacement, when it is deficient of nutrients or directly used together with the system of farm chemical spray, which is especially suitably effective for the poor soil or the soil which has strong fixation of fertilizers.

Use as the Liquid of Growth Initiation:

The liquid of growth initiation is to use diluted nutrients into the rhizosphere of transplanted seedlings to help them recover and grow quickly.

Nutrient Solution as Water Culture or Media:

Use the diluted liquid as water or media cultivation. However, the complete formula of the liquid nutrient still needs more consideration for different crops.

5.15.4 The Advantages and Disadvantages of Liquid Fertilizer

Liquid fertilizer has both advantages and disadvantages. We should use its merits and avoid its harmful factors, which is the purpose why we should know liquid fertilizer.

Advantages:

The Concentration and Proportion of Liquid Fertilizer can be Adjusted Freely:

The concentration and nutrient proportion of liquid fertilizer can be adjusted according to different nutrient formula with different necessities of crops. Especially to the some soils and different stages of crops, it can be added in some trace elements or other organic nutrients. As for the dilution of concentration, it can be adjusted according to different necessities, which means that the higher dilution with the lower the concentration can be used for more the acreage and depth of fertilization.

It can be Mixed with Pesticides Appropriately:

Liquid fertilizer is soluble and can be used with some pesticides. Before mixing, it is necessary to follow the instruction book to use the liquid fertilizer in order to lessen the pesticide effect or avoid the harms brought by pesticide or fertilizer.

Distribution can be Equal, and Fertilizer Effect is Quick:

The diluted liquid fertilizer can be equally distributed and directly contact with plant roots which is helpful to plant absorption, so that the fertilizer effect is quickly.

It can be Applied to Deep Soil:

The roots of perennial plants or fruit trees are deeply rooted into the soil. Because of that, the solid fertilizer is commonly hard to get deeper into the soil. However, liquid fertilizer uses broadcasting or pouring to fertilize in order to improve the deeper soil fertility.

Making Process is Easy:

Liquid fertilizer does not have the whole process of crystallization, pilling, dry, cooling, and packaging, which can reduce the cost of production. And also because liquid fertilizer is not powder particles, it can reduce the harm and pollution.

Both Mechanized Fertilization and Direct Fertilization are Easy, and It Can Reduce the Cost of Fertilization:

Disadvantages:
- Liquid fertilizer is likely to bring burns to plants if it does not fertilize appropriately.
- It is usually used after dilution. If there is no water source or if it is far away from the source of water, especially when it is deficiency of equipments on slope land, it costs a lot.
- When it is to be fertilized on a large scale, equipment expenses should be invested; for example, tractor, applicator, pipes, and sprayers.
- Liquid fertilizer contains soluble elements which easily combine or produce factors to form antagonism affecting the absorption effects of plants. It can be seen in the following Table 5-4:

5.15.5 Methods and Guidelines of Liquid Fertilization

Methods and guidelines of liquid fertilization are similar to methods of solid fertilizer, but we should pay attention to the followings:

Pay Attention to the Dilution Times:

The original liquid of liquid fertilizer has a high concentration; therefore, it is harmful to plants if it is fertilized directly to plants. It is commonly needed to dilute, and the dilution times should obey the instruction book of the product. When it is used to spray on leaf and the whole plant, it should has a low concentration. The concentration can be higher when it is used on the soil before land cultivation than used directly on plants, because land cultivation has the diluted effect.

Table 5-4. Antagonism and Stimulation among Plant Nutrients

Antagonism	Stimulation
N → K, B	N → Mg
P → Zn, Cu, Fe, K	P → Mg
K → Ca, Mg, B	K → Mn, Fe
Ca → K, Mg, B, Mn, Fe	Mg → P
Mg → K, Ca	
Cu → Fe, Mn	
Zn → Cu, Fe	

Pay Attention to the Soil Characteristics:

Soil physicochemical properties influence the effects of fertilization. When there are limiting factors in the soil, we should firstly improve them, and then we can increase the effects of fertilizer. For example, when alkalinity of the soil is high, the ammonium nitrogen fertilizer is likely to volatilize or lose; or, when soil has a strong fixing phosphorus fertilizer (such as laterite), the phosphates in fertilization can be fixed easily.

Pay Attention to the Characteristics of Dilution Water:

It is necessary to pay attention to dilution water of liquid fertilizer. Because,g for example, waste water from city contains high impurities or contaminations, which is tend to combine fertilizers to produce insoluble materials or form precipitate to affect the effects of fertilizer.

Pay Attention to the Time of Fertilization:

Liquid fertilizer is commonly fast response. Because of that, it is better not to use the liquid fertilizer with high proportion of nitrogen in flowering period of crops, and it should be paid more attention when it is in full blossom period. Otherwise, it would cause the flower and fruit dropping. When plants are in the vigorous period of growth, it needs more fertilizers, but the irrigation of high concentration fertilizer tends to bring burns to plant root, which should be paid attention to.

Pay Attention to the Balance of Fertilization:

It is necessary to pay attention to the balance of nutrients, and it is better not to use too much single nutrient of liquid fertilizer. There is a necessity to take the complete nutrients fertilization into consideration. Especially, too much nitrogen could be harmful to some plants. It is necessary to pay attention to the content of macro nutrients and also the content of trace nutrients.

5.16 Effects and Guidelines of Application for Slow Release Fertilizers

Plants need nutrients at anytime. According to different seasons and different growing periods, there are different amounts of absorption. Therefore, the soil or the medium needs to provide nutrients constantly, in order to maintain or promote the growth of plants. The application of soluble fertilizer should be fertilized by a small amount one time with more times, so that it would not be fixed by soil or leached away. However, frequent fertilization would cost more and increase production cost. Along with the needs and the promotion of science, slow release fertilizer is the good invention for agriculture.

118 Soil and Fertilizer

While slow release fertilizer can release nutrients slowly to get the effects that the fertilizer could provide nutrients to the plants for a longer period, soluble fertilizer dissolves quickly so that the fertilizer nutrients tend to release fast and cannot maintain for a longer period (Figure 5-4). Slow release fertilizers have different types of fertilizers, such as chemical and organic chemical slow release fertilizers.

Slow release fertilizers are useful to be applied in areas with high temperature, high rainfall, and slope lands. Characteristics of slow release fertilizer are especially useful to nitrogen fertilizer which is easily to volatilize or lose, and phosphorus fertilizer which is easily to be fixed by soils. For example, if enough attention is not paid to the application methods, fertilization time, and the amount of fertilizer for slope and mountain lands, it may lead to fertilizer leaching and water pollution. But slow release fertilizers can reduce many of these disadvantages.

5.16.1 The Advantages of Slow Release Fertilizers

Slow release fertilizer has a lot of advantages, including the followings:

Reducing the Loss of Fertilizer:

There are three ways for the loss of fertilizer, including the leaching lost, volatilization, and fixation. Slow release fertilizers are useful to prevent or avoid the loss of fertilizer.

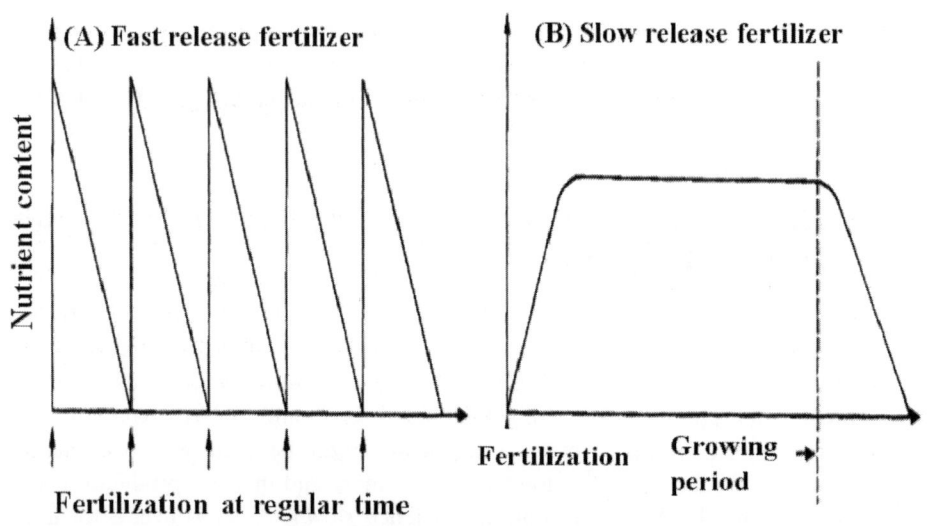

Figure 5-4. The Comparison between the Supply of Fast Release Fertilizer (a) and Slow Release Fertilizer (b)

Improving the Potential of Crop Production:

Because slow release fertilizer provides nutrients in a longer period, it is easy to get the goal of the yield and quality of the crop production. Especially, to those long-term or perennial plants or fruit trees, it can prevent the deficiency of nutrient in the later stage of crop in order to prevent the yield reduction.

Reducing Cost of Production:

The effects of slow release fertilizers can stay for a longer time, so that it can be employed with longer intervals to reduce cost and manpower of production.

Reducing the Environmental Pollution:

Excessive or inappropriate use of fertilizer could also cause environmental pollution. For example, if there are too much nutrients in the water, it would cause the growth of algae wildly and further influence the quality of water and the ecological balance. Slow release fertilizer is helpful to the conservation of nutrients and to reduce the lost under the rain and the volatilization from the air. Especially, among the slope and mountain lands, it is more important to pay attention to the environmental pollution caused by washing away of fertilizers.

Although slow release fertilizer has a lot of advantages, slow release fertilizer needs more production expenses. Therefore, slow release chemical fertilizer is commonly more expensive. In the future development, it is necessary to reduce the cost of production of slow release fertilizer in order to widely use it. And that would be the fortune of all human beings.

5.16.2 Types and Materials of Slow Release Fertilizers

The slow release fertilizer has already developed many different types. In different types of slow release fertilizers, it employs the theory of physical barrier, chemical characteristics, and bio-degradable characteristics to get the effects of slow release of the fertilizer. Usually, the slow release fertilizer is mainly made of nitrogen, because nitrogen fertilizer tends to lose. There are other slow release fertilizers containing phosphorus, potassium, magnesium, and other trace elements. Also, there are inorganic and organic fertilizers or compound organic and inorganic slow release fertilizers, such as chelate and mix organic fertilizers of humic acid and lignin or peat. There are big differences in their sizes and patterns: big ones can get to 1 cm, small ones are powder particles, and some are even in liquid form.

The slow release fertilizer is helpful to slow release with the following technical types:

Using Materials with Low Water Solubility which Only Release Fertilizers under the Condition of Chemical and Microorganism Decomposition:

Nitrogen fertilizer containing urea is the most commonly type of material. The common characteristic of urea derivatives is low water solubility. Its activity can be distinguished by degree of solubility of cold water and hot water, among which urea aldehydes (such as condensate urea of formaldehyde, butenoic aldehyde, and isobutylaldehyde) are common and popular products.

Using Insoluble Minerals:

Many minerals have low solubility which can be used as slow release fertilizers, such as the minerals of gypsum urea, magnesium ammonium phosphate, and potassium phosphate, which have the effect of slow release.

Using Fertilizers of Soluble Materials which Need to Be Decomposed:

Since being absorbed or chelated by soil particles may lead to the effect of slow release such as substance that tightly holds together iron, manganese, zinc, and copper, other mineralization is influenced by aeration.

Using Water Insoluble Products Leads to Decrease Dissolving Ability:

To deal with chemical fertilizer for inert (hydrophobic) material could use surface coating or make into films to form slow release fertilizer in granules or capsules, including three types:

- Solid membrane which decomposes depended on soil microorganism, such as sulfur coated.
- Impermeable membrane with holes where fertilizer could flow, such as plastic coated.
- Semi-permeable membrane which enables the absorbed water to dissolve fertilizer and to produce high osmotic pressure which allows the fertilizer flowing out.

The Inhibitor Based on Micro Organic Characteristics:

Some fertilizers should be transformed or decomposed by microorganism first before used by the plants. However, once the fertilizer dissolves in large amount, the fertilizer tends to lose or to be fixed. Therefore, restriction of enzyme action of the microorganism or restriction of enzyme activity accumulated by soil can help to keep the fertilizer. The common effect of microorganism is to inhibit nitrification and urease. The reduction of the formation of nitrate and ammonium are useful to keep stabilizing nitrogen forms.

5.16.3 Methods and Guidelines of Slow Release Fertilization

The slow release fertilizer is widely applied; for example, it is suitable for all kinds of grasses, orchards, agriculture, gardens, slope lands, and greenhouse culture. Methods and guidelines of fertilization for slow release fertilizer are in the followings:

Choose Appropriate Fertilizer Formula and Types:

The slow release fertilizer has many different kinds, among which nitrogen fertilizer is the most. Some of them use compound formula fertilizers, and appropriate fertilizer should be chosen based on the needs of the growth of plants; for example, some plants needs fertilizer with higher nitrogen, some needs fertilizer with higher phosphorus and potassium, and it should coordinate with soil fertility in case of arousing the negatives to plants caused by inappropriate amount of fertilizer or over fertilization. If people use organic slow release fertilizer, they not only helps to slow release nutrients, but also helps to improve soil properties.

Coordinate with Soil Characteristics:

Soil has differences in aeration with the different environment of upland and lowland. Some slow release fertilizers are easily to be dissolved in soil of anaerobic conditions, but others are likely to be dissolved by aerobic condition. It has to be known clearly and be coordinated with the soil characteristics. Besides, since sandy soil is easily to lose fertilizer, slow release fertilizer can be used to reduce the loss of fertilizer.

Coordinate with Deep Placement of Fertilizer:

For deep rooted plants, deep placement of fertilizer should be used, in which more roots are able to absorb fully nutrients. And to the whole plants, the deep placement of fertilizer is helpful to crop growth and production.

Cooperate with Enough Soil Water:

The slow release fertilizer provides nutrient constantly, and this kind of release needs enough water from soil. If the soil is too dry, then the spread of fertilizer would be a problem. Therefore, slow release fertilizer can get the fully effect only under the condition of enough water moisture. But it should pay attention to upland crops that cannot be immersed in the water for a long period.

Appropriate for Gardening and Pot Culture:

Potted plants need water frequently to provide soil water, which is a problem for the nutrient leaching. Using slow release fertilizer keeps providing nutrient and helps to the growth of potted plants in family gardening or road trees. Choose the slow release fertilizer in big particles which can hardly be washed away, or the slow release fertilizer which is easily stayed in the soil.

Using Organic Slow-Released Fertilizer to Release Slowly the Nutrients:

The compound fertilizers of meals, peat, humic acid, and lignin, the effective organic slow release fertilizers that have slight different speeds of release, are used as basal fertilizers which provide nutrients for a long period, save the costs of fertilization for many times, and also effectively improve the physical, chemical, and biological properties of soil.

5.17 Fertilization and Quality of Agricultural Products (1): Nitrogen Fertilizers

5.17.1 The Purpose of Fertilization

The purpose of fertilization is not only to increase the yield of agriculture, but also to improve the quality of agricultural products. Especially for these years, along with the development of life, the requirement for agricultural products is increasing, including the quality of agriculture, gardening, pasturage, livestock and raw materials of food. Although the quality of agricultural products is influenced by climate, crop variety, soil, cultural management, pest, and disease control, it is also largely affected by fertilization for the particular variety under particular climate and area.

It is reported that in the diseases of human beings, almost half of the diseases are directly or indirectly caused by insufficient or inappropriate nutrient. We should pay attention to the content of beneficial compounds and active compounds in foods. Appropriate fertilizer is good to the crops, but over fertilization would arouse negative effects to crops, environment, and even human beings. Therefore, it is needed to pay attention to appropriate fertilization.

5.17.2 The Concept of Quality and Factors Influenced Quality

People have different quality standards to different agricultural products. Good or poor quality has different standards according to different points of view. Usually, it can be classified into commercial quality and food quality:

Commercial Quality:

Commercial quality refers to the market quality for the sale of agricultural products; it focuses on the appearance of color and fragrance, characteristics of storage property, and quality of taste and sweetness of products. It requires the qualities of products, including oil content, sugar content, starch content, and protein content, for different food processing requirements. Good or poor commercial qualities are closely related to the prices of products.

Food Quality:

Food quality refers to the nutrient value of the food which is the inner beauty of the food. The consumer hopes to get healthy, nutritious, and nontoxic agricultural products. People and livestock want to keep healthy, so that they need agricultural products with higher nutrition. In order to get nutritious products, it is necessary to get healthy and unpolluted soil which can make sure that there is healthy and safe food. Good or poor nutrients of food is hard to judge by the appearance of food, so that it is hard to influence the price. But the nutritious value should be put more emphasizes on.

It is not possible to exactly assess the quality of nutrient of food; therefore, some concepts should be inferred. The most controversial point is the naturality of food. Naturalists believe that the food which is produced naturally is good and the food which is produced unnaturally or man-made is bad. With this point, we can infer that the food produced by using natural fertilizer (such as compost, animal manure) is good and valuable, but the food used artificial fertilizer is not good as the first one. The opponents think that the points of naturalists are not all correct. Food produced naturally is not all good and there is bad and toxic wild food.

The quality of agriculture food is affected by many factors as we have mentioned in the former part, such as climate, crop variety, soil, cultural management, pest, and disease control, all of which would influence the quality. The genetic factors of variety determine the characteristics of basic quality; climate of environment (sunlight, temperature, etc.) and elements related to the soil (supply of moisture and nutrients) will help to fulfill its genetic potential or change its results.

The skills of fertilization influence nutrients provided to plants and adjust its effects. Fertilization plays an essential role in determining the quality of agricultural products. Using inorganic and organic fertilizers can help to improve the quality of agricultural products, but it can also bring some reverse effects under the inappropriate fertilization.

For these years, there are some questions related to quality: Can the traditional fertilization method (especially organic compost) which we have used from the ancient times till now bring high quality agricultural products? As we have used chemical fertilizers on a large scale, although it has increased production, does it reduce the nutrient value of the products? In agriculture, it is increased in using chemical fertilizer in unit area of production in order to get high yield, does it reduce the quality? All of these questions are worthy to be focused on. To answer these questions, it needs to base on researches, experiments, and speculations to analyze. There are different opinions, so it needs more and deeper researches and experiments. It is commonly acknowledged that appropriate fertilization is helpful to increase quality, but inappropriate or over fertilization may lead to reverse effects.

5.17.3 Nutrient Supply of Crops and Food Quality

Fertilization provides nutrients for crops which directly affects the food quality. The amount of chemical fertilizer should be appropriate, especially when soil is deficient in the supply of nutrients; fertilization is the most effective way to increase yield and quality. When the soil could supply appropriate nutrients, fertilization sometimes can increase the quality and sometimes cannot; but over fertilization commonly leads to decrease yield or reduce the quality, and even pollute the environment or source of water. Therefore, high economic crops should not be fertilized blindly.

5.17.4 Relationship between Nitrogen Fertilizer and Quality

Nitrogen fertilizer is the main factor for controlling the crop production. Nitrogen fertilizer is closely related to food quality, and the following paragraphs will explain nitrogen fertilizer caused to the changes of food quality:

Protein Content and Values:

Protein is constituted of amino acids which contain nitrogen. Protein is the important nutritious factor in human foods. Therefore, appropriate nitrogen fertilization is helpful to compose amino acids and proteins in crops. However, over fertilization of nitrogen fertilizer tends to arouse dilution phenomenon and reduce the unit content, and the extra nitrogen is likely to stay in roots and leafs. The types and content of protein in grain affect the quality of cooking, and grain which has higher protein of gluten is helpful to increase the cooking quality.

When the Supply of Nitrogen is Excessive, There are Common Materials Contained Nitrogen in the Plants Found:

- The accumulation of nitrate tends to occur in leaf tissue of plants, especially at low sunlight, the reductive metabolisms of absorbed nitrate reduces and accumulates more nitrate in the leaf. It is not good to contain excessive nitrate in the food, and it is better not to exceed 50 mg/kg (ppm) in case of the potential danger of forming nitrite and N-nitroso compounds. Besides, if vegetables are stored in deficiency of air, the leaf contained nitrate would be formed to the harmful nitrite. When people eat them, nitrite oxidizes iron of hemoglobin to influence the ability of blood carrying oxygen.

- The increase of nitrosamines and betaine tends to occur in plant tissue. Although it commonly accumulates not so much in plant, it still needs us to pay attention to these harmful materials and their potential dangers.

Increasing Nitrogen Fertilizer Arouses the Changes of Some Materials:
- The content of vitamin B1 increases in grain cereals, but over fertilization of nitrogen fertilizer usually reduces B1.
- The most appropriate nitrogen fertilizer would increase the content of chlorophyll and carotene.
- Using nitrate fertilizer easily accumulates oxalic acid which is a harmful material. But it also should be paid attention to use ammonium fertilizer in upland soil, because most of ammonium would convert into nitrate and be absorbed by plant.
- The content of vitamin C reduces, especially when the nitrogen fertilizer is excessive.
- It increases slightly hydrocyanic acid in forage, but over hydrocyanic acid is also harmful.

In conclusion, appropriate nitrogen fertilizer is good for increasing productivity and germinating ability of crops, but over fertilization of nitrogen fertilizer would bring the former disadvantages, and it would even be likely to get worse cropping which include the increasing susceptibility to diseases and virus, excessive growth, too many branches, few flowering, and dropping of flowers and fruits of crops.

5.18 Fertilization and Quality of Agricultural Products (2): Phosphorus, Potassium, Calcium, Magnesium, Sulfur and Trace Elements

5.18.1 Relationships between Phosphorus Fertilizer and Quality

Phosphorus plays an important role in plant metabolism and also plays a "central role" in the crop quality control. Phosphorus influences quality on two aspects, which are the proportion of phosphorus content/phosphorus compounds content and the content of beneficial materials/harmful materials. Phosphorus influences the health and quality of the crop and seedlings. When the supply of phosphorus fertilizer is enough, it will change phosphate content and the constituent of phosphate compounds. The following paragraphs are the explanation:

The Total Content of Phosphorus Increases in Crops:

Apply phosphorus fertilizer in the soil, but the absorption of phosphorus is controlled by the restriction of the soil. Therefore, the increase of phosphorus content is limited, which means that it is hard to get a high accumulation of phosphorus in plant. The phosphate content of forage plants is an important quality standard. It is harmful to cattle reproduction if there is not enough phosphorus.

The Changes of Proportions of Phosphorus Compounds:

When there is no phytine (a phosphorus compound) accumulating in green part of plant, it will increase in the form of inorganic phosphorus. Phytine also accumulates in some grains, but it accumulates inorganic phosphorus in leaves and the phosphorus of nucleotide increase slightly, while phospholipid keeps no increase at all.

Appropriate Phosphorus Fertilizer is Helpful to the Changes of the Following Materials:

- Increase the content of carbohydrates (sugar, starch).
- Increase the content of essential amino acids in grain.
- Increase the content of crude protein in leaves.
- Increase the content of vitamin B1.
- Reduce the content of oxalate in leaves.
- Reduce the content of nicotine in tobaccos.

5.18.2 Relationships between Potassium Fertilizer and Quality

The way by which potassium fertilizer influences the quality of crops is unlike the way nitrogen and phosphorus fertilizer influence crops quality, for which the content of nitrogen and phosphate is the most important factor. Among the food of both human beings and animals, the balance of potassium and sodium is important. It is better not to absorb too much sodium which makes the importance of potassium becoming unavoidable.

Crops absorbing a large amount of potassium lead to reduction of the content of calcium, magnesium, and sodium. Therefore, excessive potassium fertilizer tends to arouse the deficiency of the former elements. Potassium has big effects on adjusting swelling capacity of cells and enzyme activity, and it also affects the whole metabolisms and the quality of vegetables.

Appropriately, fertilization potassium fertilizer changes crops in the following ways:

- It strengthens photosynthesis, and it is helpful to increase carbohydrate (sugar) content which means that the crop has more content of polysaccharide, starch, and crude fiber. Fertilization potassium chloride to the crop which is sensitive to chloride, the effect of potassium is to increase the polysaccharide content, but the effect of chloride is to restrain polysaccharide content in leaves which is the contrary effect.

- Appropriate potassium helps to reduce the oxalate content; however, excessive potassium would increase the oxalate content which is a harmful increase.

- It can increase the vitamin content, such as the precursor of vitamin A (carotene), vitamin B1 and vitamin C. Vitamin B1 would decrease because of excessive potassium fertilizer, which may be related to chloride ion from potassium fertilizer.

- It can increase polysaccharide indirectly to form dilution effect and then lead to the reduction of crude protein content, but it is also possible to increase protein content which may be caused by less potassium and other materials contained nitrogen. Besides, fertilization of potassium fertilizer leads to reduction of chlorophyll content.

- For root tubers (such as potato) containing starch, potassium fertilizer helps to reduce the decomposition of starch, which helps to increase starch content and also to increase potato quality.

5.18.3 Relationships between Calcium, Magnesium, Sulfur and Quality

Calcium is an important factor which takes responsibility for adjusting liquid flow and forming pectin of plant cell wall, and calcium content is crucial to affect the quality of forage. Increase the content of calcium help to reduce the concentration of magnesium and potassium, and the increase of calcium content has a limitation. A calcium deficiency influences the quality of fruits and reduces its commercial value. Especially, a calcium deficiency tends to arouse diseases such as brown patch of apple, and also tends to influence the storage of fruits.

Magnesium is an important plant constituent helpful to adjust enzyme activity. Therefore, it largely affects quality. Appropriate magnesium fertilizer helps to increase the content of chlorophyll and carotene and even increase the content of total sugar.

Sulfur is the basic element which forms the essential amino acids. A deficiency of sulfur would reduce the quality of protein. Some plants, such as cruciferae and onions which have secondary metabolites containing sulfur such as mustard oil, onion oil, etc, and a deficiency of sulfur would affect the compound of these materials containing sulfur. Sulfur is important not only for its quality, but also for increasing the disease resistance ability which could influence the quality indirectly.

5.18.4 Relationships between Trace Element and Quality

Deficiency of trace elements would lead to the problem of metabolisms which would definitely influence the crop quality. Therefore, good quality of crop at least needs appropriate supply of trace elements. Excessive supply of trace elements will lead to harmful results and poor quality.

Iron, which is included in leaf vegetables, is an important factor to the supply of human beings. But the iron content in the plant changes dramatically, which is mostly influenced by soil reaction (pH), but not fertilization.

Manganese is an important standard in the quality of food and forage. However, manganese fertilization would increase the content of manganese only when the mobility and effectiveness of soil is good. Appropriate manganese can increase vitamin content (such as vitamin C, carotene, etc.).

Fertilization copper will increase the copper content in the plant, and it is mostly obvious in forage. Copper is important to protein content and quality. A deficiency of copper tends to cause the spot problems of fruits and excessive copper would cause hazards of heavy metal.

Using zinc will increase the zinc content of plants. Over fertilization of zinc will make the zinc content of plant getting to more than digits. Zinc content is an important characteristic of crop quality.

Fertilize boron will increase boron content to get the goal of increasing crop yield. Appropriate boron would increase sugar content and increase the commercial value of the quality of fruits and vegetables; a deficiency of boron would arouse spot diseases and wrinkles in new leaves which reduce the quality and more.

Molybdenum is also an important quality standard. The effectiveness of molybdenum is related to soil characteristics. Molybdenum is a basic factor of nitrogen metabolisms; for example, reductase of nitrate affects synthesis of amino acids and proteins. Therefore, it greatly affects quality.

In the relationship among soil, plants, and human beings, sometimes there is no symptom of insufficiency in soil, but it would lead to the low content of plant and influence the health of human beings. For example, it is hard to identify the problem of green plant by its appearance. If it has low molybdenum content and the plant has no symptom, human teeth would be affected by eating these foods which has low molybdenum content for long-term diet.

5.18.5 Relationships between Fertilization, Food Quality, and Health of Animals and Human

Soil, plants, animals, and human beings are closely related to each other. The health of human beings is influenced by the food chain which is closely related to the nutritious supply of soil. Fertilization is related to the nutritious value of plants, because fertilization is related to the quality of plant contents, or related to the effect of deficiency of nutrients and the effect of food.

Quality of Food Constituents:

It is hard to assess the valuable constituent content on the whole. Because the proportions of essential substances, beneficial substances and harmful substances included in all kinds of food are different. Commonly, the valuable constituents are the followings:

Essential Amino Acids:
Eight or nine essential amino acids are important factors to form protein. Both nitrogen fertilizer and the variety of crop would influence the protein content and the formation of amino acids.

Fatty Acids:
Three kinds of essential fatty acids is important factor of forming lipids.

Vitamins:
There are about 15 kinds of important vitamins, including vitamin A, D, E, K, B1, B2-complex, B6, B12, C, H, etc.

Mineral Substances:
There are about 20 kinds of important mineral substances needed by human beings.

Beneficial Substances:
Beneficial substances and antagonist substances of color, fragrance, and taste are important factors of food quality. Antibiotics included in soil or compost is absorbed by plant, and could play a disease-resistant role. But it still needs more researches.

Effective Standard of Food:

Fertilization improves the inner quality of crops, and there are some results: in the experiments of mice, the result is to increase growth; in the experiments of babies, it helps to put more weight, increase blood cells, and strengthen resistance of diseases. At least, fertilization of crops has no obvious result of reducing the food quality.

5.18.6 Conclusion

The right way of fertilization should be established, which means that fertilization is not the more the better, and it also does not equal to the excess plant growth is the better. We should pay attention to the inner quality of plant, so that appropriate fertilizer is important. The disadvantages of over fertilization are easy for people to find out only when the reduction of yield happens, and it also reduces the quality of plants. And it is hard to assess excessive fertilizer for farmers. For example, the quality and safety of crops cultivated by water culture still needs more researches, and, we should especially pay attention to vegetables for the health of human beings.

5.19 Advantages and Disadvantages of Vegetables Cultivated by Water Culture

5.19.1 Water Culture has Great Potential, but We should Pay Attention to the Disadvantages

Water culture (hydroponics) is that liquid nutrients directly supply crops by using planting matrix of liquid state and solid state, and even cultivates in the way of hanging the root in air. Water culture has great potential in the development of agriculture. The supply of nutritious liquid has different ways of soaking root, spraying or half-soaking. The effects of different methods of water culture are different to all kinds of crops.

As soon as water culture realizes all nutritious elements of plants, it was used as a tool of pure science by botanists to study the symptoms of a deficiency of different kinds of nutrients, or the symptoms of poisons and infections. The principle of water culture is very simple. It only satisfies the basic needs of the plant which are oxygen, water, and nutrients. Especially to upland plants, we should pay more attention to the oxygen supply of roots to help the cellular respiration. But the root of hydrophytes (aquatic plants) has a special structure; for example, pore canals and air sacs transfer oxygen from over ground to root system. Few upland plants has good water resistance, because after soaking it produces aero parenchyma, and some plants has lower adaptability. In the whole root system, the method of partly soaking and the other part appearing in air can be used to fulfill the adaptable conditions of the plant.

According to different plants and different varieties, there are differences for the adaptabilities to the surrounding environment of plant roots, for example ion types, concentrations and proportions, pH, electric conductivity (EC), and temperature of liquid nutrients would cause the different degree of response. Therefore, each plant or variety has its optimum, minimum, and maximum of requirement.

In recent years, media culture has been given increasing attention, and it has great potential development on horticulture. Because the consumers care about the pesticide residues, water culture is widely used in growing vegetables. In fact, we should pay attention to all the food of cultivation on quality, safety, and nutrient balance. All these knowledge of water culture needs more researches from experts and to be understood clearly by consumers.

From the perspective of biological evolution, the existent living beings has experienced selection naturally for thousands of years. However, what do humans depend on? We all know that food plays an important role, but food contains so many substances, so, what are the roles of these substances in metabolisms of human body? The scientific knowledge gained from the researches is limited, and many questions hide their mysteries. For a long period, people depend on soil which contains inorganic and organic substances. Science community has a lot of knowledge about inorganic nutrient, but has little knowledge about what the roles of soil organism and many other trace elements or materials play in the metabolisms of humans. Because among the substances included in food, one part is synthesis by plants and animals, and the other part comes from soils. Soil provides not only inorganic nutrients but also the organic substances to plant. And the roles of these substances to human health are still a mystery that needs more researches and explorations. It helps to provide a complete and balanced formula especially to human health, rather than merely to provide basic nutrients to plants.

5.19.2 The Advantages and Disadvantages of Water Culture

To different kinds of plants and different adaptability of districts, different cultivation methods of plants are different. Water culture is exactly the same and it has its advantages and disadvantages. Taking its advantages and repairing its disadvantages is the direction and guidelines of the researches. Following paragraphs are the advantages and disadvantages of water culture:

Plants of Water Culture Grow Quickly:

This is the biggest advantage of water culture. But it needs good environment of growth; for example, it needs appropriate temperature and sunlight. And high temperature of greenhouse in summer would influence the growth of many plants. Because nutrients in water culture supply plant root successively, that is to say, nutrients in water culture contact directly to the root system without any block, which enables the plant fully absorbing nutrients, so it can grow much quickly. If all seasons could grow good crops, or people can choose appropriate plants to cooperate with the environment, it can improve the productivity in unit area.

Water Culture Reduces the Sources of Plant Pathogens:

This is also the advantage of water culture. But it does not mean that the plant of water culture would never have problems in pests and diseases, and it also does not mean that plants of water culture are not spray with pesticides. Matrix and material of water culture could reduce the plant pathogens, but pests and diseases of plants are not derived from matrix of cultivation, and it also comes from air, seed, and the surrounding environment of cultivation for a period. Therefore, in order to reduce the other sources of pests and diseases, the design of greenhouse, net house, or other facilities and equipments should be worked on, too. Commonly, water culture has less pests and diseases at the first time, but after many turns of cultivation, it needs to strengthen environmental protection; otherwise, pests and diseases would be as serious problem. Generally speaking, fast growing plants in water culture have low resistance to insects and diseases, and it even goes out of control, so at that time it needs to spray more pesticides.

Water Culture Needs to Establish New Technology and Investment Cost:

Water culture is not hard, but the technologies and guidelines of production should be fully understood. To produce plants have economic effects, we should pay more attention to nutrient matrix and seasons to cooperate the crop's various needs, and we also need to study and explore the ways to reduce investment cost, save troubles, and employ automatic equipments to reduce more labor. Especially when producing on a large scale, it has many problems which needs more understandings and cultivations; otherwise, it tends to result in failures. If it is managed inappropriately, with the dramatic changes of concentration and pH of nutrient solution, the root of plants is quickly hurt. Therefore, regulation and formula of nutrient solution need more attention.

Water Culture Needs to Pay Attention to the Growth of Algae and Microorganisms in Nutrient Solution:

The cultivation system of water culture is hard to prevent microorganisms and algae growing, and these microorganisms would absorb part of nutrients and oxygen. Algae have photosynthesis to release oxygen in the daytime, and it goes the other way round in the night. And some microorganisms even produce a great amount of poison substances and may even influence the growth of plants. If pathogens invade, it would spread more quickly.

Water Culture Needs to Pay Attention to Nutrient Formula and Safety of Production:

Nutrient solution of water culture is commonly used to help the plants to grow well and quickly, but the formula of water culture may be deficient in the substances which human needs, especially with the limitation of science knowledge, the production of water culture cannot be inferred just such as the production in soil. For years, life science has a big leap in development, but it is also lack of knowledge about disease resistance and cancer resistance of human body, and also the knowledge of the relation between health and natural substances. What kind of deficiency plants in limited solution formula will lead to cannot be estimated at once, but should be discussed and thought deeply.

Take an example as explanation. Now science has proved that Se is one of the trace elements of human body. What's more, it is reported that cancer, heart disease, and other diseases is related to the Se content of human beings. In the past, people know little about how important Se is, because the food cultivated in soil has this element. Same as other heavy metals, the excessive Se amount leads poisoning. If people do not know Se as an important trace element, who cares to add Se into nutrient solution? Eating unsound food or tending to eat unsound food for a long time would be the source of problem.

We should make it clear that plants of water culture can directly absorb nutrients which are the nutrient of plants, but not enough for the health of human body. Especially, for those plants which absorb a lot of inorganic nutrient, if it cannot be converted or accumulated, the safety should be questioned. We should control nitrate or nitrite nitrogen, and pay special attention to the problem of heavy metals included in plant nutrients. We should observe and inspect carefully and strengthen the quality of production.

In Water Culture, It should Pay Attention to the Recycle of the Matrix or Nutrient Solution to Reduce the Pollution to the Environment:

After many times of using the nutrient solution or matrix, it can hardly produce good production, because of the released exudates from plant roots, so it should be changed or discarded. Inappropriate discard may cause environmental pollution, especially the discard of the used nutrient solution. We should value the position of discard and do more researches on recycling.

5.19.3 Strengthen Basic Researches on Water Culture

Water culture has its advantages and disadvantages alike. In cultivation of non-edible food (such as flowers), water culture has great potential, and it needs more basic researches to overcome its disadvantages and to help the consumers have a better understanding of water culture. Soil is the matrix which people has been counting on for thousands of years, and soil has a lot of organic and inorganic substances which may play unknown roles in the human body. The knowledge that hasn't been recognized can only be called unknown one rather than be denied completely. Just like that we can only admit the invisibility of one certain thing rather than deny the existence of it, if it is invisible. To research the cultivation of the soil, we should not deny the valuable property of it because the knowledge mastered by people nowadays is really limited and is worth the lucubration.

The greenness of the plants does not represent that it is safe and nutrient-balanced food. It should be known that some certain special cultivating methods are not reasonable (such as water cultivation). Even though people fear the poisoning of pesticides, this immature cultivating method may bring in another problem, and it is worth the lucubration.

5.20 The Application and Problems of Fertilizer

Ever since the fertilizer has been promoted into the market, farmers can buy the fertilizer from the market. Therefore, there appear some problems about application of the fertilizer. Soil is the resource handed down from generation to generation, and is worthy attention and concern. The followings are some ideas for further discussion.

5.20.1 Never can One Fertilizer be Omnipotent

Much earlier, the farmers usually were offered nitrogen, phosphorus and potassium fertilizers, which hardly brings any problems. The application of limited fertilizers rarely causes any problems such as unbalanced fertilization and there is hardly any problem of inappropriate formula of fertilizers. The mixed fertilizers may bring in many problems, but the problems may be attributed to the misuse of the fertilizer or the long-term application of the acid-forming fertilizers rather than the property of the fertilizer itself. Nowadays, there appear many kinds of single fertilizers (incomplete) and compound fertilizers, especially the single nitrogen, potassium, phosphorus, magnesium, calcium, or compound fertilizers of two of the fertilizers above. Those fertilizers require some compatibility among each other and consideration of the property of the soil for fertilization. The emphasis on one certain single fertilizer is unreasonable, because any kind of the fertilizer cannot be omnipotent. The soil properties and fertility in different places differ from each other. When there is excessive use of one certain fertilizer to the soil, more fertilization will cause ineffectiveness and yield reduction. The more fertilizer does not get better. Conversely, the common phenomenon is that the excessive use of one certain fertilizer will cause the accumulation of the salt in the soil, and the unbalanced use of the fertilizer which is worthy of consideration.

5.20.2 Reinforcement of Farmer Education on Fertilizer Application

The toil of farmers is beyond imagination and worth our admiration. Because there is rarely any kind of the education programs for farmers, and even the existing programs usually serve as sale promotion, few farmers can master the essentials of the fertilization knowledge. Problems have accumulated during many years and are worthy of attention. Because the soil is a finite resource on the earth, misuse of the fertilizer will pollute environment and agricultural products, and it would be a loss of farmers and human being.

Lack of the knowledge about fertilizer and the fertilization will lead to disordered or over application of fertilizers. It often happens that farmers claim to be taken in by sellers, because usually farmers emphasize the notion of "witnessing the fast death of pests symbolized the efficiency of the pesticide" and hope for the fast efficiency of fertilization. Some fertilizers, such as phosphorus, potassium, and organic fertilizers, unlike the fast visible response in plant for nitrogen fertilizer, require to be mixed with other fertilizers for the better efficiency. The fertilizer efficiency cannot be exerted completely; when one fertilizer is used under the influence of some limiting factors (include some physical, chemical, and biological factors). An inappropriate use of the fertilizer will cause similar problems. Besides, there is a misunderstanding of the price of the fertilizer symbolizing its worth that makes the farmers deceived. This phenomenon is attributed to the insufficient knowledge of fertilizers and guidelines for fertilization, or misguide of the vendors and the exaggeration of the advertisements. Therefore, the farmer education on the fertilization and the improving of the soil should be reinforced to protect the farmers.

5.20.3 Concept of Application of Fertilizers and System of the Vendors

There are many business channels for the fertilizer circulation in the market ever since the fertilizer has been promoted into the market. All kinds of fertilizers are released in the market. In earlier times, when there was no establishment of the managing system of fertilizers, the vendor's management, or the fertilizer technician, nor was there enough master of the knowledge about fertilization, the problems appeared easily. Continuous application of fertilizer requires professional instructions and knowledge, while a blind application of the fertilizers without instruction of vendors will cause a lot of problems. The application and effect of fertilizer and pesticide are different from each other. Farmers usually require a try of the pesticide to see fast efficiency of it, usually with the fast one exerting function within few minutes or hours. However, the function of a fertilizer is much slower and differs from each other according to the period of plant growth. Some farmers also require having a try of the fertilizer. The fast one exerts effect within few weeks while the slow one will last longer. Few fertilizers such as the nitrogen fertilizer will present the function through the visible green color of leaves. The production and the quality of plants are not prone to be noticed, and there are a lot of factors that influence the final result of the production besides the fertilization. So, if the quality of the products fails to reach the standard, the result should be attributed to other factors such as the climate and the pests rather than to the fertilization. In this case, the trial of a fertilizer required by the buyers is unreasonable, because they think all the results should be attributed to the quality of fertilization, which will lead to a lot of misunderstanding and problems.

The fertilizer is the food of the plants. As long as the fertilizer is qualified for a product, the fertilizer will be useful. People buy the food such as oil, rice, soy sauce, vinegar, tea, vegetable, and the fruit without considering their "effect." Similarly, why should the fertilizer be tried before order and being taken home? If one fertilizer is qualified, it can be called a commodity. People's eating habit is similar to the process of fertilization. Like incorrect eating habit, unbalanced absorption of nutrient will cause a lot of problems. Therefore, considering the plant absorption, it is necessary to know the right way of fertilization.

Selling chemical fertilizers entails professional education training and the establishment of technician system. The chemical fertilizer is different from organic fertilizer. Even though disordered or over application of organic fertilizer will pollute the environment, the disordered use of chemical fertilizer will easily cause damage and problems to plants, soil, environment, and especially human beings, which is uneasy to recover or noticed and thus easily ignored by the non-professionals. Organic fertilization is similar to the natural food of human being such as the vegetable, rice, and wheat, rarely leading to problems, except for pollutants leading to some diseases. However, just as the chemicals or medicine for human being entails people's concern, the selling of the chemical fertilizer requires the obtaining of professional license and the establishment of technician system. Most importantly, the fertilizer will influence the food safety for human being and the quality of the environment; therefore, the selling and application of the fertilizers should be emphasized, and related commercial and government unions should be concerned.

5.20.4 Concern on Conservation of Soil Environment and Biological System

Throughout many years of biological evolution of the nature and the soil, there forms the mutual influence between the creatures and the environment. The mutual promotion as well as the prevention is always under operation. However, when any kind of the organisms (no matter higher and lower) is introduced in to a new place, without the mutual promotion or restraint, the phenomenon of overgrowth occurs and thus violates the natural biological balance. The illustrations of problem are the introduction of channeled apple snail (*Pomacea canaliculata*) into Taiwan and rabbits into Australia. Besides those problems that are visible to human eyes, another kind of the invisible microorganisms will cause problems, too. There is a counterbalance among the local microorganisms; however, the overgrowth of one certain kind of microorganism will appear after the introduction of it. In this case, the biological unbalance will appear and the derivative problems such as the soil deterioration and the infectious disease in waters will be caused. In order to create more benefits for our descendants, this problem merits some attention by related organizations.

The soil is used by the human being for crop production. However, the soil is just like the limited company; it requires cultivation besides utilization, unless this kind of company go bankruptcy and the soil is converted into decline after utilization. There is a group of microorganisms in the soil that are silently at work for our nature. The natural circles are made up of many kinds of microorganisms, and every circle is closely related with each other under operation. Every circle breaking down will violate the whole process and cause bottleneck and problems for the circulation of the natural elements. However, under big-scare application of pesticides and chemical fertilizers, how many microorganisms have been harmed so far? What's more, it is beyond calculation how many microorganisms in the soil cannot be cultured or researched under artificial cultivation. It is necessary to prepare a "genes bank" for preserving the information for the soil to preserve some valuable resource material for human being in the future. This kind of new idea should be paid attention to.

5.21 Common Problems and Improvement of Fertilization

5.21.1 The Inappropriate Fertilization will Cause Big Trouble

In the rainy, hot, and humid regions of subtropical and tropical area with an obvious dry season, the response of plant and soil should be noticed, even in a short period. The agriculture is a business that should be handed down from generation to generation and is a long-term career. No matter how advanced the science is, the human being need to utilize the soil for survival, and the soil should be protected rather than be impaired.

It is unrealistic to forbidden the use of fertilizers and pesticides in the agriculture. The misuse of fertilizer will impair the soil and the agricultural production, which will directly affect farmer incomes. Some problems in agricultural products such as the accumulation of the nitrate and other poisoning matters in the vegetables will not meet the consumer demand. We must take a cautious attitude toward it.

From year's observation, the problems of the fertilization frequently happen and are worthy of attention in all aspects. Considering of the benefits of the farmers, businessmen, and consumers, some problems of fertilization are for reference.

5.21.2 Recognizing the Function of Fertilizers

To apply fertilizer is to supply the nutrient necessary for the plant and improve the soil environment, and thus to reach the better soil physical, chemical, and biological situation and improve the yield as well as the quality of the product. The fertilizers applied include chemical, organic, and microbial fertilizers. The fertilizer's function is to offer nutrients for the plants and their function differ from each other. Therefore, to recognize the fertilizer is the primary job. We have to know what kind of fertilizer it is and how much nutrients does it contain. Additionally, the period, the method, the amount of fertilization, and the necessity of using other kinds of fertilizers should also be known.

Generally, farmers only regard the fertilizer useful when the fast effect is shown shortly after fertilizer application. This is a kind of misunderstanding. Only when applying nitrogen fertilizer, or only when this fertilizer is the only limiting factor in the soil, the fast effect will be shown. Other kinds of fertilizers such as phosphorus or potassium fertilizer cannot reach this fast effect. So, farmers' impression that the rock phosphate powder, calcium superphosphate, and potassium chloride are uselessness is wrong.

The fertilizer that contains nitrogen, phosphorus, potassium, calcium, magnesium, and microelements is useful; however, when they are applied to the field, the effect depends on the condition of the soil and the needs of the plant. Some fertilizers may fail to show the effect. For example, the nitrogen fertilizer presents its effect in the plant with green color and its speed of growth, but its ability of enforcing the leaf thickness and fruit sweetness of the plants, disease, wind resistance, and characteristics of enduring cannot be recognized by eyes within a short period of fertilization. Therefore, one simple fertilizer, such as nitrogen, phosphorus, potassium, or magnesium fertilizer, as well as nitrogen fixed microorganisms and phosphate solubilizing microorganisms should be supplemented with each other for the overall functions.

5.21.3 Excess or Shortage of the Fertilization

The inappropriate application of fertilization is a common problem, especially the excessive application of fertilization. Among these problems, the common one is excessive application of nitrogen fertilization. Excessive fertilization for phosphorus, potassium, boron, lime, and organic matter are also occurring. The deficiency of fertilization usually comes from the insufficiency of calcium, boron, and magnesium. The instances are listed as following:

Problems of Nitrogen Fertilization:

The Reasons:

The excessive application of the nitrogen fertilization is the commonest one, which is mainly attributed to the fast effect of the nitrogen fertilizer that attracts farmers. Actually, the excessive application of the nitrogen will hurt the flowering and fruit quality, leading to the no flowering and dropping of flowers and fruits. In spite of the good growth in plants, the yield and the quality all fail to reach the standard. The grown fruit such as melon, citrus, pear, and wax apple cannot storage and are easy to rot. Another common reason is using too much in nitrate, ammonium phosphate, calcium ammonium nitrate, bean cakes, and fish meals. If fertilizers mentioned above are mixed and used in a great quantity, the serious result of excessive nitrogen is easy to occur, and will lead to water pollution. Besides, in rainy seasons, the application of the nitrogen fertilizer makes nitrogen adsorbed quickly and is more likely to lead to the excess of nitrogen.

Ways of Improvement:

Even though the running off nitrogen fertilizer goes on quickly, once there are excessive nitrogen symptoms in plants and serious disease and pest problems occur, pesticides should be used to control these situations, and the production cost will be increased. If the no flowering occurs as symptom of excessive nitrogen, methods such spraying the diluted solution of the monopotassium phosphate or cutting away excessive branches or leaves should be taken to overturn the excessive nitrogen. It is needed to choose right way for the right method. If the dropping of flowers and fruits happens in excessive nitrogen symptoms, it is too late for overturning in this reproduction season.

Problems of Phosphorus Fertilization:

The Reasons:

The deficiency of phosphorus fertilizer is a common problem and the reasons usually are soil acidification or alkalinization. Generally, the phosphorus fertilizer is used as basal fertilization. If there isn't more phosphorus fertilizer applied within one year for perennial crops, there could be a deficiency of phosphorus. The excess of the phosphorus fertilizer usually comes from the wrong usage of single superphosphate as the lime material to improve soil acidification. The miss-guide of excessive use of the phosphorus fertilizer is another reason. When there is excessive phosphorus fertilizer, the root rot, defoliation, the dead branch, and rough stem surface will occur.

Ways of Improvement:

Except for the basal fertilizer, more phosphorus fertilizers should be used when there is a deficiency of phosphorus during growth period. If phosphorus fertilizer is of seriously deficiency, the diluted solution of monopotassium phosphate can be used by the foliar application. And rock phosphate could be used in strong acidic soils. If the excess of the phosphorus fertilizer hurt shoot and root part of plant, especially when the root is seriously hurt, a foliar application of nutrients such as nitrogen, phosphorus, potassium, calcium, and magnesium will help the rooting. For instance, the use of diluted humic acid will help enhance the rooting. No more salt fertilizers should be used for this application; it will put more osmotic stress on roots.

Problems of Potassium Fertilizer:

The Reasons:

The excess of the potassium fertilizer is also common and should be attributed to the misunderstanding in the concept that more fertilizer will be better. The excess of the potassium fertilizer will limit the growth of leaves, and even causes defoliation and magnesium deficiency. For instance, the excess of the potassium chloride will cause a hurt of chloride ion on the woody crop (such as oranges and grapes); application of the potassium sulfate for many years will also cause a serious soil acidification.

Ways of Improvement:

The improvement ways differ from each other according to the different responses of plants. The deficiency of magnesium resulting in excessive potassium should be applied with magnesium fertilizer; a foliar application of magnesium has a better effect. An excess of potassium chloride will cause chlorine poisoning, the soil acidification, and hurt of roots. A neutralizer should be used to improve soil acidification and offer nutrients from foliar application to strength the root for reducing the negative effects.

Problems of Microelement Nutrients:

The Reasons:

The deficiency of boron is the commonest one, which is mainly attributed to shortage of the knowledge of using microelements. It will lead to poor growth of leaves and harm roots. More seriously, the defoliation will happen, like that in oranges.

Ways of Improvement:

When the root are hurt and poisoned because of the inappropriate use of the microelements, a supplement of the nutrients on the leaves should be done for promoting rooting. The humic acid also can be used to improve rooting and the soil conditioning. No more other chemical fertilizers should be used to avoid the worsening of rooting.

Problems of Organic Fertilization:

The Reasons:

The inappropriate use of the organic fertilizer includes the excessive, wrong time, and wrong method of fertilization. These problems should be attributed to the shortage of knowledge for the characteristics of organic fertilizers, and the absence of instruction and the correct application of fertilization. Among all these examples, the excess of the organic fertilizer frequently happens. Usually, the credible performance of organic fertilizer at first application will cause an excessive use of organic fertilizer next year, and soil eutrophication is easy to occur. Worse still, when the compound organic fertilizer contains too much chemical fertilizer, an excess of the application will cause a trouble on root rot, defoliation, and withering. If grain meals without formulation are used before the sowing or during the flowering period, the usual results are the inhibition of germination, leaf yellowing, and dropping of flowers and fruits. Besides, other problems such as the pests and diseases and water pollution will be caused by animal feces. The animal feces sometimes contain higher heavy metals such as copper and zinc.

Ways of Improvement:

If there is an excess of the organic fertilizer, a turning of soil or an application of lime material will quicken the decomposing process. If the root is badly hurt, an offer of nutrients from foliar application should be used. If the leaf yellowing appears, nitrogen fertilizer should be used for nitrogen deficiency at the organic matter in the first period of decomposing. If animal feces and hair of organic fertilizers give off odor, a ditching or furrow should be done and covered by soil to avoid odor and the pollution caused by the running off organic fertilizer. As for problems of the excess of heavy metal in animal feces such as the copper and zinc, feed additives should be improved.

5.21.4 Appropriate Quantity of Fertilization

As soil nutrients, fertilizers usually include chemical, organic, and microbial fertilizers. The functions and characteristics of fertilizers differ from each other according to their different materials and their different strengths and weak points. The users should have the concept of "making use of the strength, to make up for the weak points" to make the best use of nutrients in fertilizer and its function of soil conditioning. Excessive or inappropriate use of fertilizer will harm the soil and plants. Worse still, it will lead to pollution to the environment. The application of fertilizer should be considered especially for the short-term crops. More fertilizer does not symbolize more benefits. Soil cultivation should be taken step by step conscientiously in a long term for conserving our soil resource.

5.22 Problems of Excessive Fertilization and Guidelines for Improvement

5.22.1 Problems Caused by Excess of Fertilization

What is excessive fertilization? When excessive application of fertilizers exerts influence on plants, soils, and environment, it is called excessive fertilization. Unlike chemical fertilizer and organic fertilizer, there are rarely problems of the excess of microbial fertilizer, because carbon sources and energy sources required for reproduction of microorganisms are limited in the soil. When there is enough carbon or energy sources in the soil without applying microorganisms from microbial fertilizer, indigenous microorganisms will reproduce in large scale and utilize most of the carbon and energy sources in the soil, while the additive microbial fertilizer only exerts its influence within the rhizosphere.

The excessive fertilization will lead to chain reactions of plant, environment, and soil. The extent of the excess can be divided into slight excess, medium excess, and serious excess. Different fertilizers will have different influences on plants. The detailed explanation is listed as following:

The Influence for Plants:

Different plants will response differently to the fertilizers. The detained explanation is listed as following:

- A serious excessive fertilization usually appears as harm to plant roots by dehydration; the high osmotic pressure will lead to the withering of the leaves in the initial stage.

- When the excessive fertilization is not serious, plant roots will be hurt partially. Even though the withering can be recovered, the plant's ability to absorb nutrients will be damaged, and its growth will be inhibited if the capability of root absorption is hurt.

- If there is only a slight excessive fertilization, the plant will absorb the excessive nutrients and lead to different responses. For example, if the plant absorbs too much nitrogen, the excessive growth and the dropping of flowers or fruits are easy to occur. If the plant absorbs too much phosphorus, potassium, calcium, and microelements, the plant growth will be retarded. The plant will be small and have dry leaves and rough stem bark.

The Influence for Soils:

The excessive fertilization will directly raise the quantity of salts and electricity conductivity value, fix nutrients and the acidification of the soil which damage the physical property and structure of the soil. If the soil biological property is damaged, it will cause retarded growth and even death of many soil microorganisms and other creatures, leading to soil degradation.

The Influence for Environments:

The excessive fertilization will directly impact the physical, chemical, and biological properties of environments, running off, volatilization and infiltration of fertilizers, leading to pollution of rivers, sea, groundwater, air, and soils. Especially, the excess of nitrogen fertilizer will cause these serious problems. The running off of organic fertilizer and phosphorus fertilizer will cause the algal bloom and problems of water systems.

5.22.2 Improving Ways of Excessive Fertilization

An excessive application of fertilizers should be prevented, because of the difficulty to solve the problems and there are definitely some side effects or harm caused in this process. The methods of improving the excessive fertilization are mainly to remove, or fix and adsorb the excessive fertilizers. The economical benefits and the convenience should be considered when choosing a method.

The ways of improving the excess of fertilizer include physical, chemical, and biological methods are listed as following:

The Physical Methods:

A dilution or adsorption of the fertilizer is suitable to remove or reduce the quantity of the excess of fertilizer. The ways include:

Soaking or Planting Aquatic Crops:
Soaking or planting aquatic crops (such as rice and taro) can be used for increasing solubility, dilution, and draining of excessive salts. Also, the underground channels can be installed to drain out salts. A soaking can lead to a denitrification for excessive nitrate in the soil.

Mixing Soil from Other Places:
The use of the soil from other places can help for a dilution of excessive fertilizers to reduce the high concentration.

Addition of Adsorbents:
The addition of absorbents such as peat, zeolite, and other porous minerals can help to adsorb high salts.

Turning Plow:
The physical method is always applied because of the low possibility of side-effects. A turning plow can be used to mix topsoil with bottom soil for diluting excessive fertilizers.

The Chemical Methods:

Chemical reactions can be used for reducing high salts. The ways include:

Change Soil pH Value by Additives:
The alkaline pH will convert ammonium into ammonia gas, if urea and ammonium contact with alkaline substances such as calcium carbonate, magnesia, and dolomite. When phosphates contact with calcium ion in the alkaline condition, insoluble calcium phosphate is formed. Additionally, application of alkaline substances for decreasing

availability of metal elements such as copper, zinc, and manganese can reduce the quantity absorbed by plants.

The Addition of Similar Chemicals:

If there are excessive cations in the excessive fertilizers, an addition of another kind of cation can replace them. And then they can be washed off. For example, the excessive sodium (Na^+) can be replaced by the calcium (Ca^{2+}) and then easily to wash off sodium.

Arranging Excessive Fertilizer by Additives:

If the excessive fertilizer is converted into insoluble solid by the arranging of additives, the damage can be reduced. For example, the phosphates can be formed into insoluble solid when reacted with calcium, iron, and aluminum materials.

When conducting all the chemical method mentioned above, the side-effect of additives should be noticed, and there shouldn't be too much additive in case of the chain effect.

The Biological Methods:

The excessive fertilizer applied can be wiped off by biological ways to reduce high salts. The ways include:

To Grow Salt Resistant Plants:

Salt resistant plants can absorb mass of excessive fertilizers, and then harvest and remove the salt resistant plants which can reduce soil salts. Usually, the best choice of the plant should be the fast growing plant.

To Apply Microorganisms and Carbohydrates:

An application of salt resistant microorganisms can help to absorb the excessive fertilizers, such as nitrogen, phosphorus, and so on. When added a small amount of carbohydrates (ca. 0.5%) to soil, the beneficial strains in the microbial fertilizers can help to dilute the salt.

All the biological ways above entails the resistance and adaptability of plants and microorganisms, when reducing excessive salts.

5.22.3 Improving Ways for Excessive Fertilization for Perennial Crops

Unlike the short-term crop, it is not easy to solve excessive fertilization in the field of perennial crops, just such as the physical, chemical, and biological methods mentioned above. Particularly, it is not easy to deal with the results of excessive fertilization such as no flowering, dropping flowers and fruits, as well as poor quality. Since methods for improving the excess of fertilizers varies from each other, in order to avoid the negative effect and secondary problems, the method should differ from each other according to different kinds of excessive fertilizers. The detailed explanation is listed as following:

The Excess of Nitrogen Fertilizer:

The excess of nitrogen fertilizer among the perennial crops is common; it can be divided into serious excess and slight excess. The serious excess can lead to the defoliation. The slight excess can lead to spindling and the exuberance of the leaves, no flowering, dropping flowers or fruits, and poor quality. The detained explanation is listed as following:

The Serious Excess:
When there is defoliation, it means that the root of the crops has been hurt. While the serious one can be dead, the slightly-hurt one will be rescued. The improving method is to irrigate water and thus wash off the excessive nitrogen fertilizer as soon as the defoliation is recognized. However, the irrigation time should be less than one hour and the drainage should be fully to avoid root rot. Once rescued, after one month, then added humic acid that can promote rooting, the crop can get a recover and grows quickly.

The Slightly Hurt One:
The most obvious symptom is the spindling and the exuberance of the leaves during a long time. The following measures should be taken to solve this problem:

- Spraying Substances for Preventing the Shoot Spindling:

The substances for preventing the spindling can be divided into the pesticide type and the non-pesticide type. The pesticide type substance should be used according to the instruction of the company. However, in order to reduce the residual effects, the non-pesticide type inhibitor can be used. A moderate spraying of liquid with high osmotic pressure, the nitrogen fertilizers or other nutrients on the leaves will take effect. For example, the mix of monopotassium phosphate, monocalcium phosphate, and borax can be used when 1,000 times diluted. But densities of different crops are different, which needs to be noticed for excess of fertilizers.

- The Application of Anti-Nitrogen Fertilizers and Root Inhibition:

Monopotassium phosphate, monocalcium phosphate, and potassium chloride may cause anti-nitrogen function to depress spindling. However, they should not be over-applied in case of serious root inhibition.

- The Application of Calcareous Materials:

The acidic soil needs the application of calcareous materials (such as lime stone, the oyster shell, etc.), which can convert the ammonium nitrogen (NH_4^+) into ammonia gas (NH_3) that can volatilize to release nitrogen from soil.

- The Flooding Process:

The water resistant fruit trees in the field can be flooding with water to wash off nitrogen fertilizers or denitrification (NO_3^- → N_2O → N_2) to reduce excess of nitrogen fertilizers.

- The Cut of the Branches or Leaves Spindling:

The crops usually absorb water and nutrients by the force of leaf transpiration and root pressure. If spindling branches or leaves are cut, the absorption of nitrogen will be lower. However, an excessive cutting should be avoided for fear that the photosynthesis of crop cannot be enough to promote the crop growth.

- The Application of Microorganisms:

To apply the microbial fertilizer with carbon source (such as 0.5% of sugars) to the root will influence nitrogen absorption by plants. The microorganisms in the rhizosphere will absorb nitrogen for nutrient to prevent the shoot spindling of crops. However, there cannot be too much carbon and other nitrogen sources such as liquid organic fertilizer or amino acids.

The Excess of Phosphorus Fertilizer:

The excess of phosphorus fertilizer is relatively rare. There are farmers who make a mistake by using calcium phosphate as lime to improve the soil condition. When there is an excess of phosphorus fertilizer, the leaves of the plants will get smaller and the bark of tree will get rough and fall off. The improving ways include the physical spraying with water, the using soil from other places, and the turn plowing. Because phosphorus fertilizer cannot be removed easily from the soil, the excessive phosphorus fertilizer only can be mixed with some elements such as calcium, iron, and aluminum to fix into insoluble compounds. The other nutrients such as iron, copper, zinc, manganese, potassium, and other fertilizers should be applied to leaves and to the soil for the nutrient balance.

The Excess of the Potassium Fertilizer:

The excess of potassium fertilizer is relatively rare. When there is an excess of potassium fertilizer, the deficiency of magnesium is obvious on leaves and get small leaves or rough bark. The improving ways are similar the physical ways above -- spraying water, use of adsorbents, add soil from other places, and turn plowing. Because potassium presents in cation (K^+) that is easy to be washed off or fixed by minerals (such as vermiculite), magnesium and calcium fertilizers should be applied effectively onto leaves or into the soil.

The Excess of Calcium Fertilizer:

The excess of calcium fertilizer occasionally occurs. It is usually caused by the excess of lime materials, alkaline soil, and when the pH value rises above 7.4. The excess of calcium fertilizer will inhibit the growth of some crops and cause the deficiency of potassium and magnesium. The improving way is to use the acid peat or sulfur to lower the soil alkalinity. Potassium and magnesium fertilizers should be applied effectively onto leaves or into the soil.

The Excess of Magnesium Fertilizer:

The excess of magnesium fertilizer occasionally occurs and is mainly attributed to the use of dolomite powder or calcium magnesium fertilizers and the alkaline soil. The excess of magnesium fertilizer will easily lead to the inhibited growth of crop, root rot, or the deficiency of potassium and the calcium. The improving way is similar to the above calcium fertilizer. Potassium and calcium also should be applied effectively onto leaves or into the soil.

The Excess of Sulfur Fertilizer:

The excess of sulfur fertilizer is rare to see and is attributed to the application of the sulfur to improve the soil alkalization. Too much sulfur fertilizer will cause soil acidification and lead to the deficiency of many elements in crops according to different kinds of the soil. Another chapter of this book, "Problems and guidelines for management and diagnosis of soil pH," will be a good reference to improve the problem.

The Excess of other Fertilizers:

The excess of many microelements such as boron, zinc, copper, manganese, and iron is commonly seen in many cases of pollutions. Usually, the physical ways can deal with it. The way of improving the most serious cases will be discussed in another chapter.

5.23 Fertilization and its Quality Influence on the Human Health

The food system of the soil-crop-animal-human is really complicated. The safety of human beings and animals is closely related to the quality and safety of crops which is influenced by the soils. The health of modern people is threatened by the soil pollution and pesticides in/on crops. Because the growth period of crops is really short and the crop's resistant capability is high, many pollutants will accumulate in crops. Among which the poor qualified or low yielding crops are not easy to be noticed by the consumers and will enter human food chain. The pollutants accumulated in crops will unconsciously threaten human health. Therefore, every possible influence of the materials applied to the soil should be considered for human health.

In order to get a better knowledge of the influence of fertilization on human health, the appraisal standard can be divided according to the content of nutrition, the influence of deficiency on the health, and the related medical standard. The detailed explanations are listed as following:

5.23.1 The Content of Nutrition and Effects of Deficiency

The quality of food needs to be assessed by the content of nutrition. It is a great difficulty because there are plenty of assessment programs and standards for food quality, or because some properties of food are uneasy to be assessed, or because the existing technology only can master some matters within a certain limited scope, including the essential, beneficial and poisoning matters in the food, but leaving out the unknown matters.

When the beneficial matter is deficiency in the food, the symptom of human and animal is obvious to be observed. However, the symptom can only be observed when the beneficial matter is seriously deficient. Therefore, there is always a possibility of latent shortage. As the report showed, the latent shortage actually is really a serious problem. The beneficial matters for the human being include:

There are more than 50 Essential Nutrients Needed in the Human Body:

Essential Amino Acids:
There are nine amino acids called essential amino acids; because the human body cannot synthesize them from other compounds, they derive from the decomposition of protein. They include leucine, valine, lysine, isoleucine, threonine, tryptophan, phenylalanine, methionine, and histidine (essential for the children). Essential fatty acids: three kinds of the essential fatty acids derive from the products resulted from the decomposition of lipids, including linoleic acid, linolenic acid, and arachidonic acid.

Essential Vitamins:
There are essential 15 vitamins, including:
- The lipid soluble vitamins include vitamin A, D, E and K.
- The water soluble vitamins include vitamin B complex (vitamin B1, B2, B3, B5, B6, B12, biotin and folic acid), vitamin C and H.

Essential Minerals:

With the exception of the organically bound elements hydrogen, carbon, nitrogen, and oxygen, there are about 20 inorganic minerals, including major minerals and microelement minerals such as calcium, phosphorus, magnesium, sulfur, sodium, potassium, and chloride. Microelement minerals include chromium, copper, fluoride, iodine, iron, manganese, molybdenum, selenium, zinc, chloride, potassium, sodium, cobalt, stannum, and vanadium.

The Beneficial Matters:

Aromatic Compounds:
Aromatic compounds are beneficial to taste sense and fragrant.

Stable Compounds:
Beneficial and stabilize for the cell metabolisms.

Special Active Compounds:
Such as antibiotics, antivirus, and detoxification.

5.23.2 Assessment According to Medical Index

The effect of food quality can be assessed by some medical standards such as the growth speed, reproduction ability, resistance to diseases, and so on. These methods can reach conclusions through the trials of nutrients in the human body or animals, and still need to be strengthened, because of the lack of the effect of soil fertilization research on the food quality. Rare results can be found so far for the knowledge on a system from the fertilization of crop to nutrients in human body and animals. The related research should be enforced for the benefit of human health.

Studies show that feeding cattle with organic forage and conventional forage, respectively, then it turns out that the former forage can result in more sperms in cattle. And rabbits that eat the forage without soil fertilization have a higher reproduction capability than those that eat the forage with soil fertilization. However, the trials of the mice show the converse result. And another trial shows that babies who eat the vegetables cultivated by chemical fertilizers present a higher performance in the weight and blood test than those who eat the vegetables cultivated in the animal manure. The definite and reliable conclusion can only be drawn by more evidence and comparison of more experiments.

5.23.3 Correct Fertilization and Healthy Food of Human

It is regretful that there is no integrated research about the soil-crop-animal-human system. The cooperation research of fundamental agriculture and medical science still merits more emphasis for benefits of human being. The increase of human population makes a higher requirement on the supply of food. The increase of crop production is the need of human beings on the earth, and it is essential to use the chemical and organic fertilizers to reach the goal. Therefore, the right soil fertilization is important and should be emphasized more under the high requirement of food quality. Is agricultural product cultivated by chemical fertilizer harmful to the human body? Is the product cultivated by organic agriculture by all means beneficial? Or, does the organic cultivation need to be applied with some chemical fertilizer? Are those compositions in crop useful or harmful to our health? How important is the many compounds absorbed by crops from soils to the human body? Is the hydroponic vegetable healthy and safe? These problems merit further research.

Table 5-5. General List of the Characteristics and Guidelines for Application of Fertilizers

Fertilizer	Characteristics	Guidelines for Application
1. Nitrogen Fertilizer (1) Urea	1. $CO(NH_2)_2$ (46% N), neutral fertilizer. 2. Strong moisture absorption, high dissolubility. Easily harm roots. Be careful when used to pot plants. 3. Decomposed by urease in the soil into ammonium (NH_4^+) and carbon dioxide. The ammonium then is converted into nitrite (NO_2^-) and nitrate (NO_3^-) to acidify the soil, with some denitrification. The decomposing rate is influenced by the soil, moisture, temperature, and pH value.	1. Can apply to any crop and soil. 2. Cannot near root to avoid the hurt when applied, especially for pot plants. 3. Easily washed off; cannot be applied too much in rainy season. 4. Except for that on the solid, applied as 0.2 ~ 2% diluted solution on the leaves, differing according to different crops.
(2) Ammonium (NH_4^+) Fertilizers	1. Ammonium sulfate, ammonium chloride, and ammonium nitrate are commonly used as acid-forming fertilizers. 2. Easily dissolved in water and fixed by the mineral in the soil. 3. Mix ammonium (NH_4^+) fertilizers with the alkaline soil or alkaline fertilizers, and ammonia (NH_3) will be produced. 4. Ammonium will be converted into nitrite (NO_2^-) and nitrate (NO_3^-) by the microorganisms in aerobic soil (nitrification). The NO_3^- is easily washed off in rainy seasons and can be prevented by the nitrification inhibitor. Not easy to be converted under 3°C.	1. Perfect apply to aquatic crops. Can be applied to uplands. 2. Similar application guidelines as the 2 and 3 of urea above.

Table 5-5. General List of the Characteristics and Guidelines for Application of Fertilizers (Continue)

Fertilizer	Characteristics	Guidelines for Application
(3) Nitrate (NO_3^-) Fertilizers	1. Calcium nitrate, sodium nitrate, and ammonium nitrate are commonly used; a physiological alkaline fertilizer. 2. Easily dissolved in water and removed, washed off from the soil. 3. The nitrate is easily to be deoxidized in the anaerobic environment and get lost by denitrification.	1. Can not appropriate apply to paddy field to avoid denitrification. 2. Similar application guidelines as the 2 and 3 of the urea above.
(4) Cyanogen Ammonium (C_2N-2) Fertilizers	1. Calcium cyanamide is commonly used; aalkaline, but a physiological acid-forming fertilizer. 2. Powder form; easily dissolved in the water. Granulous form, slow-release action fertilizer. 3. Strong moisture absorption and agglomeration ability. 4. Poisonous.	1. Perfect apply to strong acidic soil. 2. The effect is preferable when apply as basal fertilizer.
(5) Slow-Release Urea Fertilizers	1. The nitrogenous fertilize is common. Next are compound and other fertilizers, such as formaldehyde, butenoic aldehyde, isobutylaldehyde urea condensate, urea gypsum fertilizer, coated sulfur urea, guanyl urea, and so on. 2. Slow decomposing ability.	1. Appropriate apply to any kind crop with slow action. 2. The effect is preferable when applied in rainy districts and pot plants.

Table 5-5. General List of the Characteristics and Guidelines for Application of Fertilizers (Continue)

Fertilizer	Characteristics	Guidelines for Application
2. Phophorus Fertilizers (1) Water-Soluble Phosphorus Fertilizers	1. The common water-soluble phosphorus fertilizers are phosphoric acid, monopotassium phosphate, calcium hydrogen phosphate, ammonium phosphate, magnesium phosphate, single superphosphate, and triple superphoshpate. 2. Soluble, but easily fixed by calcium, iron, aluminum, and minerals in the soil.	1. Apply in small amount yet repeatedly to avoid the fixation with soil. 2. Apply in drip irrigation.
(2) Weak Acid Soluble Phosphorus Fertilizers	1. Calcium magnesium phosphate and basic slag are common. 2. Not easy to be dissolved in water; have slow-release function; slowly absorbed by the crop. 3. Slow-release function; slowly absorbed by the crop. 4. The finer the particle is, the better it is.	1. Apply as basal fertilizer. 2. The effect is preferable when applied to acidic soil.
(3) Insoluble Phosphorus Fertilizers	1. Bone meal and rock phosphate powder are common. 2. Not able to dissolve in the water and weak acid. 3. The finer particle is better.	1. Apply as basal fertilizer. 2. The effect is preferable when apply to acidic soil. 3. The effect is better when applied with an appropriate amount of organic fertilizers.
3. Potassium Fertilizers (1) Refined Potassium Fertilizers	1. Potassium chloride, potassium sulfate, and potassium nitrate are common. Others are such as potassium bicarbonate. 2. Water soluble; easy to be washed off. 3. The minerals of the soil can fix the potassium, but much less than the ammonium fixation.	1. Apply as basal fertilizer or top-dressing fertilizer. 2. Easy to cause injury; inappropriate to be applied near root. 3. Potassium nitrate is a potassium fertilizer as well as a nitrogen fertilizer.

Table 5-5. General List of the Characteristics and Guidelines for Application of Fertilizers (Continue)

Fertilizer	Characteristics	Guidelines for Application
(2) Mineral Potassium Powder	1. Kainite, fenaksite, carnallite, and other potash deposits or their products are common. 2. Potassium content is lower than purified salt refined potassium fertilizers.	1. Apply mainly as basal fertilizer. 2. The effect is better when applied with organic fertilizers.
(3) Plant Ash	1. Palm ash and wood ash are common. 2. High water-solubility; easy to be washed off; similar to refined potassium fertilizer.	1. Apply as basal fertilizer or top-dressing fertilizer. 2. Easy to cause injury. It should be used carefully as refined potassium fertilizers.
4. Calcium Fertilizers (1) Soluble Calcium	1. Calcium oxide, calcium hydroxide, monocalcium phosphate, calcium cyanamide, and calcium sulfate are common. Others such as single superphosphate and triple superphoshpate are soluble calcium. 2. Most are soluble and easy to be washed off by the rain. 3. Most are alkaline, except single superphosphate, triple superphoshpate.	1. Apply as basal fertilizer; soluble in water; can be applied by foliar spray. 2. Alkaline material is soil conditioner which has a neutralization function for acidic soil. 3. Calcium is easy to be fixed by phosphorus to form insoluble calcium phosphate. Usually calcium in the liquid manure is better as chelating type with EDTA or humic and fulvic acids. 4. Calcium sulfate is useful apply to alkaline or saline-alkaline soils to speed up sodium elution and as soil conditioner.

Table 5-5. General List of the Characteristics and Guidelines for Application of Fertilizers (Continue)

Fertilizer	Characteristics	Guidelines for Application
(2) Weak Acid Soluble Calcium	1. Calcium carbonate is common and there is other kind such as the calcium silicate (basic slag). 2. Soluble in most weak acids.	1. Apply as basal fertilizer. 2. Alkaline material is soil conditioner which has a neutralization function for acidic soil. 3. Calcium is easy to be fixed by phosphorus to form insoluble calcium phosphate. Usually calcium in the liquid manure is better as chelating type with EDTA or humic and fulvic acids. 4. The effect is better when apply with organic fertilizers.
(3) Shellfish and Its Fossil Powder	1. Oyster shell powder, other shellfish and its fossil powder are common. 2. Uneasy to dissolve in water; alkaline; contains the calcium carbonate and easy to dissolve in water after heating.	1. Apply as basal fertilizer. 2. Alkaline material is soil conditioner which has a neutralization function for acidic soil. 3. Calcium is easy to be fixed by phosphorus to form insoluble calcium phosphate. Usually calcium in the liquid manure is better as chelating type with EDTA or humic and fulvic acids.
5.Magnesium Fertilizers (1) Soluble Magnesium	1. Magnesium sulfate and the magnesium phosphate are salt with acids. Magnesium hydroxide is alkaline. 2. Easy to dissolve in water. Be careful not to use excessively to avoid injury.	1. Apply as basal fertilizer or top-dressing fertilizer. 2. Soluble in water; can be applied by foliar spray. 3. The acidic one can be mixed with acid-forming fertilizer, while the alkaline one should be mixed with alkaline fertilizer.

Table 5-5. General List of the Characteristics and Guidelines for Application of Fertilizers (Continue)

Fertilizer	Characteristics	Guidelines for Application
(2) Weak Acid Soluble Magnesium	1. The common ones are lime fertilizers that contain magnesium and calcium, such as dolomite, magnesium limestone, the bitter earth lime, serpentine powder, and so on. 2. Alkaline; slow-release function; different solubility according to different mineral sources.	1. Apply as basal fertilizer. 2. The effect is better when applied with organic fertilizers.
6. Silicon Fertilizers (1) Soluble Silicate Fertilizers	1. The common ones are silicate fertilizers. 2. Water soluble.	1. Apply as basal fertilizer or top-dressing fertilizer. 2. The grass crops are commonly used.
(2) Acid Soluble Silicate Fertilizers	1. The common ones are silicate minerals or basic slag.	1. Apply as basal fertilizer. 2. The grass crops are commonly used.
7. Microelement Nutrient Fertilizers (1) Soluble Microelement Fertilizers	1. Boron: boric acid, borax (more soluble in hot water). 2. Copper: copper sulfate. 3. Zinc: zinc sulfate, zinc chloride. Manganese: manganese sulfate. 4. Molybdenum: ammonium molybdate, sodium molybdate. 5. Iron: ferrous sulfate.	1. Apply as basal fertilizer or top-dressing fertilizer; can be applied by foliar spray. 2. The amount of the microelement is decided according to the need. Excess may lead to poisoning in plants.
(2) Acid Soluble Microelement Fertilizers	1. Manganese: manganese melted from basic slag or minerals. 2. Zinc: zinc melted from basic slag or minerals.	1. Apply as basal fertilizer. 2. The amount of the micro element is decided according to the need. Excess may lead to poisoning in plants.

Chapter 6

The Concept and Application of Foliar Fertilization

6.1 The Concept and Application of Foliar Fertilization (Part I)

6.1.1 The Correct Concept of Foliar Fertilization

The leaves of plants not only can absorb or exchange gases (such as carbon dioxide, oxygen, etc.), but also absorb a variety of nutrients and other chemicals. We all know that the root of plant is the main organ responsible for nutrient absorption from the soil; leaves also have the same function. As a matter of fact, foliar fertilization has been widely applied to modern agricultural production. Along with great successes in short- or long-term crop cultivation techniques in recent years, the concept of fertilization has gradually expanded from soil to foliar fertilization.

Plant nutrients include macro elements (nitrogen, phosphorus, potassium, calcium, magnesium, sulfur), and micro elements (iron, manganese, zinc, molybdenum, copper, boron, chlorine, etc.). Amongst these nutrients, macro elements are mainly obtained from the root and cannot be totally dependent on foliar fertilization. Because foliar absorption is under the hindrance of leaf cuticle, weather conditions, and solution characteristics, foliar fertilization can only supply part of plant nutrients. Foliar fertilization can be very effective especially for trace element deficiency.

The applications of foliar fertilization are broad. Even with the lack of symptoms, foliar fertilization can supplement inadequate or unbalanced root absorption. Foliar fertilization can be used to adjust tree growth and flowering time, improve quality, and other special purposes. Other than trace elements, all nutrient ingredients could also be applied by foliar fertilization. Although foliar fertilization is widely applied, the technology and correct concepts are still required, otherwise would have adverse effects. In short, foliar fertilization is a high potential fertilization method that requires tests and correct concepts in practice.

6.1.2 The Absorption of Foliar Nutrients

Except for coniferous leaves, the stomata numbers and structures on both sides of plant leaves are more or less different. The extracellular surface layer of leaves is commonly known as the cuticle. Cuticle is a semi-hydrophilic structure that protects leaves and allows the penetration of little water and solute. The cuticle is an amorphous polymer that contains hydroxyl fatty acids. Figure 6-1 illustrates the structure of the leaf surface which is divided into different layers. The outermost layer is called the outer cuticle wax; the second layer is the primary cuticle, a substance which does not contain cellulose or cell wall; the third is the secondary cuticle; the fourth is the inner cuticle that consists mainly of pectin; and the final layers are the cell wall and membrane. Therefore, the solutes in water enter cells through the leaf surface first through stomata or water pores, or alternately through these cuticle layers, and finally reach the cell wall and membrane. Through this

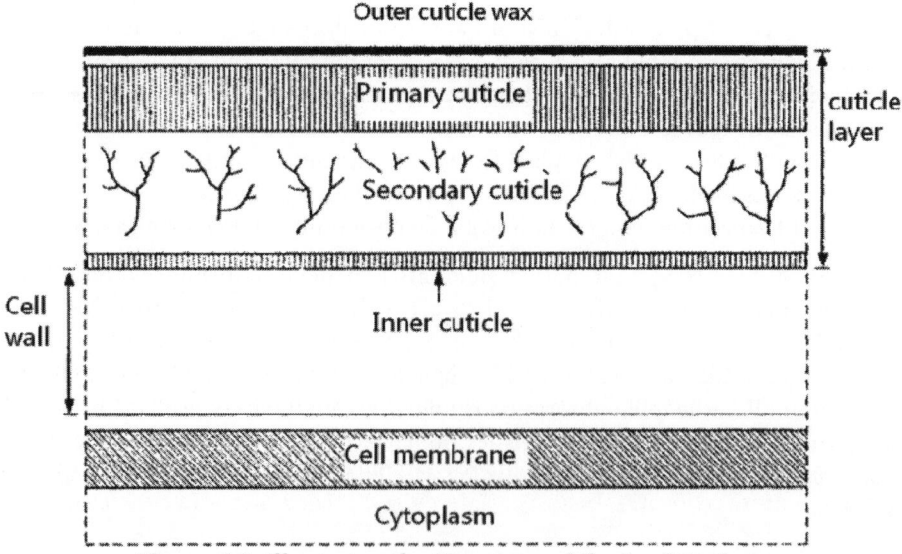

Figure 6-1. Illustrates the Structure of the Leaf Surface

progress, the solute can be absorbed by the leaf and transformed from epidermal to other cells for absorption and utilization.

Generally, foliar absorption of nutrients or substances via the cuticle wax pathway is very important. Because of the poor hydrophilic property of the cuticle wax, surfactants are often added with foliar nutrients to reduce the leaf surface tension, and produce a thin film on the cuticle wax to promote absorption. In addition to the cuticle wax route, nutrients can also enter plants through stomata, water holes, and other apertures. When gas exchange is active in stomata, the surface tension of water in the air spaces will render the aqueous solution entering the stomata, thus surfactant is necessary.

When the stomata is closed, foliar spraying will have minimal nutrient absorption through this route. The opening and closing time of stomata are different in various plants. Generally, stomata are open during daytime and close at night; but many succulent or desert plants (CAM plants), in contrast, close their stomata during daytime and open at night, such as pineapple, cactus, agave, cattleya, etc.

The absorption of foliar fertilization is not only a physical phenomenon, but it also includes physiological and biochemical phenomena, i.e., the absorption and transportation of nutrients consume energy. Overall, the absorption rate of foliar fertilization is mainly controlled by the diffusion of the nutrients in water on the leaf surface. The cuticle and cell wall act as barrier structures; therefore, the thicker the cuticle, the worse the absorption. In other words, new leaves have thinner cuticle and would lead to a better absorption rate via foliar fertilization; otherwise, aged leaves with a thick cuticle is bound to slow absorption and poor effect.

The distribution of stomata and water holes on both sides of plant foliar are different between crops. The average plant has more stomata on the lower epidermis of a leaf. As such, applying on different sides of plant leaves would result in different effects.

6.1.3 Advantages and Disadvantages of Foliar Fertilization

The fact that traditional agriculture applied diluted urine on vegetables and seedlings suggests that we have already known the use of foliar fertilization. As of today, urine is replaced by fertilizers. Currently, foliar fertilization has been applied to agronomy and horticulture in developed agricultural regions worldwide. To enhance product quantity and quality, tests to adjust or cure foliar fertilization are widely applied. It is necessary to follow the accurate technology and methods, and recognize the pros and cons of foliar fertilization to avoid the occurrence of shortcomings when applying foliar fertilization with flexibility. Listed below illustrates the advantages and disadvantages of foliar fertilization applications:

Advantages:

Fast Effective:
The efficiency of foliar fertilization is higher than soil fertilization, especially for crops that lack recognizable symptoms. Urea or trace elements should be applied to plants which are deficient in nitrogen or trace elements, and especially throughout the growing season.

High Nutrient Recovery Rate:
The recovery rate of foliar fertilization by the plant can reach 50 ~ 85%, while soil fertilization is relatively lower at about 10 ~ 30% and at most at 50% with different fertilizers. From an economical point of view, foliar fertilization is more competitive.

Spraying Commonly with Pesticides:
It is inevitable to use pesticides in agriculture. When applying pesticides, addition of appropriate amount of foliar fertilizers concomitantly could save manpower. However, improper mixing could result in plant injury; thus, care should be taken.

Improving Production with Quantity and Quality:
Several reports have stated that appropriate and accurate application of foliar fertilization can improve both the quantity and quality of crop products, including fruit color and sweetness.

Other Special Purposes:
When applying foliar fertilization with plant hormones, plants could efficiently ripe and be harvested earlier, especially for crops that need to be harvested after the drying of leaves, such as cereals and beans. Other special purposes include the adjustment of harvest date, the prevention of excessive growth, the inhibition or control on physiological disorders, etc.

Disadvantages:

Apt to Cause Foliar Injury:
It tends to cause fertilizer injury when the concentration of solution applied on leaves is too high, or when the solution, left dried on leaves after sprayed, results in high osmotic pressure. These high concentrations of salts are easy to cause leaf burns or even death of plants. On the other hand, urea is an organic compound, not a salt; at lower osmotic pressure, it is a good spraying solution.

Need the Effective Application with Multiple Coordination:
The effect of foliar fertilization is unstable due to reasons such as the growth status of the plant, the weather condition, and the kinds of solution. For stable outcomes, skilled technology, methods, as well as efforts should be considered.

Supply Limited Amount of Nutrients:

As we already know, leaf surface acts as a barrier that limits the amount of available nutrients to be absorbed by plants. The plant must absorb major nutrients from the soil, e.g., macro elements including nitrogen, phosphorus, potassium, etc.

Easy to Cause Over-Application or Unbalanced of Leaf Nutrients:

The required amounts of trace elements in plants are low; overdose or improper spraying will result in toxicity issues, and appropriate amount or frequency of use should thus be considered.

6.1.4 Conditions for Foliar Fertilization

Although foliar fertilization is widely applied, we must be sure of our target before applying it on plants. Each application suits different targets and is dependent on the plant nutrient situation. An understanding of the condition of foliar fertilization is necessary to get the best results. Listed below are several conditions under which foliar fertilization should be applied:

When Crop Leaves or Shoots Show Lack Symptoms:

When leaves show symptoms, we can apply foliar fertilization immediately on plants. Its effect shows best on new leaves and worst on aged ones.

When the Soil is too Acidic (pH < 5) or too Alkaline (pH > 7.5):

When the soil is too acid or too alkaline, the development of crop root system is poor and will result in poor absorption efficiency. Furthermore, acidic or alkaline soil will decrease the absorbing efficiency of many nutrients. Under this condition, a foliar fertilizer spraying or direct soil improvement is necessary. An acidic soil lacks elements such as magnesium, calcium, potassium, phosphorus, boron, molybdenum, etc., while an alkaline soil lacks iron, zinc, manganese, phosphorus, etc.

During Drought or the Soils Has Poor Drainage Ability:

If the soil is under poor physical condition such as too wet or too dry, absorption of nutrients by the plant root system will be impaired. In addition to soil improvement, the application of foliar fertilization in short-terms can immediately relieve this problem.

Pests and Pathogens Show on Roots or Shoots:

Soil-borne diseases or nematode diseases seriously affect the root absorption function. In order to compensate the loss in crop production, foliar fertilization is a good way to reduce the damages caused by pests or diseases.

When Crop Develop Physiological Disorders:

The reasons for crop's susceptible disease may be pathogens, nutrient deficiency, or unbalance. When physiological disorder occurs on plant shoot, it indicates that the crop requires nutrient application. This sort of symptom is a physiological sickness and cannot be solved by using pesticides. Adjustment of crop harvesting date in the modern advanced agriculture often attempts to change the flowering time of crops by making harvest earlier or later. During flowering season, supplying the crop with excessive nitrogen will cause less flowering or premature dropping of flowers and fruits. Under this circumstance, we can supply phosphorus and potassium as foliar fertilizers to solve this problem by reducing the overgrowth and improving early flower differentiation.

6.2 The Concept and Application of Foliar Fertilization (Part 2)

6.2.1 The Timing for Foliar Fertilization

Crops' demand for different nutrients varies at different growth stages. At vegetative stage, it requires large amounts of nitrogen, phosphorus, and potassium. At pollen development stage, it needs sufficient amounts of copper. Therefore, the timing to apply foliar fertilization must meet the crop need, and we also need to consider the inadequate nutrients in the soil. Table 6-1 is summarized from the 1985 First International Symposium on Foliar Fertilization. The nutrients listed in the table for each crop does not necessarily mean that we must apply all of them; instead, we should match both the crop's need and the deficient element in the soil.

6.2.2 The Selected Concentration for Foliar Fertilization

Different purposes affect our determination of applying foliar fertilization. When foliar fertilization is necessary, the concentration and amount should be determined as different crops require different concentrations. The amount of foliar fertilization to be supplied not only differs by each crop but also depends on the soil fertility. Generally, one hectare requires 100 to 200 liters of spray solution, sometimes up to 400 liters is possible when the crop size and leaf surface characteristics are considered. A higher concentration is easy to cause harm to new leaves; thus, reading the manual carefully before use is important. For sensitive crops, special care is needed, such as performing spraying tests of various concentrations in small areas. With the test records and experiences, we can avoid serious fertilizer injury. Table 6-2 illustrates the recommendations of common concentrations and quantities when applying foliar fertilization.

Table 6-1. Recommended Timing for Applying Foliar Fertilization for Different Crops (Extracted from the 1985 First International Symposium on Foliar Fertilization)

Crop	Element Needed	Recommended Timing for Application
Grape	Zinc	Two weeks before full flowering; may be extended to flowering and fruit growing
	General nutrients	Stem bud growth period, continuous apply every 10 days; before flowering and during fruit growth
	Phosphorus and potassium	During the period of fruit growth and coloring
	Potassium	Beginning of fruit growth period
	Boron	Flowering and fruit growth period
	Magnesium	The beginning of fruit growth The beginning and after ripening
	Iron	10 days before flowering and after fruit growth
General Fruit Trees	Nitrogen	Before and after flowering; after harvest (leave is still green)
	Potassium	The 2nd, 4th and 6th week after flowering
	Magnesium	Every two weeks after the flower fell
	Iron	3 ~ 4 weeks after flowering
	Potassium	3 ~ 4 weeks after flowering; after harvest (leave is still green) (prune, pear, and apricot)
	Manganese	After flower fell and 4 weeks after flower fell
	Copper	After sprouting and after harvest
	Boron	Before flowering
Orange	Zinc, manganese, Iron, magnesium	Growing season in spring (leave is in 2/3 growth period and before get hard)
Corn	Phosphorus, zinc, manganese	Leaf stage 5 and 7 and 14 days after 4 and 8 leaf stage
Potato	Nitrogen, phosphorus, potassium, magnesium, Manganese and other trace elements	Stem elongation Bud formation and flowering
Lettuce Red Pepper	Calcium Nitrogen, phosphorus. Potassium and trace elements	Before the ball forming, once or twice a week 8 leaf stage; early flowering and late of large number Flowering

Table 6-2. Useful Concentrations and Quantities of Foliar Fertilization

Element of Deficiency	Fertilizer	Concentration (%)	Amount (kg/ha)	Crop
Nitrogen	Urea	8 ~ 16	32 ~ 65	Grain (the high tolerance)
		0.5	2 ~ 4	Vine, fruit, vegetables
Phosphorus Potassium	Monopotassium phosphate	0.2	8	Used to control the over N application
Potassium	Potassium sulfate	1	4	Used to control the over N application
Calcium	Calcium nitrate, Calcium chloride	0.3 ~ 0.5	3	Fruit
Magnesium	Magnesium sulfate	1 ~ 2	8	Fruit and grain
Iron	EDTA-Fe 5%	0.1 ~ 0.5	0.4 ~ 0.8	Fruit, vegetables and vine
Manganese	Manganese sulfate	1 ~ 2	4 ~ 8	Grain
		0.5	2	Horticultural crops
Zinc	Zinc sulfate	0.5	2	Grain
		0.2	0.8	Horticultural crops
Copper	Copper sulfate	0.2	0.8	Horticultural crops
Boron	Borax	0.2 ~ 0.5	2	Fruit, vegetables
Molybdenum	Ammonium molybdate	0.1	0.4	Cabbage, half the amount Applied on fruit

6.2.3 Factors Affecting Foliar Fertilization

The success or failure of foliar fertilization is determined by several internal and external factors. Many scholars found that the main factors that affect the efficiency of foliar fertilization include: the quality of the spraying solution, the correct weather environment, and the plant characteristics. All the three factors should be in coordination; otherwise, the efficiency of foliar fertilization will be low. For example, if the solution is good but sprayed on plants while the temperature is relatively high, the liquid sprayed on leaves will soon dry up and lead to a poor foliar absorption efficiency. Table 6-3 shows the factors which affect the efficiency of foliar fertilization.

6.2.4 Guidelines for Applying Foliar Fertilization

Guidelines for applying foliar fertilization are based on the following consideration: Integrate the factors which affect the absorption efficiency to make an

Table 6-3. The Factors Affecting the Efficiency of Foliar Fertilization

Spraying the Solution	Environment and Climate	Plant Characteristics
Concentration	Temperature	Cuticle wax
pH	Humidity	Leaf surface wax
Surfactant	Wind	Age of leaf
Carrying agent	Sunshine	Stoma, gas pores
Carbohydrate	Time	Surface humidity
Salts or chemicals	Rain	Growth duration
Nutrient proportion	Drought	Nutrients condition
Adhesion	Osmotic pressure	Leaf turgor
Hygroscopic	Nutrient stress	Variety
polarity	Leaf morphology	
Applying application		
Applying technology		

optimal spraying solution, apply it in the most suitable climate, and then apply the foliar fertilizer on a suitable crop. The most adequate application is to be used on the most suitable and most appropriate crops. The following illustrates the essentials for foliar fertilizer application:

The Spraying Solution:

The Optimal Concentration:
Choosing the concentration for foliar fertilization is very important. High concentration will result in fertilizer injury, while low concentration will lead to limited absorption. In practice, we can apply lower concentration solution multiple times on the same crop, but it is not economically viable. According to different varieties of crops, their age, cropping season, and soil condition, the concentration of solution may be different. We can firstly refer to the common spray concentration, either increase or decrease it, and test it on a small field spot.

The Optimal pH:
The pH of the spraying solution affects crop growth. Too acidic or too alkaline may show damage, while neutral is the best. Too acidic or too alkaline not only lead to plant injuries, but also cause nutrients to precipitate, thus lowering the effectiveness of elements. When the solution is dried on the leaf surface, white sediment will form and block leaf water pores or stomata for gas or nutrient exchange.

The Right Formulae of Nutrients or Salts:

Each nutrient element comes with its own unique molecular structure, and thus their efficiency through foliar absorption is different. For example, one report indicated that leaves absorb more potassium nitrate than potassium sulfate do, while magnesium chloride is better than magnesium nitrate and magnesium sulfate. In addition, urea is the best nitrogen source with high efficiency. The solubility of the selected salt should be noted; insoluble salts are not applicable.

Adding Surfactant or Adhesive Agent:

The leaf surface usually possesses a cuticle wax layer that renders the solution to form a water film when sprayed on the leaf surface. Therefore, surfactants or wetting agents is applied to reduce the surface tension and assist the solution to be attached to the leaf surface. Warning! Do not choose toxic surfactants.

Adding Chelator or Carrier Agents:

Nutrients are prone to be fixed or inactivated in extra- or intracellular space. Chelators can reduce the sedimentation or fixation of nutrients, while adding carrier agents such as carbohydrates, proteins, amino acids, humic acid, fulvic acid, or other acids can improve nutrients absorbability, transportation, and utilization. Moreover, adding vitamin or water soluble hormone may also improve the efficiency of the spraying solution.

The Climate and Environment when Spraying:

Weather Conditions:

The water content of leaves can affect foliar absorption efficiency. Under high temperature, strong sunlight, and windy conditions, the sprayed solution on the leaf may easily evaporate and limit the absorption of nutrients. Rain shower also reduces the concentration of foliar fertilizer nutrients. Humidity is the main factor affecting absorption efficiency; proper humidity keeps the sprayed nutrient solution in liquid form. Maintaining moisture prolongs the time and increase the efficiency of nutrient absorption by leaves. In a day, morning and evening are the better timing for spraying than noontime; in four seasons, the warm spring days are better than hot sunny summer days.

Soil Nutrient Environment:

Soil environment affects the effectiveness of nutrients, and in turn affects the uptake by crops. When the soil is too acidic, it can be sprayed with magnesium, calcium, phosphorus, nitrogen, potassium, etc. When the soil is too alkaline, it can be sprayed with iron, zinc, boron, phosphorus, etc., to quickly and effectively supply nutrients. When nutrients in the soil are unbalanced, foliar fertilization can show obvious effects.

Adjusting Seasonal Production:

The adjustment of seasonal production is an urgent subject in modern agriculture. To achieve the purpose of changing the production season, we need to do more research to understand the plant nutrients status and improve our skills on flowering and the time of bud differentiation by foliar fertilization. Foliar fertilization can be very effective to help achieve early flowering.

The Plants:

Leaf Age:

The cuticle thickness and structure of new leaves are different from those of aged leaves; thus, the absorption efficiency differences will lead to differences in the effectiveness of foliar fertilization. In general, new leaves have better responses to foliar fertilization than aged leaves do. The cereal new leaves have better absorption efficiency than the aged leaves do.

Growth Stage:

Crops need different nutrients at different growth stages; therefore, foliar fertilizer application should correspond to the crop's nutrients demand. Generally, most crops require substantial nitrogen, potassium, and phosphorus at early growth stages, while potassium and phosphorus at flowering stages. Because of the differences in flowering season between perennial fruit trees, coordination with their growth stage should be considered when applying foliar fertilizers. At different growth stages, the crops need specific and appropriate formula.

Keep Stoma Function Well:

Stomata are the gas exchange channel essential for photosynthesis and respiration. Improper foliar fertilization could lead to the blockage of the stomata and render gas exchange. This would harm crop growth and should be noticed.

Other Precautions:

Nitrogen Solution Should Not Be Sprayed under these Circumstances:
- Cereals:
  After sowing to the third leaf and the floret development, spraying nitrogen fertilizer solution should be avoided in case of injury.

- Grass Land:
  Within one day after cutting, it is not suitable to spray liquid nitrogen fertilizers, or grass growth will be poor.

Careful Choosing Mixtures of Agricultural Chemicals:

Appropriate mixing of foliar fertilizers with agrochemicals is economical. However, it should be noted that inappropriate mixing will result in fertilizer injury or chemical harm to the plant.

The potential of foliar fertilization is high and has positive effects on crop yield and quality. Coordinating the needs between soil, environment, and crops is the key to effective foliar fertilization. Moreover, further research and tests are required to establish a database for efficient methods.

Chapter 7

Guidelines of Improvement of Fertilization and Soil Management

7.1 Guidelines to Improve the Effect of Fertilization

7.1.1 Causes for the Decline of Fertilization Efficiency

Fertilization is an indispensable way to increase production and improve quality in modern agriculture. In order to balance nutrients in soil and supply abundant nutrients, we need to fertilize the soil based on soil condition and crop needs. The effect of the same quantity's fertilizer varies in different soils, because the capability of adsorbing, fixing and preserving in soil varies according to various kinds of fertilizers. Therefore, improving physical, chemical, and biological properties of the soil enhance the utilization of fertilizers directly or/and indirectly.

Fertilizers that supply to soil cannot be absorbed immediately or completely by the crop. So the properties of soil play important roles in the efficiency of fertilization. Soil can adsorb, preserve, and even convert fertilizers. Soil supplies them to crops for absorption. However, when the soil is fertilized, it often occurs to the degradation of fertilizer efficiency. The inclusions of common causes are listed below:

Occurrence of Insoluble Matter after Fixation:
Phosphorus, calcium, and trace elements are the most prominent.

Loss of Fertilizers:

Nitrogen and potassium are the most prominent.

Volatilization of Fertilizer:

After conversion with the denitrification of nitrogen, it becomes the most prominent.

Fertilizer is Fixed and Absorbed by Large Numbers of Organism:

Nitrogen and trace elements are the most prominent.

Lack of Water or Too Much Water in the Soil:

Therefore, the fertilizer is hard to be absorbed or the root system of crop is easy to be hurt.

The Fertilizer Becomes Difficult to be Absorbed:

Under inappropriate mixed condition.

The Root System has Difficulty in Nutrient Absorption:

The efficiency of fertilizer degrades due to the fact that crops are affected by the adverse soil environment, pests and diseases.

To reduce the loss and waste of fertilizers, it is essential to improve the fertilization efficiency. The targets of today's skilled agriculture and the improvement of productivity are to increase the return rate of agricultural investment and achieve highest productivity and quality by minimizing fertilizer.

We will discuss ways to improve the fertilization efficiency in different sections of this article by providing information about the occurrence of problems and relevant solutions, so that people can put into use, take as references for solving problems nimbly.

7.1.2 How to Reduce the Loss and Volatilization of Fertilizers

Occurrence of this Issue:

We need to pay attention to fertilizer loss in high rainfall areas. Many fertilizers, such as urea and potassium chloride, are easy to dissolve in water. Some fertilizers take the form of fine powder, so they are easy to be washed away by water. We need to take more precautions of fertilizer loss, especially for those with difficulties to preserve fertilizer soil efficiency; for example, in sandy soil, slopes and hilly areas. Otherwise, there would not only be individual loss, but also cause reservoir, stream, river and headwaters pollution. Fertilizer loss includes both from being washed away by water and volatilized to air. Among all, the nitrogen volatilization is the most severe. There are two kinds of nitrogen loss in the gas state: the volatilization and denitrification of nitrogen. Urea and ammonium sulfate in alkaline soil are very easy to volatilize to gas state. After being fertilized into soils, urea can be quickly denatured by urease enzyme and release ammonium ion (NH_4^+) or ammonia (NH_3) in alkaline soils. The ammonium ion is also easy to volatize under high temperature and arid conditions. If ammonium ion can be converted by microorganisms to nitrate (NO_3^-) nitrogen, it would not volatize easily. However, nitrate nitrogen is easy to be washed away by water or to be denitrified and become nitrogen gas or nitrous oxide under anaerobic circumstance.

Strategy and Solutions:

Cover in Soil or Deep Fertilization:

Fertilize the soil, as long as manual labor permits, the deep placement efficiency is the best for deep-rooted crop, especially in orchard. We can reduce the loss by applying large granular urea or other chemical fertilizer in the deep layer of paddy field. Urea can be decomposed into ammonium ion and reduce the occurrence of nitrification. Ammonium ion will not be denitrified in the deep layer of paddy field, because denitrifiers act on nitrate under anaerobic condition, instead of ammonium ion.

Do not Fertilize on the Surface Abundantly when It Rains Heavily. Doing So can Help Reduce Washing-Away of Fertilizers by Rain:

- Cover Grass Cultivation:

  Cover grass refers to all sods on the soil surface except trees. At first glance, the cover grass would absorb and compete with fruit trees for the fertilizer; but in the long run, it is beneficial to the perennial crop. Cover grass cultivation works especially well on orchards which are cultivated on slopes and hilly areas. The selection of cover grass is critical. It should be noted that the best ones are those that do not grow tall and fail to thrive in winter. We can mow the cover grass, mulch it on the surface soil, and increase the organic matter content to reduce the occurrence of fertilizer loss and surface soil runoff. Long-term water and soil conservation is rather important for orchards. Cultivating of cover grass not only improve the fertilizer and soil conservation but also enhances water retention property of the soil. Cover grass is multifunctional. As it can loosen the soil, water can permeate into the soil easily.

- Increase Soil Organic Matter:

  Organic matters in the soil are rather efficient in conserving fertilizer and water. The capacity of soil to conserve fertilizer is closely related to the quantity and quality of soil organic matters. The more soil organic matters contain, the better capacity conserves the fertilizer. The characteristic of coarse texture, good aeration and high leaching ability of sandy soil can increase the decomposition rate and loss of organic matters. Therefore, the proportion of organic matters in sandy soil is quite low, so fertilization should be more frequent but at a lower concentration. The application of organic fertilizers can help the conservation of fertilizer and water. For example, asparagus is often cultivated on sandy soil and farmers tend to apply more chemical fertilizers than it is required. This would result in waste. It is better to apply more organic fertilizers instead.

- Reduce the Volatilization of Ammonia:
  It is better to use nitrate nitrogen fertilizers on strong alkaline soil, since nitrogen in urea and ammonium fertilizer is easy to volatilize. The vegetation on the surface soil can adsorb and preserve part of nitrogen fertilizers, while nitrogen tends to volatilize more on non-covered soils.
- Use More Coarse Grained Fertilizer to Reduce the Fertilizer Washing-Away by Rain.

7.1.3 Pay Attention to Soil pH Value

Occurrence of this Issue:

The pH value of soil can affect the availability of fertilizers after they are applied into the soil. As in Figure 7-1, the availability of various nutrient elements differs

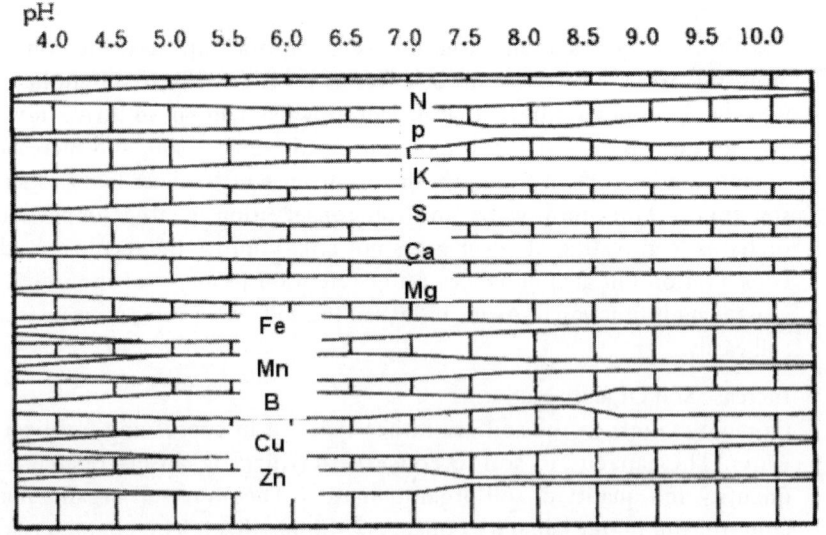

Figure 7-1. Relationship between pH Value and the Availability of Various Nutrients

CH7 Guidelines of Improvement of Fertilization and Soil Management   173

under different pH conditions. The availability of both macro- and micro-elements will be influenced by the pH value of the soil. For instance, it will be difficult for crops to absorb phosphorus when the availability of phosphate decreases under too acidic (pH < 6.0) or alkaline (pH > 7.5) soil. Large proportion of tropical and subtropical area is acidic soils. The acidity of red and yellow soil is most prominent, which can result in many issues concerning cultivation, production and quality of crops.

Micro-element is also easy to be affected by pH value. In the alkaline soil, pH value is around 8.0. The common leaf yellowing shown in the new leaves of plants, even large amount of chemical fertilizer application cannot mitigate the problem. Lack of micro-elements, such as iron, manganese, boron and zinc is the main reason causing new leaf yellowing in alkaline soils.

Inappropriate methods to measure pH value can also lead to fertilizer efficiency misjudgment. We should not insert a kind of metal electrode directly into the upland soil to measure pH value. Instead, we should collect small amount of soil sample, put it into a container, add water (1:1 ~ 5 weight ratio), and stir it into slurry or let stand for 30 min. When measuring pH value, we should stir it again and insert the electrode into the slurry and read pH value.

Strategy and Solutions:

Utilize Neutralizer:
- We can apply alkaline matters (e.g., agricultural limes, magnesia limes, dolomite, oyster shell powder, etc.) for acidic soil. The amount of lime depends on pH value and should apply progressively over years.

- We can apply acidic matters (e.g., sulfur powder, dilute Sulfuric acid, etc.) for alkaline soil. Be sure not to mix neutralizers with chemical fertilizers, or it will reduce the availability of chemical fertilizers gravely. Otherwise, mixing neutralizer with organic fertilizer is a good option. Utilizing neutralizer into the soil is a way to improve the fertilizer efficiency. Yet, we cannot overuse it, or the soil may become too acid or too alkaline.

We can apply organic fertilizer in alkaline or acidic soil, which is helpful to the availability of various nutrient elements. It improves fertilizer efficiency, mobilization of cations, and helps reduce the toxic effect of aluminum in acidic soils.

Immediately Supply Nutrients by Foliar Fertilization:
When pH value cannot be improved in time, we can apply fertilizer on foliage as a way for fertilizer deficiency remediation. This approach cannot cure the problem permanently; thus, the soil pH value improvement is still the ultimate approach. Foliar fertilization is a rather effective method to provide immediate nutrients for the short-term. We normally

supply micro-elements and nitrogen as foliar fertilizers when there is an urgent need or deficiency. Yet, we should not overuse, or it will antagonize our goal and produce high-quality products.

7.1.4 Improve the Fertilization Technique

Occurrence of this Issue:

The aim of fertilization is to provide enough and balanced nutrients to crops. The nutrient elements that crops absorb from the soil includes both macro-elements and micro-elements, such as nitrogen, phosphorus, potassium, calcium, magnesium, sulfur, sodium, iron, silicon, zinc, copper, manganese, molybdenum, chlorine, cobalt, etc. Apart from inorganic nutrients, crop can directly absorb or utilize many kinds of organic compounds from the soil, such as various organic acids, plant hormones, humic acid and fulvic acid. All these are important to the quality and the health of crops. The followings are some common causes for fertilization unavailability.

Unbalanced Fertilization:

Unbalanced fertilization will lead to antagonistic effects between the nutrients, impede the availability of elements, or interrupt with plant metabolisms and affect the availability of fertilizer directly or indirectly. We should not overuse any macro-elements, such as nitrogen, phosphorus, and potassium. If we overuse nitrogen, crops will tend to grow tall, be bushy liable to diseases, be hard to blossom, and drop fruits. If we overuse phosphorus, crop will grow slow. Overusing potassium would result in symptoms of magnesium deficiency.

Inappropriate Mixing of Fertilizers:

All fertilizers contain certain amount of effective constituents; therefore, mixing fertilizers inappropriately will decrease the availability or lead to volatilization. For example, mixing phosphorus fertilizer with calcitic limestone will fix the phosphate and become insoluble forms. Mixing ammonium fertilizer (e.g., ammonium sulfate) with alkaline fertilizer will lead to the volatilization of nitrogen.

Fertilization Disaccording with the Soil Characteristics:

Inappropriate fertilization on alkaline soil (pH > 7) or acidic soil (pH < 5) is a common mistake. Usually, it is not proper to apply alkaline fertilizers, limes or overuse soil additives into the soil with pH value higher than 7. Otherwise, it will result in leaf yellowing or stunted growth of crop which is caused by the lack of iron, zinc, boron and copper. It is not proper to apply acid-forming fertilizers on soil with pH value lower than 5 for a longer time. Otherwise, it will result in more soil issues and also problems in magnesium, calcium and phosphorus utilization. Fertilization disaccording with the soil characteristics not only is a waste, but also triggers soil problems. So we have to choose fertilizer carefully which is in accordance with the nature of the soil.

Strategy and Solutions:

Soil Diagnosis and Plant Analysis:

It is our primary task to understand the characteristics of the soil. At least we should know its pH value and more by understanding its physical, chemical, and biological properties. Plant analysis is a significant indicator for the problem soil. We can refer to the plant analysis results and apply suitable fertilizers, which is beneficial to the availability of fertilization. Fertilizing blindly may cause serious harm to the soil.

Understand the Properties and Functions of Fertilizers:

Different crops and stages of plant require different rate and amount of nutrients. The lack of any important element can lead to unavailability of other elements. So we must have an integrated concept on fertilization. In order to achieve the availability, macro- and micro-elements supply in the soil should be sufficient. For instance, if soil lacks micro-elements, plant growth and metabolisms will suffer, even applying large amounts of other elements would be useless.

Pay Attention to the Mixing, Storage and Delivery of Fertilizers:

We need to prevent fertilizers to become moist and caking, and avoid mixing different pH fertilizers and storage. We shall immediately apply mixed fertilizers, or it might affect the fertilizer efficiency. We should also avoid exposure of fertilizers to sun, rain, high temperature and high humidity during deliver and storage.

Appropriate Application of Organic Fertilizers:

Applying organic fertilizers can bring many benefits, such as improving nutrient efficiency. The nutrient elements that the soil supplies can influence the organic matter content, especially in upland or orchard, compared to in paddy field. After fertilizers are applied to the soil, they cannot be absorbed by the current crops completely. The proportion of nutrients absorbed by crops may differ from varieties and soil conditions. For instance, after nitrogen or potassium fertilizer is applied to the soil, maybe only < 50 percent can be absorbed by the current crops, and less for phosphorus fertilizers with only < 20 percent can be absorbed. The application of organic matters can improve the absorption efficiency of various elements.

Other Means of Soil Management and Fertilization Techniques:

Improving the natural condition of soil is beneficial to fertilizer efficiency. Deep plowing, soil turning, extirpating weeds, covering, irrigation and draining are all helpful methods to improve fertilizer efficiency in different ways. Sufficient supply of water is rather important to efficient fertilizer. Water acts as an important factor in absorption efficiency, as it promotes the movement and diffusion of nutrients in the soil and supply the root with sufficient nutrients.

7.1.5 In Accordance with the Need and Characteristics of Crops

Occurrence of this Issue:

The absorption and response to fertilizers differ between crops, even amongst different cultivars. To enhance the fertilizer efficiency, one should meet the need of crops. If the crop or cultivar needs more fertilizers, we shall supply more in order to increase production and improve quality. For some crops, it is a waste to overuse fertilizer. We should establish the concept that we should not overuse fertilizers.

Common Cases of Discordance are Listed Below:

Unseasonable Fertilization:

The requirement of fertilizers between annual and perennial crops differs a lot during their growth and maturation process. Crops need large amounts of fertilizers in their vegetative stage, during which the fertilizer could be very effective. Applying phosphorus and potassium fertilizers are important for the crop quality in later stage of growth and fruit-bearing period. The need for various fertilizers in these periods is quite different from that in the vegetative period. Hence, the kinds of fertilizers may strongly affect the fertilization efficiency. In warm spring, crops flourish and the availability of nitrogen fertilizers is rather remarkable. If there is not sufficient sunlight or the weather is very cold, or there is too much or too little rainfall, crops grow slow and the fertilizer is ineffective. Therefore, we should not apply too much fertilizer, or it might injure the root or cause root-rot and the fertilizer would be wasted.

Inappropriate Amount of Fertilizer:

The response of crops to the amount of fertilizers also differs. There is much difference among the capacity of root for tolerating fertilizers. For instance, the root of melons is easy to be injured by fertilizers, while the root of grass family can adapt to higher fertilizer concentration. Large amounts of fertilizers do not always increase the production. Thus, it is very important to optimize fertilizer amount.

Solutions:

Understand the Features between Crop Cultivars:

There is always a handbook for the suggested amount of fertilization after a new variety is available in the market. Cultivation will become much better if we understand the feature of the crop before applying fertilizers. Or it would be too late when fertilize damage occurs. For example, we should apply more organic fertilizers to crops which prefer growing in sandy soil, such as asparagus and watermelon. Crops which are rich in starch and fibrous materials require more potassium fertilizer. Legume are capable of nitrogen fixation, thus we do not need to apply too much nitrogen fertilizers.

Fertilization should be in Accordance with the Growth and Age of the Crop:

The nutrients for a crop are different in various seasons: all differences in the vegetative, blooming and fruit-bearing periods. Generally speaking, for perennial fruit trees, we should fertilize based on the growth trend of trees. For weak trees, we should not overuse chemical fertilizers, or it might have adverse effects or injuries. Instead, we should apply slow-release organic fertilizers with chemical fertilizers. We can also supply fertilizers by foliar application regularly.

Fertilization should be in Accordance with the Rotation System:

Fertilization is rather important for short-term vegetables. Since we have to cultivate and harvest the crops in only a few months, fertilizer availability is closely related to the production and quality of these crops. Yet, it is common that continuous cropping can lead to lower production and quality. Relying on fertilization is difficult to recover the continuous cropping problem. For crops which suffer from continuous cropping problems, fertilization should be in accordance with the rotation system, so that the availability of fertilizers could be improved. For instance, rotation between vegetables and paddy is quite beneficial to some fertilizers efficiency.

Other Means of Management that are in Accordance with the Crop:

Fertilizer efficiency is likely to decrease due to diseases and pests. So eliminating the occurrence of diseases and pests is a necessary means to enhance fertilizer efficiency. The recently-developed facility agriculture aims to improve the microclimate of crops for their optimal growth and reduce in the occurrence of diseases and pests to ensure that healthy roots can guarantee effective utilization of fertilizers.

7.1.6 It is Easy to Reach the Essence of Matters, if You are Aware of the Important Sequence of Development

All the above-mentioned methods to enhance fertilizer efficiency are either direct or indirect approaches. We can enhance fertilizer efficiency by applying several approaches and methods may differ among soils. The best way is to focus on the improvement of the major weaknesses of the soil and the crop, so that we could double our efforts. Diagnosing the deficiency of the soil can help us put forward solutions which can produce an immediate effect. We also need to fertilize based on the need and growth of crops, prevent diseases and pests, take managerial measures, pay attention to environment and climate, improve soil fertility, enhance fertilizer efficiency, avoid pollution and avoid excessive continuous cropping. Hence, our soil, the treasure for generations to generations, can be preserved.

7.2 Nutrients and Soil Management for Winter Crops

There are various kinds of crops in the subtropical and tropical zones, and their capacity to adapt to the climate differs considerably. Some crops can adapt to high temperature in summer, but can hardly bear the low temperature in winter. The low temperature and frost caused by cold weather is rather dangerous to many perennial and annual crops. Besides the strengthening of the cultivation management of crops in winter, enforcing of soil and fertilizer managements can also protect crops against cold, benefit their growth, and maintain soil protection.

7.2.1 Characteristics of Winter and its Relationship with Crops

The key to soil management in winter is to know the characteristic of the climate and its relationship with crop growth and development. Then, we would know crop needs for nutrients and their capacity to absorb nutrients in winter. We can double our efforts, only when we know crop needs and soil supply of nutrients accordingly. Besides, we should apply these guidelines wisely.

Low temperature is the most prominent characteristic of the climate in winter. Low temperature can influence crop cell metabolisms and directly affect its growth and development. The reason why winter crops can adapt to low temperature is that their cell membrane and metabolic enzymes can tolerate at the low temperature. Many crops can bear low temperature, but not necessarily can tolerate freezing temperature and the common damages caused by frost and hail. Considering this point, we should avoid the water freezing on the soil surface to hurt the root system.

Besides low temperature, winter is also characterized by the differences of sunlight, humidity, rainfall, and the soil ability to provide nutrients. For example, some areas are quite rainy in winter, while others are dry. This obvious contrast of the climate can also affect the amount of water in the soil. For upland crops and fruit trees, flooding and poor drainage of the soil can inhibit the growth of the root system.

The duration of sunshine is short in winter, so it is a good season to promote the blossom of short-day crops. It is also a season suitable for vegetative growth for long-day crops which can also adapt to low temperature. Crops can adapt gradually if the climate changes gradually for the law of nature. However, if the climate changes abruptly, such as low temperature in the period of cold weather, crops are prone to injury. Winter is not entirely harmful to perennial crops, because it can reduce the occurrence of pests and diseases. Many perennial crops from the temperate zone will be dormant. Growth after dormancy is helpful to economic production. Asparagus is a good example.

7.2.2 Nutrient Requirement and Absorption Capacity of Winter Crops

Fertilization management should depend on the crops needs of nutrients and their absorption capacity. Crop growth and activity of root system are key factors in absorbing nutrients. When temperature is low, the respiration of root decreases. While it takes energy to absorb nutrients, so when energy is not acquired sufficiently, the absorption capacity decreases naturally. Moreover, the decline of activity of cytoplasm in root cells and the transport ability of ion leads to the decreased capacity of transporting nutrients to aboveground. Crops that grow slow need fewer nutrients than those that grow fast; therefore, the need for macro nutrients is reduced.

Low temperature can affect the absorption of different nutrients. The absorption of nitrate nitrogen is affected more by low temperature than ammonium nitrogen. The impact of low temperature on phosphorus and potassium fertilizers is also quite severe, whereas calcium and magnesium fertilizers are less affected by low temperature. Reports shows that the micro-element ferrous absorption tends to be affected seriously by low temperature. When the temperature drops below 10°C, the absorption of various elements is limited and the root system suffers from cold injury. The cellular membrane of root will lead to phase change (e.g., from liquid-crystalline phase to gel phase). That is to say, in normal conditions, the chilling injury is a transition from a liquid-crystalline phase to a gel phase in the cellular membranes, which is less mobile. Therefore, it cannot prevent small molecular weight matters to enter the cell and loses the function to regulate cellular membrane transportation. This will cause the unbalance of ions and result in cell death. This phenomenon usually happens to those chill-sensitive crops.

The capacity to absorb water also decreases in low temperature. Since crops absorb nutrients while absorbing water, less absorption of water represents less absorption of nutrients.

7.2.3 The Availability of Soil Nutrients in Winter

The availability of nutrients in soil can be affected by various factors, including physical, chemical and biological properties of the soil. As long as the climate factor in winter can influence the above-mentioned properties of soil, it can influence the availability of nutrients. For example, microbial activity is obviously inhibited by low temperature; thus, it is more difficult for many elements (e.g., nitrogen, phosphorus, micro-elements, etc.) to be released from organic matter, and the supply of soil nutrients decrease. The transformation of nitrogen fertilizers, i.e., from ammonium nitrogen to nitrate nitrogen, is severely affected by low temperature, because nitrifying bacteria are inhibited by low temperature. If we apply large amounts of ammonium nitrogen fertilizer or urea, ammonium nitrogen is likely to accumulate by low temperature. Those crops which do not favor ammonium or cannot bear too much ammonium nitrogen are likely to suffer from excessive ammonium. If we do not cultivate crops, or if we view from the conservation of fertility, the existence of ammonium nitrogen is beneficial for the conservation of nitrogen fertilizer in soil. Ammonium nitrogen is less likely to run off than nitrate nitrogen. Paddy field or anaerobic conditions are especially beneficial to the preservation of ammonium nitrogen. In addition, nitrogen fixing microorganisms can also be affected by low temperature, which will reduce the supply or fixation of nitrogen.

Phosphorus fertilizer is less fixed in low temperature, which means the availability of phosphorus is high. The higher temperature and the higher probabilities of rainfall, the more phosphorus will be fixed.

The availability of potassium is influenced by its exchange capacity. Its exchange capacity will decrease in low temperature, and higher temperature benefits its exchange capacity. However, some research shows that the freezing and thawing of soil can help release more effective potassium.

The source of many micro-elements is related to the decomposition of organic matter; thus, the supply of micro-elements would be affected at a lower temperature. It is particularly true for those micro-elements that are not rich in the soil. In winter, there is less absorption and less supply. Insufficiency of these elements is likely to happen. For example, deficiency of manganese is likely to happen at a low temperature and in a poor drainage. When the drainage gets better or the temperature becomes higher, the symptom will disappear. The drought in winter can also lead to boron deficiency.

7.2.4 Guidelines for the Management of Soil and Fertilizer in Winter

The features of winter climate will exert different impact on various kinds of crop and soil. The management of soil and fertilizer should be in accordance with the condition and needs of crops. We should also adjust the application rate of fertilizer based on the soil capacity of nutrients supply. The general guidelines will be elaborated as follows:

Fertilization Periods and Methods:

Some crops can adapt to low temperature in winter and continue to grow and develop, while others cannot. We should apply fertilizers to winter resistant crops. For those crops which cannot bear the cold winter, the aim of fertilization is to improve the soil fertility, so that the soil can supply sufficient nutrients when spring arrives.

Perennial fruit trees of temperate zone usually grow slow or stop growing for some periods in winter, so it is rather important to control the timing of fertilization. It is not correct to fertilize too early or too late. Crops require little fertilizers in winter.

If we fertilize too early, the fertilizer is easy to be fixed, run off or volatilized, causing the loss of fertilizer and decrease of efficiency. If we fertilize too late, there would not be sufficient nutrients for spring growth or new bud formation. For instance, some organic fertilizers require two weeks for decomposition before supplying large amounts of nutrients. In the beginning of organic matter decomposition, some nutrients may be fixed. Thus, we need to estimate the time for organic fertilizer to supply nutrients.

The deeper the soil we supply nutrients or improve, the better it will be for tree crops. Hence, we should consider deep fertilization which can enhance the fertilization efficiency. This is particularly important for crops whose root system is deep and wide. Winter is the best season to improve the deep soil, because many crops grow slow in winter. Even though the root is injured when we are digging the deep soil, it would not affect crops too much.

Kinds and Amount of Fertilization:

We should refer to the handbook of cultivation when deciding the kinds and amount of fertilization for winter crops. For perennial fruit trees, the primary task is to improve acidic or alkaline problem of the soil. We can apply limy or alkaline fertilizers to improve acidic soil, and sulfur powder or acid-forming fertilizers for alkaline soils. Although the amount of fertilizer to improve the pH value differs, we can adjust gradually and annually. That is to say, we can set one metric ton per hectare of lime as a standard and adjust the amount based on the pH value of soil.

Winter is also the best season to apply organic fertilizer to improve the physical, chemical and biological properties of soil. It is important to apply organic fertilizer as a basal fertilizer and is a factor that cannot be ignored when deciding the production and quality of perennial fruit trees. As mentioned in Section 1, we should not apply fertilizer too early or too late. The deeper and wider of rhzosphere we apply the fertilizer, the better it will be.

Regulate the Harvest Season of Fruit Trees and Avoid Dropping of Fruits:

Regulation of the harvest season violates the natural plants growth. When weather changes in the winter season, fruits-dropping is likely to happen. Fine management of soil and fertilizer can prevent massive fruit droppings. If we overuse fertilizers, especially for nitrogen fertilizers, or if the organic fertilizer releases too much nitrogen, fruits-dropping is likely to occur. Therefore, we need to pay attention to the balance of nutrients and the amount of phosphorus, potassium, calcium and magnesium we added. Good aeration and moderate humidity of the soil are all beneficial to the crops health and their resistance to low temperature.

Foliar Application to Supply Nutrients:

Adjust and supply nutrients to the soil are the essence for crop production. We usually take foliar application of fertilizers as a supplementary way, when the nutrients are insufficient, or when we adjust the unbalanced absorption of nutrients. In winter, the absorption capacity of soil decreases mainly because of the adaptability of the crop. So it is useless, even harmful, to overuse fertilizer under this condition. For crops which are suitable for growing in winter, the absorption can still be inhibited by low temperature or by nutrient deficiency in the soil. Under this condition, foliar application of fertilizers is a feasible way. However, the amount of fertilizers should not be excessive so as to avoid harvesting after foliar application in a short period.

Take Precautions against Cold Weather:

Cold weather will cause damage to crops in many areas. We need to pay attention to the management of plant breeding and cultivation practice, as well as the management of the soil. For crops that cannot resist cold or frost, it is easy to observe the chilling injury above the ground of plants, while we often ignore the chilling injury underground.

We can regulate the soil moisture and irrigation to prevent injury caused by low temperature. Different crops need different control methods. For some crops, the root might get rotten because of the high moisture. There is no universal principle. It differs from one crop to another, from one kind of soil to another. In principle, how to improve soil temperature and escape the cold is the primary task. For instance, in vegetable bed, we use water current or irrigation to reduce injury caused by cold. We should also avoid applying fertilizer before and after a cold weather, and reduce any fertilizer application that would add the osmotic pressure to root and foliage.

7.3 Soil Environment and Management in Rainy Season

7.3.1 Soil Environment in Rainy Season

Causes of Oxygen Deficiency in the Soil:

During rainy season, the aeration pores in the soil are filled with water, so diffusion of air into soil reduces, and result in more consumption of oxygen than supply. Plant roots and microorganisms gradually consume the oxygen and finally lead to oxygen deficiency of soil (Figure 7-2). The severity of anoxia depends on the drainage capability of the soil. Anoxia is not severe in soil with fine drainage, while it is severe in soil which is flooded (Figure 7-3).

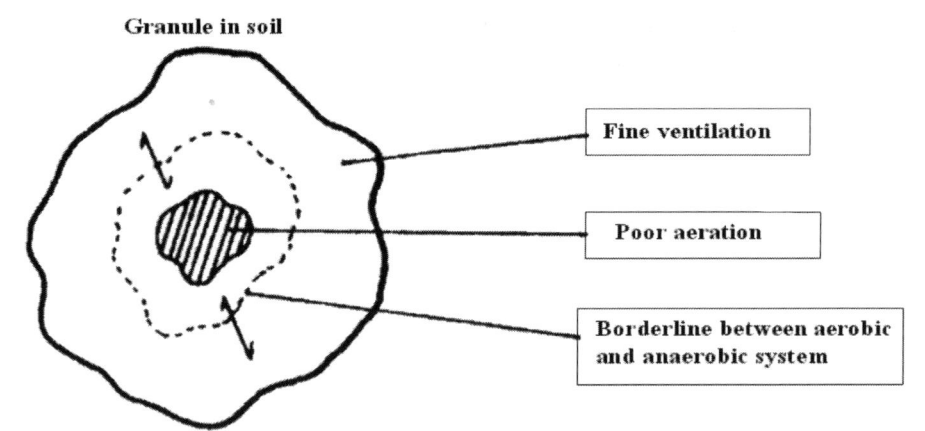

CH7 Guidelines of Improvement of Fertilization and Soil Management   183

Granule in soil

- Fine ventilation
- Poor aeration
- Borderline between aerobic and anaerobic system

Figure 7-2. When Soil Granule is Soaked in Water, the Aeration is Poor. If the Flooding is Serious, the Poor Aeration Zone May Expand

(A) Moderate flooding   **(B) Severe flooding**

- Root System
- Soil Granule
- Poor aeration

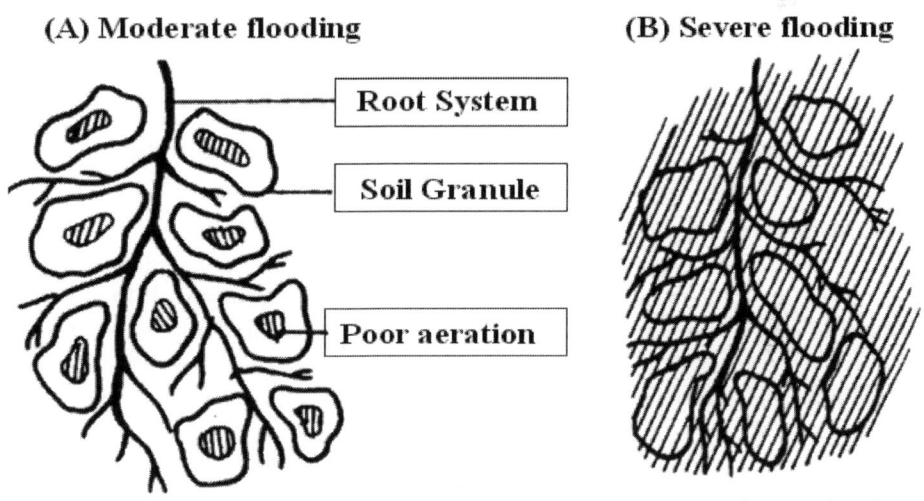

Figure 7-3. Root System in Rainy Reason: (A) Fine Drainage, Moderate Flooding and Few Areas of Poor Aeration; (B) Poor Drainage, Severe Flooding, Holes between Granules are Filled with Water, so that there is Severe Depletion of Oxygen in the Root System

The anaerobic soil in rainy season makes chemical property of soil become reductive and transforms microbial activity from aerobic to anaerobic. Therefore, there are considerable changes of soil physical, chemical and biological properties during rainy season, and the main reason is oxygen deficiency of the soil caused by rain.

Soil Microbial Activity under Anaerobic Condition:

Rain affects the physicochemical properties of the soil. When the redox potential is between +530 mV to +420 mV, nitrate nitrogen reduced to form nitrite nitrogen. In addition, when the redox potential is between +640 mV to +410 mV, manganese tends to transform from manganese (IV) to manganese (II). When the amount of oxygen continues to decline, ferric (III) transform to ferrous (II). In addition, sulfuric acid will be reduced to hydrogen sulfide, hydrogen and methane.

7.3.2 Soil Problems in Rainy Season

From the above description, we concur that oxygen deficiency in soil during rainy seasons leads to a higher soil reduction and more anaerobic activities of microorganisms. Paddy field is less affected by rainy reason, while there might be serious problems if the drainage system of upland is poor. Common problems of soil in rainy seasons are listed below:

Loss of Soluble Nitrogen Fertilizers:

Soluble nitrate nitrogen is easy to be lost in rainy seasons. Two main reasons are listed as follows:

Loss Caused by Leaching:

Nitrate the easiest nitrogen fertilizers to move from soil. It may be lost by the runoff from rainfall, especially it may be severe in hillside fields.

Loss Caused by Denitrification:

In soil where drainage is poor or oxygen is deficient, denitrifying microorganisms can transform nitrate to nitrogen gas and nitrous oxide, and nitrogen fertilizers will be lost due to volatilization. Research has shown that the loss of nitrogen fertilizers through denitrification may reach twenty to forty-five percent of soil nitrogen. Therefore, the loss of nutrients caused by denitrification in rainy seasons should be one of important concerns.

Formation of Toxic Compounds:

Table 7-1 illustrates toxic compounds produced by microbial activities under anaerobic soil condition and their potential to harm crops. These compounds are listed as follows:

Organic Acids:

Most of these are the volatile fatty acids. For instance, when the concentration of acetic acid is about 10^{-2} M, it is toxic to rice. Formic acid, propyl- and butyl- acids are also common organic acids. In addition, aromatic acids with ring structures such as phenolic compounds are also common.

Table 7-1. Activities of Microorganisms and Relevant Reaction in Anaerobic Soils

Activity of Microorganisms	Chemical Reactions in Anaerobic Soil
Respiration of Nitrate Nitrogen	Organic matter + nitrate nitrogen → nitrite nitrogen
Dissimilation of Nitrate Nitrogen	Organic matter + nitrate nitrogen → ammonium nitrogen
Reduction of Ferric and Manganese	Organic matter + ferric (III) and manganese (IV) → ferrour (II) and manganese (II)
Denitrification	Organic matter + nitrate nitrogen → nitrogen gas + nitrous oxide
Fermentation	Organic matter → organic acids
Reduction of Sulfate	Organic matter or hydrogen + sulfuric acid → sulfide (hydrogen sulfide)
Reduction of Carbon Dioxide	Hydrogen + carbon dioxide → methane, acetic acid
Decomposition of Acetic Acid	Acetic acid → carbon dioxide + methane
Reduction of Hydrogen Ion	Fatty acids and alcohols + proton (hydrogen ion) → hydrogen gas + acetic acid + carbon dioxide

Hydrocarbon Gas:

Methane is the most common product produced under anaerobic condition. Ethylene can also be found to reach a concentration of 10 ppm. Usually, the growth of root system will be inhibited when ethylene reaches the concentration of 2 ppm.

Carbon Dioxide:

When rain inhibits the gas exchange ability of soil, the amount of carbon dioxide in the soil tends to increase to high concentrations that are toxic to crops.

Sulfides:

Hydrogen sulfide is rather toxic. Poor drainage soils with large amounts of organic matter have the formation of hydrogen sulfide to injure crops.

7.3.3 The Influence of Rainy Season on Root System

The root system of various crops has different adaptive capacity to rainy season. Oxygen deficiency and the production of toxic compounds in soils are common in rainy seasons. The growth of root system requires oxygen for respiration. Aquatic crops and water-tolerant crops can tolerate oxygen deficiency, because these crops can transport oxygen from the stem to the root or by spongy tissues, and aeration ducts to roots. Thus, roots soaked in water will not suffer from anoxia. However, most upland crops are not tolerant to flooding by water.

Oxygen deficiency and the existence of toxic compounds in rainy season can cause severe harm to crops. The most obvious phenomenon is the inhibition of root growth. The normal absorption and transportation of nutrients and water are also affected. It may even lead to the death, rotten root or withering. The reason why these problems occur is that the root is affected by the following factors:

Change of Root Respiration:

Under anaerobic condition, aerobic respiration of the root cells cannot proceed and result in the decrease of energy sources. The crop under anaerobic condition can only conduct anaerobic metabolisms and produce ethanol or ethylene, which may do harm to crops.

Change of Root Permeation:

Oxygen deficiency and the existence of toxic compounds will reduce the permeability of root cellular membrane. The absorption of nutrients requires energy and an unbalanced absorption of water can cause plants to wither.

Changes in Synthesis and Transportation of Hormones in Root:

Gibberellins, cytokinins and other hormones are synthesized in the root. When oxygen is deficient and there are toxic compounds, the capacity of root synthesize and transport hormones to stems and leaves decreases. This can cause poor growth.

Influence of Nitrogen Fixation in Root Nodules of Legumes:

When root nodules of legumes are flooded by water, nitrogen fixation activity would be inhibited due to the inhibited gas exchange.

7.3.4 Guidelines of Soil Management in Rainy Season

Importance to Soil Drainage Ability:

The drainage ability of uplands in rainy season should be strengthening to avoid water accumulation. The primary management task is to drain excessive water by making ditches or concealed conduit. The faster the draining is, the better it will be. However, we should also avoid soil being washed away by water.

Cultivate in High Plot:

In areas where groundwater level is high, it is hard to drain large amount of water during rainy season. So cultivation in a high plot can prevent submergence of crops. The root system, which is above the ground, acquires sufficient oxygen to maintain the normal absorption of nutrients and water. However, the fertilizer we apply on high plot is easy to be washed away or be eluted, so we shall pay attention to the amount of fertilizers we applied.

Selection of Fertilizers:

In rainy season, soluble fertilizers and nutrients are easy to be washed out by water. The extent of the washed-away is influenced by many factors, such as characteristics of the soil and fertilizer. Nitrate fertilizers, potassium, calcium and magnesium ions are usually the easiest to be lost. Amongst all kinds of soil, sandy soil is the easiest to lose nutrients. Therefore, we should pay attention to the loss and supply of nutrients. Covering with materials or soil for the applied fertilizer can help reduce nutrients loss caused by water washing-out.

Cultivation by Protected Agriculture:

In order to avoid too much rainfall in rainy season, we can use rain shields and other facilities to help reduce the occurrence of oxygen deficiency and the formation of toxic compounds.

Reduce the Amount of Organic Fertilizer Applied in Poor Drainage Soils:

If we overuse organic fertilizer in areas with poor drainage, it will cause the formation of toxic compounds and result in severe rotten roots.

Reduce Plowing:

Plowing softens the soil. However, during rainy or windy seasons, the soft soil would weaken the plants and lead to the lodging of tall crops. Under normal soil conditions, no tillage or zero tillage farming can be used for bringing advantages to increase yield because of higher accumulation of organic matter, higher water infiltration and storage capacity, and less erosion. Covering grass cultivation for perennial tree crops minimizes soil loss in rainy seasons, especially for orchard soils and increases the organic matter content in soil in turn.

Apply Nitrification Inhibitors:

In rainy seasons, nitrogen fertilizers can be nitrified into nitrate nitrogen. Because nitrate nitrogen is easily washed away by rain; therefore, limiting the formation of nitrate nitrogen is beneficial to the preservation of nitrogen fertilizers. Therefore, the application of nitrification inhibitors in rainy seasons is more efficient then when applied in dry seasons.

Apply Plant Hormones:

The transportation of hormones from the root is not effective in rainy seasons. There are reports indicating that applying plant hormones can promote the growth of stem and reduce the withering of leaves. Further research and development is required for the promotion of crop growth in rainy seasons.

7.4 Guidelines for Fertilization under Climate, Disease and Pest Stresses

There are various stresses in agricultural cultivation, both natural and artificial. Common environment stresses include physical, chemical, and biological stresses. These environment stresses will lower crop production and quality, and may even become a negative factor. That is to say, even we have the best crop cultivars and cultivation methods, the production will be severely affected under harsh condition. Hence, it is a rather important task to strengthen crops capacity to cope with environmental stress. Ways to cope with environment stress, including physical, chemical and biological means. This section will illustrate favorable fertilization guidelines to cope with unfavorable climate, diseases and pests.

Under favorable conditions, fertilization is a factor to improve production and quality of crops. On the other hand, under environmental stress, we should pay attention to guidelines of fertilization. Effective or correct control of fertilization will help crops endure and reduce damages under environmental stresses.

7.4.1 Climate Stress and Fertilization

All farmers expect optimal climate for agriculture. However, the climate is always changing. Common environment stresses such as drought, too much rain, low temperature, blaze, heat-flash, violent wind and rainstorm will exert huge influences upon crops. Just as the old saying goes "the agriculture is dependent on the climate." Although it is hard to resist harsh conditions, sufficient management of cultivation and fertilization will minimize the loss caused by environmental stresses.

There are many methods to cope with unfavorable climate conditions. In terms of fertilization, we should pay attention to the crops protection. We should strengthen fertilization management before unfavorable climate was occurred, so that the ability of crops to resist environment stress will be enhanced. Or it will be too late to fertilize when unfavorable climate occurs. I will illustrate several common climate stresses (drought, cold, mechanical injury) and relevant guidelines of fertilization in the following sections:

Drought:

Occurrence of this Issue:
Due to the uneven rainfall, drought also takes place in areas where there is large amount of annual rainfall. When the irrigation system works well in farmland, there will be no drought. However, drought might be severely influence crops in hillside or mountain areas, especially harmful to shallow rooted plants. Though fruit trees are deep-rooted, the quality can be affected when drought occurs. When it occurs, not only the

water will be limited, the ability to absorb nutrients for crops will be also inhibited. When severe water shortage occurs, plants will exceed their withering point and die. The capacity of different crop cultivars to tolerate drought differs. Thus, we can breed or select cultivars that are more resistant to drought.

Guidelines of Fertilization:

It has been previously stated that fertilization during environmental stress would cause more harm. Potassium can increase the turgor pressure of plants and reduce transpiration and moisture loss. Hence, we can provide appropriate amount of potassium fertilizer before drought to increase crops capacity to tolerate drought. Big foliage, excessive vigorn growth and small root system are all characteristics of intolerance to drought. Supplying more phosphorus fertilizer can help the root system become longer, deeper and allow better deep-water absorption. Moreover, we should avoid excessive nitrogen fertilizers that cause excessive vigor growth and big foliage to reduce the harm to short-term drought.

The withering of crops may not necessarily related to severe water shortage. Sometimes, the lacking of inorganic nutrients could expand the harm caused by drought. For example, lack of manganese during drought can lead to leaf yellowing, which is similar to the symptom of water shortage. Therefore, precise management of fertilization before drought can reduce or prevent harm caused by drought. However, we should be more careful about fertilization during drought period, and the amount of fertilizer should be at the minimal amount as suggested. The increased amount of fertilizer salts applied during drought will expand crop injuries.

Heat-flash can cause similar harm as drought, for heat-flash will increase the respiration and transpiration of crops. If there is moderate supply of nutrients, crops will not be affected by the hot weather and the photosynthesis of crops will not be affected either. Usually, C4 crops (e.g., corn, sorghum, sugarcane, tropical pasture, etc.) are more resistant to heat; CAM crops (e.g., pineapple, cactus, cattleya orchids, etc.) are resistant to drought or heat, while C3 crops (e.g., paddy rice, legumes, etc.) usually cannot resist heat too much.

Cold:

Occurrence of this Issue:

The occurrence of cold current always hurts many crops. Low temperature reduces metabolisms of plant cells and severe low temperature will cause plant cells to freeze and injure the cell membrane. Severe low temperature not only injures the aboveground plant parts, but also causes serious frost injuries of the root in surface soil, and this will drive plant injury or even death. Plants of temperate zone are more capable to resist cold than those of tropical zone. This is closely related to the composition of the cell membrane of plants. Researches on breeding and biotechnology will help produce cold-resistant crop cultivars.

Guidelines of Fertilization:

Supplying moderate phosphorus and potassium fertilizers before winter will help crops resist cold. Applying excessive nitrogen fertilizer is not helpful to resist cold. The lack of micro-elements (e.g., manganese, copper, etc.) may expand the harm caused by sudden drop of night temperature. This indicates that these elements can improve the capacity of crops to resist low temperature. The fertilization of various crops differs. However, as long as the water movement through cell membrane reduces, it can help crops resist cold. In short, apart from improving cultivars, moderate supply of nutrients can also reduce or prevent the harm caused by cold. But even the best fertilization method cannot help crops to resist severe cold climate. It is especially true when the temperature drops abruptly, so that crops cannot adapt to it immediately.

Mechanical Stresses:

Occurrence of this Issue:

We cannot ignore mechanical stresses exerted by the environment. Common ones include strong wind, rainstorm or heavy now, which can cause lodging or injuries to crops.

Guidelines of Fertilization:

It is necessary to sustain a balance nutrient for crops to develop abundant cellulose and lignin to resist mechanical injuries caused by strong wind and rainstorm. Moderate supply of phosphorus and potassium fertilizer can enhance the capacity of crops to resist these environmental stresses. Excessive nitrogen fertilizers can soften plants tissue, so that plants are easy to be injured by strong wind and rainstorm. Lodging is closely related to the excess lusher growth of plants with high nitrogen fertilization. The healthy of plants is closely related to its silicon content. The application of silicate materials can strengthen branch and stem so as to prevent lodging.

7.4.2 Resistance to Diseases and Fertilization

Occurrence of this Issue:

Disease is one of factors that limit crop production. Pathogens can invade crops either from aboveground or underground of plants. Therefore, prevention and treatment of diseases is an important part in agricultural production, and special attention should be focused on soil borne diseases. There are many pathogens that can cause diseases and their pathway to infect crops differences. Therefore, ways to prevent diseases are also different. We can eradicate pathogens, infectious mediators of diseases, and breed for cultivars that are tolerant or diseases resistant. Diseases of crops are severe in subtropical and tropical zones where there is heavy continuous cropping. It is difficult to observe pathogens by our eyes. Thus, sometimes it is too late when we discover the diseases, and huge injury has already been done. To prevent and cure diseases are difficult, so we should conduct more efforts in doing research.

The reason that plants can resist diseases in the natural environment is due to either internal or external factors. Internal factors are related to the biochemical metabolisms and the structural characteristics of plants. For instance, when pathogens infect resistant plants, the plants can produce phytoalexins, so that the pathogen cannot hurt plants. External factors indicate special protection from the environment, such as suppressive soil, which can protect crops from pathogens. Some soils are able to suppress diseases and the reasons might include pH value of soil, and the antagonistic effects between microorganisms, inorganic and organic materials. There are several reports indicate that soil conditioner can be used to improve the capacity of soil to resist diseases. The health condition of crops is also closely related to the soil capacity for resisting diseases. The relation between nutrients and disease-resistance can be illustrated as follows:

- Disease resistant may decrease due to excessive or insufficient nutrient supplies. When crops lack nutrients, they become weak and are prone to suffer from diseases.
- Sufficient and balanced nutrient supplies will help crops grow healthily.
- Direct or indirect fertilization is beneficial for crops to resist diseases.

Therefore, disease resistant of crops is more or less related to the supply of nutrients. Fertilization can help crops improve disease resistant. Yet, some pathogens are difficult to control, and cannot be suppressed completely.

Guidelines of Fertilization and Management:

Optimal management of fertilization is beneficial for crops growth and can easily compensate for the loss caused by unfavorable factors. The following illustrates several common relationships between malnutrition and disease resistant of crops:

- Excessive nitrogen fertilizers will weaken crop tissues and allow easier infection of viruses, bacteria, and fungi. However, crops which lack nitrogen are also prone to infection.
- Crops that lack phosphorus are prone to be infected by fungi. This may be caused by inappropriate nitrogen/phosphorus ratio.
- Lack of potassium can reduce carbohydrate content in crops and induce thinning of cell wall and weakening of leaf trichome. Thus, render crops susceptible to parasite invasion. In addition, intermediate products of metabolisms are produced in excess, which means more sugar will be produced and replace starch. Hence, crops are prone to aphid attack and increase the chance of virus infection.

- Lacking of calcium would weaken plants and render them more susceptible to fungi infection.

- Lacking of silicon would decrease the silicalization of leaf epidermis. Thus, crops are susceptible to insect invasion and disease infection.

- The relation between lacking of micro-elements and disease resistant of crops is not clear, but we can infer that their relation is quite close. For example, the virus infection symptoms of orange are similar to symptoms caused by lacking of micro-elements. For instance, virus infected foliage shows similar symptom as zinc deficiency, and the analysis result of the infected leaves, indeed, shows zinc deficiency. Whether the lacking of zinc resulted in infection of virus or infection of virus lead to lacking of zinc, this requires further research.

- It is not clear whether absorption of antibiotics can improve disease-resistant of crops. Antibiotics produced by soil microorganisms can play an important role in crops protection. Fertilization can promote nutrients crops absorption, increase soil microbial antibiotic production and improve the capacity of crops for resisting environmental stress.

- The relation between nutrient supply and disease resistant of crops has already been confirmed. Plants are more susceptible to be invaded by parasites under the circumstances of unbalanced or insufficient nutrient supplies, and under toxic compounds presented in soil. For instance, when the soil is too acidic, the availability of many nutrients will decrease and there will be evident diseases. If we can elevate pH value, then we can improve disease resistant of crops. Applicable materials (lime, slag, silicate, oyster shell powder, etc.) not only can elevate pH value of soil, but also increase nutrients, such as calcium, magnesium and silicon in the soil. There are several cases that show nematode infection leading to the crops death. The disease is not due only to nematode infection but other factors, such as unbalanced nutrient supplies and toxic aluminum, will also accumulate to the harm. Hence, it is rather important to improve malnutrition of the crops along with parasite prevention to achieve effectiveness.

7.5 Problems of Soil pH and Guidelines of Diagnosis and Management

7.5.1 Understand the Features of Soil pH

Either too acidic or too alkaline of the soil will affect crops production and quality. Among various crops, their response to the soil pH is quite different. Most crops are not tolerant with much acidic or alkaline soils. Yet, some crops can resist acidic or alkaline soils. For instance, tea and pineapple are quite tolerant to acidic soils. Too acidic or too alkaline soils will exert huge influences on the physical, chemical and biological properties of the soil. The rigidity, salinity, availability of nutrients and diseases of soils are all issues that we should pay attention to.

In order to prevent soil becoming too acidic or too alkaline, first of all, we should understand pH value. pH is defined as the decimal logarithm of the reciprocal of the hydrogen ion activity in a solution. pH value means the capacity of a material for releasing hydrogen ion in aqueous solution. Hydrogen and hydroxide ions can influence many functions, such as metabolisms, absorption and growth of cells. They can also influence physical and chemical factors, such as solubility of compounds, chemical reactivity and availability of elements. Generally speaking, those substances that can release hydrogen ions are acidic materials, while those that can release hydroxide ions are alkaline materials. To measure the pH value of the soil is to measure the hydrogen ion concentration in soil solution. Therefore, water is an essential part when measuring the pH value of soil. Some farmers measured their soil pH value directly insert a metal pH sensor into dry soils, and that is an incorrect way.

7.5.2 Sources of Soil pH

Only by understanding sources of soil pH, we can effectively reduce or prevent soil from becoming too acidic or too alkaline. The sources of agricultural soil pH mainly include natural characteristics of soil, the influences of human activity, and the environment. Since it is hard to change natural characteristics of the soil, we should strengthen the control of the acquired factors to prevent soil from becoming too acidic or too alkaline. I will illustrate the source of acidity and alkalinity in the following sections:

7.5.3 The Source of Acidity in Soils

Natural Characteristics of Soils:

Sources of acidity in soils includes hydrogen ions released by several materials, such as soil organic matters and the weathering of primary and secondary soil minerals, i.e., clay fraction of alumina silicate, hydrated forms of aluminum and iron, exchangeable aluminum, iron, manganese, soluble salts and other acidic matters. The source of various soils may be different.

Influence of Human Activity and the Environment:

The acidity of soils can be influenced by human activity and the environment.

Applying Acid-Forming Fertilizer or Agrochemicals:

There are lots of acid-forming fertilizers mainly consist of acidic groups or strong acids, among which sulfuric acid, nitric acid and hydrochloric acid are the strongest. We should avoid continuous application of fertilizers with strong acids.

Absorption of Cation by Crops:

When large amounts of cations, such as calcium, potassium, magnesium, and sodium ions are absorbed by crops, hydrogen ions will be released from plant roots into the soil, and increase the acidity of soil.

Soil Leaching:

When soil is leached by rain or irrigation, cations will be lost and replaced by hydrogen ions, making the soil more acidic. Since there is much rainfall in the area, the leaching in hillside regions are quite severe and the soil is quite acidic.

Decomposition of Organic Matter:

Large amounts of organic acids will be released when organic matter is discomposed, which makes the soil more acidic.

The Dissolving of Carbon Dioxide in Water:

The metabolisms of microorganisms and crop roots will produce large amounts of carbon dioxide. When carbon dioxide is dissolved in water, it forms carbonic acids and make the soil becomes acidic.

The Weathering of Sulfide:

Nature or artificial applied sulfide in soil will dissolve in water, or will be utilized by microorganisms. It produces sulfuric acid or other acid matters make the soil acidic. When it becomes severe, the soil pH value may fall to around 3.

Acid Rain:

When industrial pollutants enter the atmosphere, it will add acidic matters in rain and make the soil acidic.

7.5.4 The Source of Alkalinity in Soil

Natural Characteristics of Soils:

A common alkaline soil in some alluvial soils contains large amounts of lime materials (e.g., calcium carbonate). In terms of saline-alkaline soil, saline soil and saline soil with sodium are more common than saline soil without sodium. These soil pH can reach up to 8.5 and contains high proportions of salts.

Influence of Human Activity and Environment:

The aims of applying lime material or sodium salt in acidic soil is to neutralize the pH value of soil. However, excessive application may make the soil alkaline. In addition, incorrect, excessive, or long-term application of salt will make the soil saline and injure crops. Therefore, we have to pay attention to this.

Inappropriate Irrigation:
Farmland alongside the ocean often lacks water. If the quality of irrigation water is poor and high electric conductivity (> 4 dS/m), it may lead to the salinization and alkalization of the soil. In addition, inappropriate irrigation of seawater can also lead to the salinization of soil, which is not worthwhile. So, we should consider the conservation of soil seriously.

High Salt Content in the Groundwater:
The evaporation of water in soil will transfer salt from underground to the surface soil. Plus, less irrigation, it will make the salinization of soil even worse.

7.5.5 How do We Know whether the Soil is too Acidic or too Alkaline

The answer to the question is to measure pH value of soil. Common ways of measuring pH value of soil include electrode measure, colorimetric measure with measure paper and titration measure. Since measuring pH value equals measuring the concentration of hydrogen ions, we do not know the concentration of hydrogen ion by plant responses or by our eyes. Experienced personnel could only estimate pH value, but direct soil measurement is still required. The concise methods and notes are listed as follows:

Electrode Measure:

Soil and water should be mixed and stirred sufficiently. We should stir it again after standing for 0.5 to 1 hour and measure pH value with an electrode pH meter. We should not insert the meter directly in dry soil for measurment. Before using the electrode pH meter, we should calibrate with buffer solutions of pH 4 and 7. This correct method is most commonly applied in the laboratory and fields.

Colorimetric Measure with Litmus Paper or pH Strips:

Put one gram of soil into a measure tube, add 25 grams of water, shake one minute, use measure paper to absorb the water solution sufficiently, and observe the change of color. Because there are various brands of measured papers and their sensitivity also differs, we need to use a specific standard reaction color kit for each type for checking and comparing. This method is quite simple and convenient and it has been extensively used in agriculture.

Titration Measure:

The reaction between the acid-base indicator and extracted soil could allow us to compare with the colorimetric standard. This is a common, simple and quick method and only need to follow the manual.

Other Methods:

Add a small drop of 6 N hydrochloric acid (50% diluted concentrated hydrochloric acid) on soil sample on a porcelain container. Then, observe the bubble formation (CO_2) to estimate the lime content in soil. The more bubbles there are, the higher content of calcium carbonate is in the soil sample. This method cannot estimate the precise pH of the soil. If the bubble is light, it may indicate the proportion of calcium carbonate is about 2%. If the bubble is obvious, it may indicate about 2% to 5% calcium carbonate. If the bubble is drastic emission, it may indicate over 5% calcium carbonate.

7.5.6 The Harmful Effects of Extreme Acidity or Alkalinity of Soils

The Harm to Extreme Acidic Soil:

Different crops or cultivars will respond differently to extreme acidity of soil (Table 7-2). Those which can resist acidity are less likely to be injured. For crops which cannot resist extreme acidity (pH < 5), their growth will be limited, and may show symptoms related to the lack of phosphorus, calcium, magnesium, molybdenum, etc. They may also show symptoms of aluminum and manganese toxicity. For some crops, there might be abnormal stripes and spots on the foliage and/or under extreme conditions. This might lead to crops death. Severe diseases are likely to occur on crops cultivated on acidic soil, so we should pay special attention to it.

The Harm to Extreme Alkalinity of Soils:

Different crops or cultivars will response differently to extreme alkaline soil (pH > 7.5). In alkaline soil, the availability of iron, zinc, manganese, copper, and boron ions will decrease, and thus severe symptoms may show. Common symptoms in alkaline soil include the yellowing of peanut foliage and leek, and especially for new developing leaves. For crops that cannot adapt to saline alkaline soil, their growth may be inhibited, and the root system may rot.

7.5.7 Guidelines to Neutralize Extreme Acidity or Alkalinity of Soils

Apply Neutralizer:

Conditioners for Acidic Soil:

Apply limes or alkaline materials, such as agricultural lime, magnesia lime, dolomite powder, slag, and oyster shell powder, to the soil. The amount of these materials should depend on pH value of soil. Since the capacity of different material to neutralize pH value differs, we should be careful of the application amount (Table 7-3). We can neutralize gradually over years for perennial crops. We should not apply too much lime on soil with poor buttering capacity, such as sandy soil, sandy loam and other soil in which the

Table 7-2. Optimal Range of pH for Crops

Crops	pH value
	4.5　　5.0　　5.5　　6.0　　6.5　　7.0　　7.5
Pineapple	4.5–5.5
Tea	4.5–6.0
Rice	4.5–7.0
Tobacco	5.0–6.0
Strawberry	5.0–6.0
Watermelon	5.5–6.5
Potato	5.5–7.5
Sugarcane	5.5–7.5
Radish	6.0–6.8
Legume	6.0–6.8
Soybean	6.0–6.8
Peanut	6.0–6.8
Cotton	6.0–6.8
Green pepper	6.0–6.8
Cucumber	6.0–6.8
Eggplant	6.0–7.0
String bean	6.0–7.0
Pumpkin	6.0–7.0
Maize	6.0–7.0
Wheat	6.0–7.5
Barley	6.0–7.5
Tomato	6.0–7.5
Ginger	6.0–7.5
Orange	6.0–7.5
Grape	6.0–7.5
Papaya	6.0–7.5
Mango	6.0–7.5
Loquat	6.0–7.5
Pear	6.0–7.5
Apple	6.2–7.5
Carrot	6.0–7.0
Cabbage	6.2–7.5
Banana	6.3–7.0
Celery	6.3–7.0
Cauliflower	6.3–7.0
Brussels Sprouts	6.3–7.0
Lettuce	6.3–7.0
Spinach	6.3–7.0
Pea	6.3–7.0
Asparagus	6.3–7.5
Onion	6.3–7.5
Leek	6.3–7.5

proportion of organic matter is low. Lime materials should not overuse, otherwise the disadvantage will show. The degree of alkalinity and capacity of common lime materials to neutralize acidity differs a lot and it is illustrated in Table 7-4.

Conditioners for Alkaline Soil:

Applying acidic materials (e.g., sulfur powder, dilute acids, acidic peat powder) is the fundamental way to neutralize alkaline soil, but we should not overuse it.

Table 7-3. The Application Amount of Lime Material to Improve the pH Value of Acidic Soil to 5.5 ~ 6.0. If the Amount is over 2,000 Kilogram per Hectare, It Should Be Applied Gradually and Annually. For those Soils in which the Proportion of Organic Matter is Low, We Should Apply Lime Material with Organic Fertilizer as to Prevent Soil Injury.

Type of Soil	\multicolumn{5}{c}{pH Value of Soil}
	3.5	4.0	4.5	5.0	5.5
	\multicolumn{5}{c}{Application Amount of Lime Material (kg/ha)}
Sandy Loam	1,400 ~ 1,700	1,000 ~ 1,400	1,000 ~ 700	400 ~ 700	0 ~ 400
Loam	2,400 ~ 2,800	1,700 ~ 2,000	1,000 ~ 1,500	500 ~ 1,000	0 ~ 400
Clay Loam	2,900 ~ 3,500	2,000 ~ 2,600	1,300 ~ 2,000	600 ~ 1,200	0 ~ 600

Table 7-4. Common Lime Materials and their Quality

Name	Major Chemical Composition	Capacity to Neutralize Acidic Soil[1]	Alkalinity[2]
Quicklime	CaO	179	100
Hydrated Lime	$Ca(OH)_2$	136	76
Limestone Powder	$CaCO_3$	100	56
Magnesia lime	$CaMg(CO_3)_2$	90 ~ 105	53 ~ 59
Silicic Slag	$CaSiO_3$	60 ~ 80	34 ~ 45
Lime Slag	$CaSiO_3$	65 ~ 85	36 ~ 48
Crab Carapace Powder	$CaCO_3$	38 ~ 45	21 ~ 26
Oyster Shell Powder	$CaCO_3$	92	51

p.s:1. Take the alkalinity of limestone powder as 100, and then the relative alkalinity of every material is its capacity to neutralize acidic soil.

2. Alkalinity = CaO% + MgO% × 1.39.

When applying the above-mentioned soil conditioners, we should not mix them with chemical fertilizers. It is suggested that we apply fertilizers one or two weeks after applying soil conditioners. By this mean is to improve the fertilizers efficiency after pH value of soil is neutralized, or the efficiency will be reduced due to extreme acidity or alkalinity of the soil. Only after eliminating the unfavorable factors can the fertilizer become effective.

Apply Organic Matters:

Organic matters can improve the availability of nutritive elements, especially when applying together with soil conditioners. If we only apply soil conditioners, the physical property of soil may become poor, and this might lead to adverse effects. Organic matters produce organic acids that improve the cations mobilization (e.g., Ca_2+) and reduce the harm caused by aluminum.

Pay Attention to the Amount of Fertilizers as to Prevent Acidification and Alkalization of the Soil:

Chemical fertilizers may be acidic or alkaline. In terms of fertilization, it is not proper to apply acid-forming fertilizers on acidic soil for long period and vice versa for alkaline fertilizers. We should apply proper fertilizers as to prevent unfavorable soil reactions.

Foliar Application under Emergent Condition to Supply Nutrients:

When we cannot change the pH value of soil in a short time, fertilizing on foliage can help temporally to a certain degree for the plant growing in the soil without neutralization. However, the fundamental way is to neutralize the soil pH.

Apply Insufficient Nutrients:

If we can supply nutrients of low availability, it will help offset extreme acidity and alkalinity issues, e.g., application of phosphorus fertilizers. For acidic soils, it is important to supply nutrients, such as calcium, magnesium and boron. For alkaline soil, it is important to supply nutrients, such as copper, iron, zinc, and manganese.

Pay Attention to the Irrigation Quality and the Drainage:

Utilizing poor quality irrigation water for a long period on a farmland will influence buffer capacity, pH, or physical, chemical and biological properties of the soil. Therefore, we should be cautious when utilizing various waste water and groundwater, and prevent it from hurting soil resource for our generations of mankind. It is easier to eliminate unfavorable factors in soil with fine drainage than with poor drainage. This is particularly important in the saline land.

7.6 Problems of Soil Electric Conductivity, Salinity and Guidelines of Diagnosis and Management

7.6.1 Understand Problems of Soil Electric Conductivity and Salinity

Soil electric conductivity (EC) is the ability of soil to conduct electric current, and is measured from extracted saturated soil solution. Soil EC is used as an indicator of the degree of salinity in soil. The standard unit of soil EC is dS/m or mmhos/cm at 25°C. 1 mmhos/cm = 10 S/m = 1 dS/m. It can also be used to indicate the osmotic pressure in soil (OP = 0.39 EC; OP means osmotic pressure). The higher the degree of soil salinity, the higher its soil EC value is, and the higher its osmotic pressure. Therefore, soil EC can be used to indicate the concentration of soluble salts in the soil solution. The concentration of soluble salts in the soil solution is directly related to the plants growth.

Method to measure soil EC: for agricultural application, we put certain amount of soil (200 ~ 400 g) into a plastic cup, with ca. 1:2 soil-water mixture. After 0.5 hours of shaking and make it to pasty form, place the saturated soil solution on a filter paper and use a pump to collect the filtrate for measurement using an EC meter (with temperature compensation ability). Simple and portable EC meter is available in the market.

According to the American Institute of Saline Soil, the sensitivity and tolerance of various crops are different (Table 7-5). Most crops can resist soil EC lower than 4 dS/m (mmhos/cm). While salt-tolerant crops, such as plants alongside the seashore or in the desert, these can resist soil EC between 8 to 16 dS/m.

7.6.2 The Source of the Soil Electric Conductivity

There are many kinds of dissociated ions and saline matters in soil. These saline matters will dissociate to positive and negative ions to a certain degree based on

Table 7-5. Relations between Soil Electric Conductivity, Salinity, and Growth of Crops

Soil EC (dS/m)	Salinity	Crops Growth
0 ~ 2	Very low	Common crops grow normally
2 ~ 4	Low	The growth of sensitive crops is inhibited
4 ~ 8	Medium	The growth of many crops is inhibited
8 ~ 16	High	Only salt-tolerant crops can grow
> 16	Very high	Extremely salt-tolerant crops can grow

different environmental factors and lead to the EC of the soil solution. The more ions there are, the higher the soil EC will be. Pure water without ions will not conduct electricity. The saline matters in soil may derive from natural minerals, salts, drought, irrigation water, invasion of sea water, excessive and inappropriate application of fertilizers, and pollution.

During the formation and development of soils, if ions are eluted by water for a long period, the soil EC will decrease. In agricultural activities, almost all fertilizers are acidic salts, such as sulfate salts (e.g., potassium sulfate and calcium sulfate), chloride salts (e.g., potassium chloride and calcium chloride), phosphate salts (e.g., potassium phosphate and calcium phosphate), and nitrate salts (e.g., potassium nitrate, and calcium nitrate). Animal excrements that contain salt, such as animal manures can also lead to high salt content in soils. The more fertilizers we apply, the higher soil EC will be. Long term excessive application of fertilizers will result in high contents of salts in soil and inhibit the crops growth, especially for short term crops (e.g., vegetables, cash crops) in protected (facility/shelter) cultivation. The accumulation of salt will be accelerated in areas where leaching is not sufficient, and where the temperature is high, and where the evaporation is severe. Irrigation water is also a source of concentrated salts.

7.6.3 Problems of Soil EC and Saline Soil Damage

If the soil EC is rather low (< 2 dS/m), then there will be no harm caused by the salts. However, this does not necessarily indicate that nutrients in soil are insufficient. Some people believe that low soil EC indicates that the soil is insufficient of fertilizers, so they decide to apply more fertilizers. This is a wrong concept. Long term application of optimal amount of fertilizer will not lead to the massive salt accumulation in the soil. Usually, if soil EC is greater than 2 dS/m, the issue of salt accumulation is evident. The accumulation of salts is common for vegetable farmlands and often lead to poor growth, or even seedling death, or withering when severe.

We can observe white crystallization of salts on the surface of top soil where there is severe salt accumulation. Excessive salt issue is common in fields for continuous cropping of vegetables and in facility or shelter cultivation. It is also common in soils that have been immerged by sea water.

The crops absorption in high saline soils will be affected, owing to the antagonistic effects between different nutrients. Plant absorption of excessive macro-elements (e.g., nitrogen and phosphorus) may also lead to micro-element availability decrease (e.g., iron, copper, zinc, and manganese). The symptoms of salt injury include: (1) Crops grow slow and stunted, or seedlings dead when severe; (2) Abnormal growth of crop portion and seedlings death; (3) When sunlight is strong and the weather is hot, foliage will suffer from lacking of water and wither; (4) When it becomes severe, leaves will turn small and dark, and leaf tips and margins may show burns; and (5) The root system grows poorly and the root tip appears sticky.

7.6.4 Methods to Improve Saline Soil Damage

The main method to deal with salt injury is to improve saline soil and management of crops cultivation. Saline soil improvement includes removing salt from soil and reducing the addition of salts. Details will be elaborated as follows:

Saline Soils Improvement:

Methods to Reduce Salt from Saline Soils:
- Salt Leaching Methods.

- Improve the Drainage Saline Soils Capability:
  The method to leach salts is to use irrigation water or rainfall to wash away salts from saline soils. Therefore, we should improve the drainage capability of the soil by utilizing deep drainage by open ditch or deep ditch to drain water. We can also establish bedding cultivation to allow easier leaching of salts by water. In addition, if there is a hard crust layer under the plough depth, we should break this layer to enhance water drainage and wash off of excessive salts.

- Establish Sprinkling and Irrigation Systems:
  Sprinkler irrigation is the most effective method to deal with salt injury in vegetable fields. Sprinkler irrigation will allow salt to leach each time when water runs from the top to the bottom part of the soil. We should establish long term effective irrigation system in the saline land alongside the seashore and use fresh water to leach salts.

- Apply Organic Matters and Microorganisms:
  Organic matter, such as peat, compost, animal manure and green manure, and microorganisms can increase salts exchange and water permeation into soil to wash off excessive salt.

- Apply Soil Conditioners to Improve Salt Exchange:
  The cause of saline and alkaline soil is due to the ability of soil particles to adsorb salt cations, such as sodium ion of sea water and other salty water. These ions can cause salt injury of plants. We can use calcareous materials, sulfuric acid, sulfur powder, organic acids, and bioagents to neutralize saline soils. For instance, gypsum contains calcium and can be used to replace sodium and leach out excessive sodium in seashore farmlands. Sulfate can be made into calcium sulfate (gypsum), and then applied to replace sodium and

reduce soil pH value. Under alkaline conditions, organic acids will become negative ions and release hydrogen ions to increase the exchange and leaching rate of salts.

Methods to Reduce Salts Entering Saline Soils:

It is not wise to use irrigation water that contains high salt content, especially in areas where the amount of evaporated water is higher than rainfall; otherwise, the salinization may be accelerated. We need to construct dam or trench alongside the seashore, so that sea water will not flow into the soil. We should also plant sand breaks to prevent saline wind entering the farmlands. Avoiding excessive fertilization is a significant principle in agricultural management. The fundamental way to deal with salt injury is to prevent abundant external salts entering into the soil.

Methods to Improve the Management of Crop Cultivation:

Cultivate Highly Salt-Tolerant Crops:

Various crops have different capacity to resist salinity (Table 7-6). We should cultivate highly salt-tolerant crops in saline soils; otherwise, sensitive crops will not grow well

Table 7-6. The Capacity of Different Crops to Resist Salt Injury

Sensitive Crops (EC < 4 dS/m)	Medium Resistant Crops (EC = 4 ~ 6 dS/m)	Sub-strong Resistant Crops (EC = 6 ~ 8 dS/m)	Strong Resistant Crops (EC = 8 ~ 12 dS/m)
All legumes	Soybean	**Grains:**	**Grains:**
Vegetables and	**Grains:**	Oats	Barley
Fruits:	Corn	Wheat	Rye
Carrot	Rice	**Special Crops:**	Asparagus
Celery	Sorghum	Cotton	Bermudagrass
Cucumber	Sweet corn	Sunflower	Sugarbeet
Kidney bean	**Vegetables and**	**Vegetables and**	Palm
Pea	**Fruits:**	**Fruits:**	*Phoenix dactylifera*
Strawberry	Green pepper	Cabbage	
Fruit Trees:	Lettuce	Cauliflower	
Apple	Muskmelon	Spinach	
Avocado	Onion	Tomato	
Cherry	Pumpkin	Watermelon	
Citrus	Sweet potato	**Fruit Trees:**	
Lemon	**Fruit Trees:**	Fig	
Plum	Apricot	Olive	
Pear	Grape	Pomegranate	
Pomelo	Peach	**Green Manure:**	
Walnut	Castor Bean	Alfalfa	
White clover	Flax	Clover	

in highly-saline soils. In order to save the cost for the saline soils improvement, it is important to choose appropriate crops to cultivate. Crops which can clear away salt in soils and absorb large amount of fertilizers, such as corn, can be used in facility or shelter cultivation to reduce the accumulation of salts.

Cultivate Cover Crops or Mulching:
The evaporation of soil can transfer salt from underground part of soil to aboveground part through capillary activity of water, and result in the salts accumulation on the surface soil. Therefore, cultivating cover crops or utilizing mulching can effectively reduce the rising and accumulation of salts.

Choose Good Cultivation Method:
When we cultivate crops on highly-saline soil, we can utilize bedding cultivation. If the crop is not cultivated on the top of ridge due to high salts but on the slope of ridge, it will also reduce the injuries caused by excessive salts.

For saline-alkaline soils alongside seashore, we can choose cultivate aquatic crops mixed with or without fish. We can cultivate paddy crops (such as rice), which are highly tolerant to salinity in soil and fish-farming in water. We can also cultivate crops on high beddings and fish-farming in ditch.

7.7 Characteristics of the Saline-Alkaline Soil Formation and Improvement Guidelines

7.7.1 Causes of the Formation of Saline-Alkaline Soil

About one billion hectares of land on Earth are saline-alkaline soils. In recent years, global warming and the surge of oil and grain prices raise force people pay attention to soils improvement. This would make full use of vast areas productivity of saline-alkaline soils and to reduce the decline of soils.

The major cause of saline-alkaline soils formation is the climate and the environmental conditions of soils, which includes less rain, large amount of evaporation, sea shore, poor water quality, and poor drainage in low lands. Materials that would accumulate salts (e.g., Ca^{2+}, Mg^{2+}, Na^+, and SO_4^{2-}) and their ways of accumulation can all result in the characteristics and amounts of salts, soil pH value and crop injuries.

In terms of climate and environmental conditions, saline-alkalinization of soil is prone to occur in areas where there is little rainfall and where soil drainage is poor on low land aside seashore. In addition to poor irrigation water quality and severe evaporation-condensation effect, this will accelerate saline-alkalinization of the soil.

Little rainfall areas defines as the annual rainfall lower than average annual evaporation rate. In these areas, large amounts of ions released due to mineralization and weathering of soils. Also, the inefficient leaching of salt will result in the soil accumulation. During rainy seasons, salt ions in higher grounds will be washed out into the groundwater. Because of the poor drainage in low lands, the accumulation of salts will increase in the groundwater. During dry seasons, the salts will raise to the surface soil along with the capillary water capability, which will be in a repeated cycle. Areas with little rainfall rely on irrigation water. When evaporation is severe, irrigation water will concentrate and accumulate more salts. Long term application of chemical fertilizers is also another cause of the salts accumulation in soils.

There may be various kinds of ions in saline-alkaline soils, such as Ca^{2+}, Mg^{2+}, Na^+, K^+, Cl^-, SO_4^{2-}, HCO_3^-, CO_3^{2-}, NO_3^- and other ions. Borate may also exist in some areas. Salts will influence physical, chemical and biological properties of soil, such as the amount of soluble salts, pH value, EC, organic matter characteristics, the movement of exchangeable cations, the state of soil colloid, and the activities of microorganisms and roots of plants.

7.7.2 Types of Saline-Alkaline Soils

Saline-alkaline soil is the summation of a type of soil. Base on the amount of salts and the degree of saturation of exchangeable sodium, saline-alkaline soils can be divided into three types:

Saline Soil:

Saline soil contains abundant soluble salts which is sufficient to affect the normal growth of plants. However, the saturation of exchangeable sodium is not high and the soil pH is usually below 8.5.

Alkaline Soil:

Alkaline soil contains little soluble salts, while the saturation of exchangeable sodium is high enough to inhibit the plants growth. Its pH value is between 8.5 and 10 and the soil colloid is in a dispersive state.

Saline Alkaline Soil:

Saline alkaline soil has the characteristics of the previous two. It contains large amounts of soluble salts and high saturation of exchangeable sodium which can affect the plants growth. The pH value of this soil is usually below 8.5 and the soil colloid is in a flocculent state.

7.7.3 Guidelines to Improve Saline Alkaline Soils

In terms of cultivation on saline alkaline soils, the priority is to choose the right crops. We should choose those crops that have high economical value that can resist salt, drought, low nutrients, cold, heat and have strong adaptability (Table 7-6). In order to make crops grow better and to improve and sustain soil productivity, we must improve saline alkaline soils. The guidelines include reducing salinity and alkalinity, supplying balanced nutrients, etc. Details will be given in the following section:

Guidelines to Reduce Salinity:

Reducing salinity is the essential method to improve saline alkaline soils and reduce soil EC as a good indicator. In the previous section, we have already presented the methods to deal with salt injury. The basic method is to use salt leaching, dilution, adsorption, absorption and coverings. Details will be given in the following section:

Salt Leaching:
- Use Physical Methods, such as Concealed Conduit or Open Trench to Remove Salts:
  This method is closely related to the groundwater level. If the level of groundwater is high, it will be difficult to remove water and pump out water, and it will be necessary to use deep ditches drainage systems to assist in removing salts in water at a quicker pace. Open ditches should be at least 2 meters deep to prevent salts being transferred to soil surface through water capillary activity in dry seasons. We can also loosen the soil structure to break the water capillary activity in dry seasons, which can reduce the rise and accumulation of salts in surface soil.

- Use Chemical Methods such as Applying Additives:
  Additives mainly consist of acidic matters, such as inorganic acids, organic acids, acids were produced by organic matters, or gypsum (calcium sulfate). These matters can promote the dissolving of salts and the exchange of sodium, so as to accelerate salt leaching. In addition, acidic matters can improve pH value in soil.

- Use Biological Methods, such as Applying Acids Produced by Microorganisms:
  Some microorganisms are able to produce acids, which can promote dissolving of salts and accelerate salts leaching.

Dilution:
This method mainly uses water to soak or adding material of low EC (e.g., compost) to dilute salts in the soil. We can use sprinkler or dripping irrigation method, so that water will move from the top to the bottom and reduce salts in the rhizosphere of crops.

- Adsorption:
  Adsorption method mainly uses additives that can adsorb salts, such as acidic peat, charcoal, bamboo charcoal, lignin, etc.

- Absorption:
  Absorption method implies planting crops that are highly tolerant to salts and can absorb and reduce excessive salts in soils.

- Covering:
  Covering method implies utilizing any covering materials (e.g., plastic cover film, straw mat, straw, etc) to reduce evaporation rate and allow crops to grow well. However, we need to make sure that the plastic cover film after cropping for the next crop should be taken away, rather than be ploughed into the soil; otherwise, excessive plastic residues will accumulate and affect normal root system growth over time.

Guidelines to Reduce Alkalinity:

Reducing alkalinity is the basic method to improve saline alkaline soils. We can take the pH value reduction as an indicator. We have already presented the guidelines to improve extreme acidity or alkalinity of soil in previous sections. The basic point is to apply soil conditioners that contain acids or which can produce acids. Details will be given in following sections:

Apply Soil Conditioners with Acids:
This method means applying acidic matters, such as sulfur powder, dilute sulfuric acid, peat soil and other materials with strong acidity. It is better to apply these soil conditioners gradually, rather than applying large amounts of soil conditioners in one time. Large amounts of soil conditioners inhibit the microorganisms' activities and increase organic matter's decomposition to decrease soil organic matter content.

Apply Organic Matters or Organic Acids:
Organic matters can produce organic acids in the process of decomposition, which can improve alkaline soil.

Apply Microbial Fertilizers that can Produce Acids:
Several microorganisms inhabit the plant rhizosphere and produce large amounts of organic acids, which improve and neutralize the alkaline soil.

Guidelines to Supply Balanced Nutrients:

Similar to fertilization on other types of soil, before we apply fertilization on saline alkaline soil, we need to test the soil and apply fertilizers in accordance with the need of crops, and prevent excessive or insufficient fertilization. Because phosphate in saline-alkaline soil is prone to be fixed and its availability declined, we should apply phosphate solubilizing microorganisms to improve this issue. Both ammonium nitrogen and urea are easy to release ammonia under high alkaline conditions. We should better apply small amounts of fertilizers with higher frequency.

Because of the extreme alkalinity in saline alkaline soils, crops are prone to lack micronutrients or consume excessive sodium ions, resulting in antagonism of other cations, especially for calcium, magnesium and potassium ions. In terms of micronutrients, crops easily lack zinc, iron, boron, manganese, copper and other micronutrients in saline alkaline soils.

Among the nutrients supplied to crops, besides the important N, P, K, Ca, Mg, S, Fe, Zn, Cu, Mn, and B, the proportion of different crops need to be concerned for Ca/Mg, Ca/K and Mg/K ratio. Various fertilizers have been supplemented or added should be mainly applied into the soil; otherwise, foliar application can also be conducted immediately to insufficient or unbalanced fertilizer supplement. Generally, it should be soluble to supply nutrients, when foliar application is adopted. The guidelines for foliar application can be referred to Chapter 6 of this book.

7.8 Guidelines of Soil Fertilization and Management in Slope Lands

Due to the features of landscape, agricultural soil managements of slope lands are quite different from that of the flat farmland. Reclamation and cultivation of the soil on slope lands require correct exploiting concepts more so for ordinary flat fields; otherwise, improper reclamation, cultivation and management of slope lands would bring out more aftereffects, such as the existing problems of the water and soil conservation, and the aftereffects of unfertile soil in slope lands. Higher investment costs will be required, if cultivating the soil after turning to poor quality.

In subtropical and tropical regions, high temperature and rainfall, slope lands management should be one of the important concerns in these climates. During early stage of using a slope land, the usual way of reclamation would chop away trees and bushes on the surface, and cultivate crops, such as cassava, lemongrass, pineapple, ginger and fruit trees. Later, the slope lands are developed into horizontal stair-type landscape where fruit trees are cultivated. This practice will loosen the soil on the slide slop. Moreover, without sufficient surface soil protection, this will lead to mud erosion during heavy rain seasons. Soil loss due to this practice can be easily determined by observing rivers for being washed out. Improper use of slope lands would bring about the soil fertility loss, and may flush away our precious good surface soil that had been accumulated for several thousands of years. Particularly, leaching large amounts of valuable surface soil and the reduction of stable organic matters in the soil (for example, the humus) cannot be replenished artificially in a short time. Therefore, reclamation and utilization of the slope land should be dealt with care.

7.8.1 Common Improper Ways of Reclamation and Cultivation

Although disadvantages of the improper or incorrect reclamation and cultivation of slope lands sometimes could not be observed in a short term, it is still worth our attention or the losses will outweigh the gains. The common improper ways of reclamation and cultivation are listed below:

Bare Cultivation:

This is the most common improper way of reclamation. Most farmers thought that weeds can compete with fruit trees for nutrients absorption. But after weeds have been completely weeded out, the surface soil in bare land is easy to be washed away and is also easy to raise temperature under sunshine. Moreover, when the soil is moisturized, the organic matter will decompose rapidly due to higher soil temperature. These losses of soil organic matter are much more severe than nutrient competition of weeds. Thus, these are the disadvantages of weeding out of all weeds or burning the mountain.

Horizontal Stair-Type Landscape Cultivation with Poor Water and Soil Conservation on Side Slopes:

To make horizontal stair-type fields (terraced fields) is allow convenient fertilization management for the slope lands. But the way to bulldoze the surface soil should be concerned. The surface soil should be filling into platforms with surface soil, instead of bulldozing onto the slope or deep pits. If the landscape exposes its subsoil for planting, there will be negative influences on the fertilization and management after crops have been planted. If protective vegetation is not cultivated on the side slopes, there would be severe losses of the soil. The setup of drainage ditches and the cultivation of vegetations in the ditches must be established to coordinate with from preventing soil washed off by rain.

Improper Fertilization:

The surface soil on the slope land is often bare in many places. After fertilization, a heavy rain would cause great fertilizer losses. Therefore, this will bring about pollutions to streams, reservoirs and water sources. Sometimes the fertilizer is applied too close to the base of fruit trees and may lead to fallen flowers and fruits. The deep application of fertilizers and improvement of the subsoil should be considered.

7.8.2 Soil Types and Limiting Factors for Slope Lands' Crops

The soil type of slope lands differs from regions. Because soil types and characteristics are different, the limiting factors influence crop growths are different as well. For agricultural management, we should first know what the limiting factors are in the land. Crop varieties and cultivars can be selected to avoid harms of limiting factors. In addition, we can directly improve soil limiting factors so as to ensure trouble-saving, labor-saving and money-saving productions of high outputs and qualities. The following are explanations of the limiting factors in various soils of slope lands:

Red Soil:

Red soil is very common in slope lands. The limiting factor of this soil is its strong acidic character, which is a negative factor in most crops. Apart from the cultivation of acidic-resistant crops (such as tea and pineapple), strong acidity of red soil should be neutralized. Due to the strong acidity and the properties of red soil, large amounts of phosphorus will be fixed, and it is a severe problem for common red soils. Therefore, red soil often has rather low content of organic matter, weak in the water and fertilizer conservation capacities.

Yellow Soil:

Yellow soil is named by its color. It is similar to the red soil as it is also strongly acidic, thin, and show high capacity of phosphate fixing capability, which made the low availability of phosphorus fertilizers. Particularly, when there is heavy rainfall, much of the nutrients in yellow soils will be leached out a great deal of nutrients, and organic matters should be provided to the soil. The acidity of the soil should also be adjusted for broader application purposes.

Stony Soil:

After surface soil on many slope lands has been washed away by heavy rainfall, only stony soil on the surface remains. The thin soil layer, as well as, the poor and weak buffering capacity of this kind of slope land is main limiting factors of crop growth. This kind of soil tends to become acidic and has extreme weak capacities for the water and fertilizer conservation.

Colluvial Soil:

Colluvial soil is formed from the accumulation of soil and shattered stones. The soil property is loose and unstable. Due to the above properties, the soil buffering and water/fertilizer conservation capacities are rather weak. It thus needs particularly the supply of organic matters and nutrients for improvement.

Others:

Red and yellow Spodosols soil or the typic Spodosols is common in high altitudes (greater than 1,500 m) with much rainfall and leaching, while the black soil is similar to cultivated soil. Their limiting factors differ due to their different parent materials.

7.8.3 Guidelines of Fertilization for Slope Lands and Orchard Management

From the above mentioned soil limiting factors of slope lands, these slope land soil properties are generally low organic matter content, high acidity, and low water/fertilizer conservation capacities. Since slope lands are actively leached by heavy rain, focusing on fertilization and management of slope lands are different from those of the flat lands. Each focus and essential part is explained as follows:

Strengthen Grass and Cover Cultivations in Orchards:

Long-term exposure of surface soil would lead to leaching and loss of nutrients and the reduction of organic matters, which is bound to influence crop growth in orchards. Grass cultivation has many advantages for the long term cultivation. For instance, rain water running out of orchards with grass cultivation is clean without muddy. Shorter plant of grass species should be chosen for the cover grass crop cultivation in orchards, as they are weak competitors to fruit tree for nutrients, and have good benefits in soil and water conservation. Short legume plant of green manures can be selected, and cover grasses can serve as a cover, and help increase organic matter content in soil. During rainy seasons, cover grasses present special function in preventing the erosion of surface soil.

Correct Fertilization:

Fertilizing chemicals on the slope land should avoid applying on the soil surface, especially, when there is much rain. The loss of fertilizers would increase, and it will result in environmental pollutions. However, the loss of fertilizers would reduce, if there is grass cultivation on slope lands. When labor and cost permits, deep fertilization in soil should be adopted as it can reduce the loss and the application rate of fertilizers. Furthermore, if deep injection of fertilizers could be conducted, the effect of fertilization would be even better. If the soil has fixed considerable amount of nutrients, the fertilizer can be divided into several times, or apply slow-released fertilizers. Supplement of fertilizers can be applied coordinately with foliar application for regions of micronutrient deficiency. Correct fertilization should coincide with shoot and root growth to achieve high production and high quality.

Strengthen Organic Fertilizers Application:

There are many good effects for organic fertilizers, e.g., slow-released nutrients, promote fertilizer efficiency, and increase water and fertilizer conservation capacities. Organic fertilizers should be the key point for production management of agriculture on slope lands and soil conservation. Organic fertilizers can be classified into ready to be decomposed or hard to be decomposed organic materials, and matured or non-matured composts. All of them have the effect on improving soil fertility of slope lands. The non-matured organic composts can be applied in advance, or applied along with chemical fertilizers to avoid failing to supply nutrients in time for perennial crops.

Neutralize Poor Soils:

Many slope lands are commonly acidic soils, except the low rainfall area. Neutralizing materials, such as limes, dolomite, slag and oyster shells, etc., should be used for acidic soils neutralization. The concept for soil neutralization is to apply neutralizing materials concomitantly with organic fertilizers. Otherwise, the neutralizing effect would be weak and need to pay attention to.

From the above guidelines, it can be understood that methods in the leaching reduction and erosion of slope lands, the management practices for the improvement of organic matter content and water/fertilizer conservation, and the neutralization of strong acidic or strong alkaline soils are essential part to agricultural production of slope lands.

7.9 Conservation of Forest Soils and Use of the Soil Microorganisms

7.9.1 Soil Conservation in Forest Lands

The fertility of forest soils is one of important factors to determine the tree growth. Since the ancient times, the soil in mountain area has not been used by human. The nutrients of trees need for their growth is supplied by the recycling ways of decomposition return from fallen leaves to roots. The trees would be growing at a limited speed under the circumstances lacking of artificial management.

After human have increased the needs for trees, how to accelerate the growth of trees into useful timber has been the purpose for the forestry in recent years. There

CH7 Guidelines of Improvement of Fertilization and Soil Management   213

Table 7-7. Mechanisms and Methods to Strengthen the Soil and Water Holding Capacity in Slope Lands

Item	Structure	Mechanisms of Water Conservation	Methods to Strengthen the Water Conservation
1. Vegetation:	(1) Forest vegetation (2) Farmland vegetation (3) Non-farmland vegetation	(1) Temporary conservation by plants (2) Influencing environmental factors (lower temperature and wind speed, etc) to reduce water loss (3) Apertures of plants and increase water Permeability in soil (4) Reduce the erosion of water and soil (5) Increase soil organic matter content	(1) Forestation (mixed forest is better) (2) Reduce deforestation and reclamation (3) Cover grass cultivation, grass ditch, straw cord (4) Application of covers and green manures (5) Planting perennial crops (6) High density in crop plantation and contour planting (7) Cover grass planting along the roads, and vegetation to protect the slope erosion
2. Soil Surface:	(1) Plant residuals (2) Surface of gravel and carbon dust (3) Covers (4) Ponds and pools	(1) Temporary conservation of water (2) Reduce erosion of water, soil, and wind (3) Reduce water loss (4) Increase water permeation in the soil	(1) Increase water conservation and reduction by water evaporation with mulches and covers (2) Increase water permeation in the crevice of surface soil (3) Reduce the establishment of impermeable structures (e.g., cement) and application of water permeable materials (4) Build ponds and waterways to intercept water (5) Windbreak and cover on the slope land (6) Avoid damage of natural landscape and hydrology through over development of the slope land

Table 7-7. Mechanisms and Methods to Strengthen the Soil and Water Holding Capacity in Slope Lands (Continue)

Item	Structure	Mechanisms of Water Conservation	Methods to Strengthen the Water Conservation
3. In Soil:	(1) Texture, gravel and rock layer (2) Structure of soil profile (3) Density and porosity (4) Organic matter content	(1) Increase soil humic substance contents and other water conserving materials to increase water conservation capacity and water amount (2) Increase water permeation into deep soil to reduce runoff of water (3) Reduce soil and wind erosion	(1) Increase the content of water conserving substances (e.g., organic matter and humic acid) (2) Increase soil porosity (e.g., vegetation approach) (3) Increase soil organisms (e.g., earthworm) (4) Reduce the decomposition of organic matters and the disintegration of soil structures due to excessive plowing and turning of soil

are quite a lot of methods to promote the growth of trees. The main work is just to start from varieties of trees and the soil. In terms of the forest trees, introduction and trees breeding should be strengthened. Also, use varieties that can grow fast and resistant to pests, diseases and animals, and keep the ecological balances. On the other hand, soil conservation in forest should be strengthened. Consideration of these two aspects, it could achieve better goal for improving forest lands' productivity.

No matter in the remote mountain, the shallow mountain or the slope land, after trees were cut down, the exposed land would accelerate decomposition from accumulated organic matter in soils and fallen leaves. They decompose more quickly particularly in regions of subtropical and tropical climates with high temperature. The greatest cause of soil fertility loss from forest lands is by leaching in rainy regions. However, the faster forest recovers itself, the less the nutrient loses.

A lot of soils used for forestation in shallow mountains or slope lands, because they often contain poor soil conditions or poor soil fertility. In order to accelerate growth of the forest trees, soil management in forest should be strengthened for supplying sufficient nutrients in the forest soil. In the wide and bumpy forest lands, soil management is a big problem. How to improve soil fertility of forest lands in the most economic and effective way is an important task.

The soil limiting factors in forest lands refer to poor physical, chemical and biological properties of soils. The common problems are poor soil pH, thin soil layers, much gravel rock and insufficiencies of nutrients. Only the insufficiencies of nutrient among above limiting factors may be easily improved in mountain areas or slope lands. Other limiting factors are hard to be improved.

The soil nutrients include nitrogen, phosphorus, potassium, calcium, magnesium, iron, copper, zinc, manganese, sulfur, chlorine, boron, molybdenum and silicon, etc. Not a single can be omitted. Among them, nitrogen is the element which gets lost and moves most easily in the nutrient recycling in forest lands, while phosphate easily transforms into ineffective phosphates. Therefore, improvement of nitrogen and phosphorus is extremely important to conservation of soil fertility. Accumulation of organic matter in forest lands can be the standard for soil fertility. Generally, the temperature is rather low in remote mountains and regions of high altitude, so that organic matter decomposes rather slowly and accumulates easily; the contents of the soil organic matter in the shallow mountains or slope lands are low. Especially, the "clear cutting" way of reclamation would lead to decrement of the organic matter content in soils. If the soil fertility is low, trees would grow rather slowly. Although they are fast growing trees, good soil should be provided to ensure good performances.

7.9.2 Strategies to Improve Soil Fertility of Forest Lands

Strategies for improving soil fertility of forest lands can be started from improving soil physical, chemical and biological properties. However, since the feasibility and difficulty are different, attention should be paid when it is begun to make the improvements. Especially, since soil limiting factors are different in different forest lands, improving methods also differ. Generally, the habits and needs of plant types should be taken into consideration in soil improvement. The limiting factors of the forest land should be first found out from soil analysis. But not all of limiting factors can be solved in the forest land probably, because of the limitations of labor power and funds. Therefore, strategies for improving soil fertility of forest land should take economic benefits into consideration, focus on the enhancement of plant nutrients and coordinate with the principle of nutrient recycling. In other words, it is a must to increase the sources and availability of nutrients, and coordinate with the function of fallen leaves and organic matters' nutrient recycling in forest lands.

Different approaches can be used to achieve goals in light of the above strategies. For example, for increasing the sources of nutrients, when economical benefits permit, aircraft or artificial fertilization can be used. But its pollution to the environment should be evaluated. Increasing the sources of nitrogen to forest lands and promoting absorption of phosphorus fertilizer, together with nutrient recycling of forest lands could achieve the goal of the forestry growth.

Sources of Nitrogen Increment -- To Use Nitrogen-Fixing Microorganisms and Nitrogen-Fixing Plants:

Apart from fertilization, nitrogen fertilizers of forest lands' sources are nitrogen-fixing microorganisms and nitrogen-fixing plants. Since artificial fertilization has a high cost, and nitrogen fertilizers being applied is easily leached, pollution often happens to reservoirs, water sources and streams. Nitrogen fixing microorganisms and nitrogen fixing plants could fix nitrogen gas (N_2) from air and transform it into the ammonium which could be used by plants. This process happens within microorganisms, so it has the function in minimum environmental pollution.

Types of nitrogen fixing microorganisms include symbiotic, associative and free non-symbiotic prokaryotic microorganisms. Except the symbiotic nitrogen fixing microorganisms, the other two types have fewer applications. Among symbiotic bacteria, the rhizobia have been widely researched and they can form root nodule and actinomycetes. Among the forest lands, major and minor forest trees can obtain a benefit from nitrogen fixing plants grown as intercrops between the trees for those shortage of nitrogen in the soil.

The fallen leaves of nitrogen fixing plants have higher nitrogen content. After fallen leaves have been decomposed, they can supply large amounts of nitrogen for trees. Therefore, to increase nitrogen-fixing plants in the field is a conducive approach for increasing nitrogen source. In the long term, it is the most effective way to increase nitrogen in the field. Nevertheless, more researches for discussing the nitrogen-fixing plants increase with higher activity in the field need to be strengthened.

Promotion of the Absorption of Phosphorus -- To Use Mycorrhizal Fungi and Phosphate-Solubilizing Bacteria:

Phosphorus is one of elements which do not move easily in the soil. Insufficiency of phosphorus is often a limiting factor of plant growth. Decomposition of organic matter could supply phosphorus. And phosphorus in the soil is also often fixed into insoluble phosphates, such as, phosphate fixation with iron, aluminum, and calcium, etc. These insoluble phosphates could not be directly absorbed and utilized by plants, while mycorrhiza fungi, and phosphate-solubilizing bacteria could help crops to absorb phosphorus from soils. Good mycorrhiza fungi or phosphate-solubilizing bacteria could be applied in the cultivation of seedlings to achieve the inoculation goal.

Mycorrhiza fungi are symbiotic with roots of trees, including endo-, ecto- and endoecto-mycorrhiza fungi. The endo- and ecto-mycorrhiza fungi have been applied into the production of trees. Besides, phosphate-solubilizing bacteria can be in the rhizosphere or in non-rhizopshere to increase solubilization of insoluble phosphates. It also needs to be researched and developed.

The above mentioned is to increase nitrogen sources and promote phosphorus absorption through soil microorganisms. Application of soil microorganisms has a lot of advantages, including increasing nutrient availability for trees and the growing seedlings, improving soil fertility, reducing application of chemical fertilizers, low production cost, self control in the environment, and reducing pollution and occurrences of diseases. However, application of soil microorganisms has its disadvantages as well. For example, the inoculants have poor conservation of the bacteria, which could influence the survival and effects of bacteria. The bacteria have limited adaptabilities, so that the inoculants should be used in combination with various strains of bacteria so as to overcome the disadvantage of limited adaptabilities.

The forest is the focus of ensuring the territory safety, people's livelihood and water resources. Conservation of forest soil is regarded as the foundation of tree growth. In the area, with high temperature and quick rain, in order to promote trees growth, soil fertility of forest should be improved. And the selection and breeding of fast-growing trees should be strengthened. The above measures could ensure a promising future for the forestry development.

7.10 Soil Problems and Solutions of Continuous Cropping

7.10.1 Occurrences for Continuous Cropping

In the agricultural cultivation, continuous cropping is referred to continuous cultivation of same variety or same type of crops on the same land. Continuous cropping often leads to poor growth or seedling death of crops. And, even application of fertilizer cannot make thorough improvements. Common problems are withering of seedlings and rottening of roots, abnormalities of growing shoots and new branches and leaves, or their incapable growth. Some can also cause problems such as pests and diseases, and then losing plants.

Soil problems in continuous cropping are different due to differences in crop varieties, soil, climate, and cultivation management. Poor growth would happen if some crops have only been cultivated continuously for one cropping, such as ginger, mung bean, watermelon, and solanaceae (such as, green pepper, tomatoes), perennial asparaguses, peaches, apples, etc. Particularly, if younger seedlings of peaches and apples are cultivated again in the same sites of old died plants, serious seedling deaths would happen. For other crops, soil problems could only be found after several times of continuous cropping (for example, vegetables such as Cruciferae). After several years of continuous cropping, soil problems would become obvious.

For agricultural regions where various crops are cultivated during whole year round, and for lands where twice or more times of productions have been made in the same field during the whole year. Thus, problems of the continuous cropping are very common. For regions which have small agricultural acreages, there is also necessary to crop continuously. When advanced cash crops are produced in net houses or under specific awnings of the facility agriculture, in order to increase the production per unit within whole year, intensive continuous cropping would more easily bring about problem soils. The problems would be more severe, so that some specific crops could hard to be cultivated in that soil.

7.10.2 Problems of Continuous Cropping in Soil

Every crop has its special nutrient needs and their metabolism features. According to different crops, soil nutrient proportion and types, water and oxygen being needed differ; in terms of the metabolism, there are somewhat differences between all kinds of crops and synthesize different products of metabolisms. Then secretions of crop roots and decomposed products of residuals have their unique characteristics. As a result of the above needs and the metabolism features, continuous cropping may bring to that last crops affect the crops growth at later stage. Its causes may include:

- The crops have absorbed much of nutrients, which lead to insufficient nutrient supplies.
- The crops have absorbed much of cation nutrients, which lead to the increment of soil acidity.
- Due to the high carbon-nitrogen ratio, when large amounts of residuals from the decomposition of previous crops, large quantities of microorganisms would reproduce. Nitrogen would then be absorbed by the microorganisms and nitrogen would be insufficient.
- The crop roots would produce toxic substances (allelochemicals). When toxic substances are released by the decomposition of the residuals or when microorganisms reach a certain critical concentration in the soil environment, they would cause other toxic effects to crops.

The above causes have confirmed that, for a lot of crops, toxic substances of previous crops are left over in the soil, and then affect growth of later crops; for example, asparagus, sorghum, wheat and mung bean, etc.

The continuous cropping leads to soil problems. Apart from the problems of the special needs and metabolic products of the crops, there are also problems in changes of soil physical, chemical and biological properties by continuous cropping. The following are the main changes brought by continuous cropping.

Reduction of Organic Humus in Soil after a Long Period of Continuous Cropping:

According to the findings in Ohio of the United States, organic carbon in field soil, which had been cultivated corn for a continuous 30 years, would decrease by 65%; organic carbon in wheat would decrease by 38% continuously for 30 years. These severe problems may be due to facts that soil granules were gradually broken, so that organic humus is exposed and then decomposed. It may also be the reason that large amounts of residuals had been added into the soil which leads to the excessive growth of soil microorganisms and this makes more decomposition of organic substances. The continuous cropping of upland crops easily accelerates the soil organic humus decrement. It is worth of the attention, since stable organic humus is the foundation for the good characteristics of soils.

Pests and Diseases Increment after a Long Period of Continuous Cropping:

The pathogens and the pest have more or less preferences or concentrated cognizance for the specific crop. For example, the pest damaging tomato plant may not damage other vegetable plants. If the group of pests increase gradually in the continuous cropping or increase heavily after only once cultivation, the later crop would then be done harm by the pests which have increased. The pests may exist in the soil or the residuals of the crop. And then they may cause problems for continuous cropping.

Improper Cultivation Management and Fertilization in a Long Period Time of Continuous Cropping:

As a result of improper cultivation management, excessive irrigation and drainage will accelerate the leaching of soil nutrients, and the loss of soil. Or the excessive application of chemical fertilizers causes commonly salt accumulation in the field with short term cultivation of vegetables. The result makes white powder salts being formed on soil surface. These are also cause for continuous cropping problems.

7.10.3 The Control of Soil Problems Caused by Continuous Cropping

Generally, soil problems would somehow appear during continuous cropping of crops. If problems are severe, the fertilization management cannot cure them. If problems are not serious, the fertilization management can be conducted remedy for parts of losses. Therefore, in order to prevent occurrences of continuous cropping problems, there should be a concept that the prevention is more important than the remedy in terms of the control. The following are explanations of guidelines for problem controls in continuous cropping:

Adopting Cultivation of Crop Rotation:

In order to prevent continuous cropping problems, the most effective way is to abandon continuous cropping. To adopt the crop rotation system is an important work of preventing continuous cropping problems. In crop rotation, upland and paddy field take turns, or legume and non-legume take turns, or big crop and small crop take turns. All of these are best choices for crop rotation. To grow variously, diverse crops are also one of guidelines for soil conservation. Crops which have large amounts of fertilizer absorption (such as corn) could be grown to absorb the unbalanced and excessive nutrients from soils.

Applying Neutralizing Materials:

Application of calcareous materials into acidic soils could improve pH value of soil and reduce the activity of pathogens. It has the similar effect of disinfection and increases the decomposition of toxic substances as well, and improves the acidification and the environment of soil microbial phases. It really can answer multiple purposes. Application of the neutralizing medium can achieve better effects in coordination with the use of organic matters.

Applying Organic Fertilizer or Cultivating Suitable Green Manures:

Organic matter can decompose and supply nutrients. It can also supply carbon source or energy for the microorganisms' activity, which is conducive to the decomposition and adsorption of toxic substances. Advantages of green manure cultivation are crop rotation and organic fertilizer. Some even can control part of pests and diseases. It is also an approach to kill two birds with one stone.

Supplementing the Insufficient or Unbalanced Nutrients:

The continuous cropping may cause insufficiencies or unbalances of nutrients, so nutrients need to be supplemented. Application of nutrients into soils or onto leaves can only solve slight problems for continuous cropping.

Applying Beneficial Microorganisms:

Detoxifying bacteria can be applied to eliminate or decrease concentration of toxic substances brought by continuous cropping. It is the direction of agricultural research in the future.

Applying Methods such as Water Saturation and Sun Dry:

Toxic substances, and pests and diseases can cause harms to crops, when they reach a certain concentration or population. Both concentration of toxic substances decrease or population of pests and diseases reduction are all contributed to solve the disadvantageous environment of continuous cropping.

Cleaning away for Crop Residues from the Field:

Some crop residues are sources of toxic substances, or pests and diseases. Therefore, cleaning away for crop residues from the field could reduce the sources and harms of toxic substances, or pests and diseases.

The above guidelines for the control of problems in continuous cropping can be used complementarily. Use of more methods of the control could reach a more effective achievement for temporary and fundamental approaches. Summaries of causes and solutions to problems for the continuous cropping could be briefly listed in Table 7-8. Listed are the references of various problems and their respective solutions.

Table 7-8. Causes and Solutions to Problems for Continuous Cropping

Causes of Problems	Phenomenon for Problems	Solutions to Problems
1. Soil Biological Property:		
(1) Increment of harmful organisms.	Increment of pests and diseases.	Turning over the soil, exposing, soaking, and mulching with plastic cover during the rotation and application of soil conditioners.
(2) Decrement or imbalance of beneficial organisms.	Rapid increase or rapid decrease of soil microbes.	Application of conditioners, beneficial organisms, organic fertilizer or detoxifying bacteria.
2. Soil Chemical Property:		
(1) Decrement or imbalances of nutrients.	Insufficient nutrients or low content of organic matter.	Supplement of nutrients and organic matter which is insufficient or unbalanced.
(2) Deterioration of soil pH.	Soil acidification.	Application of soil conditioners.
(3) Nutrient fixation due to excessive crop residues.	High C/N ratio of residues and insufficiencies of nitrogen and microelements.	Supplement of nitrogen fertilizers and cleaning away residues from fields.
(4) Increment of toxic substances.	Toxins to plants and high soil EC.	Using dilution (soaking and leaching), adsorption (application of peat), decomposition (application of organic fertilizers, exposure and detoxifying bacteria) or polymerization (application of organic fertilizers) to make detoxification.
(5) High salinity	Soaking, application of organic matter, gypsum ($CaSO_4$) can be applied to the alkaline soil with too much sodium salt.	Cultivation for regularly (2 to 5 years) to plant crops (for example, corn) which are salt resistant and can absorb large amount of salts.
3. Soil Physical Property		
(1) Poor soil areolation	High soil compaction.	Plow and loosen the soil.
(2) Poor soil drainage	Existence of plough pan and soil compaction.	Plough deeply, turn up the soil and drain off water from the field.

7.11 Guidelines for Management and Fertilization of Coarse Soil

7.11.1 Differentiation of Soil Texture

Most soils will be composed of a variety of different particle sizes. According to standards, soil particle sizes can be classified into sand particle (the diameter is between 2.0 mm to 0.02 mm), silt particle (the diameter is between 0.02 mm to 0.002 mm), and clay particle (the diameter is below 0.002 mm). And in the cultivated land, there are even bigger gravels (the diameter is above 2 mm) or stones. Soil texture is a qualitative classification tool used for the relative proportion of soil particle sizes and to determine classes for agricultural soils based on their physical texture. According to percentages of sand, silt and clay content, soils could be classified into 12 classes of soil textures, including sand, sandy loam, loamy sand, loam, silty loam, silt, sandy loam, clay loam, silty clay, sandy clay, silty clay loam, and clay. For the convenience of classification, the above 12 classes of soil textures could be summarized briefly into coarse soil texture, medium soil texture and fine soil texture. Coarse soil texture includes sandy soil and sandy loam in above 12 classes. These kinds of soil contain more than 70% of sands; fine soil includes sand clay, silt clay and clay. These kinds of soil contain more than 35% of clay; soil includes the soil except the above five textures of soil. The coarse soil texture of agricultural lands also includes soil with more gravel (Table 7-9).

Soil texture influences soil physical, chemical and biological properties. With regard to fertilization and management of crops, flexible management should be conducted for differences of soil textures in order to achieve twice the result with half the effort.

7.11.2 Characteristics of Coarse Soil

There are advantages and disadvantages in coarse soil for different crops. Therefore, the characteristics of coarse soils should be known to cultivate proper crops in the proper soil or to improve the disadvantages of coarse soils. The following are explanations for the characteristics of coarse soils:

Table 7-9. Classification of Soil Texture

Classification of Texture	Types
1. Coarse Soil Texture	Sand, loamy sand
2. Medium Soil Texture	Sandy loam, loam, silty loam, silt, sandy clay loam, clay loam, silty clay loam
3. Fine Soil Texture	Sandy clay, silty clay, clay

Physical Property:

The coarse soil contains much sand, and the interspaces between soil particles are greater, so that it has quite good aeralation and drainage. If the water table of the land is high, it would also lead to poor drainage of the land. This is related to the regional landscape and water level, but has nothing to do with the soil drainage. In other words, if the land is a flooding area or it is a land having a rather high water table, even the coarse soil could not drain away water from the region. Texture of good drainage has relatively poor water holding capacity. Sandy soil is just a best example. Its water evaporates very quickly and it is easy to lose water. Under the sunshine, the temperature is also high on the soil surface.

Chemical Property:

The surface properties of sand in coarse soil differ due to differences in their constituents. The sand often contains many original minerals, including quartz and aluminosilicate (for example, mainly feldspar). The most obvious coarse soil has a rather low amount of organic matter, since coarse soil has good aeralation and decomposes organic matter very fast. The organic matter content is low, so that the soil has weak capacities in the fertilizer and water conservation. Its buffering capacity also decreases. Many factors are given to change a land to a coarse soil; for example, loss of fine texture on slope lands would also result in coarse texture of surface soil. Sandy soil or gravel land is commonly formed by water alluviation. The sandy soil along the seaside has a high amount of salt, which is always a limiting factor of crop growth. The coarse land is insufficient of nutrients, which should be paid special attention to in terms of the fertilization management.

Biological Property:

The coarse soil has mainly in poor physical and chemical properties. Particularly the soil with low organic matter content has rather low activity of soil microorganisms. But the soil is suitable for aerobic organisms to grow in. The coarse soil may easily bring about nematodes or aerobic fungi or the injuries of soil animals.

7.11.3 Guidelines for Cultivation Management and Fertilization of Coarse Soil

To Choose Proper Crops to Cultivate:

"Proper crops cultivate on proper lands" is an important cultivation concept. If the soil is of coarse texture, suitable crops for growing should be carefully chosen. Crops suitable for the coarse soil possess the following features: (1) deep-rooted; (2) tolerance to drought; and (3) having a hypertrophic storage root. Deep-rooted crops are crops such as woody plants (for example, fruit trees), vines (for example, melon) or taproot and large fibrous root crops. Drought-tolerant crops generally have smaller in water evaporation (such as the Crassulaceae and cactus, etc). Moreover, having large fibrous root crops could help conserve more water and nutrients (for example, asparagus). Some plants have preferences for the sand and thus prefer to grow in sandy soil. Crops would grow poorly in the heavy soil, such as asparagus, peanuts and melons, etc.

To Adopt the Cultivation of Surface Mulching and Moderate Irrigation:

There is poor capacity of water conservation in the coarse soil. In sunny and windy days, there is a great amount of water evaporation, so that crops would easily be short of water. Use of surface mulching could reduce the evaporation of surface soil. A lot of materials could serve as the mulches. Mulching used by grass and synthetic plastic film, together with proper ways of irrigation could be seen as water sources for the soil. However, excessive amount of irrigation would wash away soil nutrients and result in harmful effects. Therefore, it should be dealt with care.

To Apply More Organic Matter for Improving Properties of Coarse Soil:

Extremely poor capacities of water and fertilizer conservation exist in the coarse soil. More application of organic matter could improve soil physical and chemical properties. The organic fertilizer is better in the form as thoroughly-decomposed or long-term effective organic matter. The organic matters decompose quite quickly that have short-term effects. For example, peat, humic acid or the thoroughly-decomposed organic fertilizer belong to those rather long-term effective in improving coarse soil.

Constant cultivation of green manure could also effectively improve the disadvantages of coarse soils. Apart from supplying nutrients, it has good effects on the soil physical, chemical and biological properties. Even the mulching of green manure has quite satisfying effects as well.

Application of Chemical Fertilizer Based on the Principle of "Apply Lighter and Apply Fertilizations More Frequently":

The coarse soil has extremely poor capacity of fertilizer conservation and poor buffering capacity. It is suggested not to apply too much chemical fertilizer; otherwise, it would easily be washed by the rain or it would damage crop root systems. Therefore, application of chemical fertilizer into coarse soil had better be done based on the principle of "apply lighter and apply fertilizations more frequently." Application of nitrogen fertilizers should be paid special attention to. Since it disappears most quickly, the management should be strengthened.

To Adopt the Crop Rotation for Improving Continuous Cropping:

Problems of continuous cropping often happen to agricultural crops growing in the coarse soil. It may be related to the characteristics of crops, toxic substances, pests and diseases, and unbalance of soil nutrients. Up to now, the most effective way for improving the continuous cropping has been adopting thecrop rotation. The crop rotation has quite a number of benefits, such as an indispensable essential in the soil conservation.

7.12 Relationships between Management of Soil Fertilization and Crop Diseases

7.12.1 Soil Fertilization is Close to Diseases

For those regions in subtropical and tropical areas or where continuous cropping is practiced heavily, there are quite a lot of severe diseases of crops. Diseases in the soil cannot be detected by our eyes. When we finally find out the diseases, they have not been cleaned up. These harms to crop production are really serious. Prevention and care for diseases are also of great difficulties, so those researches and developments should be emphasized. There have been reports that crops health is also closely related to their capacity of resisting diseases. Disease resistance may decrease due to excessive or insufficient supply of nutrients. Especially, when nutrients are insufficient, crops are weak and more susceptible to the attack of the pathogens. Direct or indirect fertilization are conducive to the crop resistance to diseases or restraint of the pathogens. If nutrients are supplied sufficiently and proportionately, the crops can grow normally. Excessive nitrogen fertilizer could increase soft tissues of plant. The plants would thus be more vulnerable to the virus attack, bacteria and fungi. Insufficiency of nitrogen can also make plants easily infected with pathogens and parasites. Plants that short of phosphorus would easily be attacked by fungi and the possible cause may be the disadvantageous ratio of nitrogen/phosphorus. Insufficiency of potassium could lead to carbohydrates decrement in plants, weakness of the cell wall, thinness of leaf hair, and the plants would be easily attacked by aphids, increasing the infection of the parasites and virus; insufficiencies of calcium and silicon would weaken plants, and fungi would easily intrude into plants. Insufficiency of the microelements is closely connected to crop resistance to diseases. Many researches have reported that diseases results in soil acidification. Improvement of the soil fertility and health is the basic work for reducing crop diseases.

7.12.2 Soil is Related to Plant Diseases

Plant disease is one of limiting factors for agricultural production. Varieties of pathogens causing plant diseases differ from each other. So do their approaches of attacking plants. Therefore, the control of diseases is quite different as well, including eliminating pathogens, preventing and killing vectors, crop breeding for resisting diseases or strengthening the crop capacity of resisting diseases. Crop diseases are rather severe in subtropical and tropical area. Diseases in soils cannot be detected by our eyes. When we finally find out the diseases, they have often been irremediable. Their harms for crop production are really serious. Prevention and care of the diseases are also of great difficulties, so researches and developments should be emphasized.

Although we do not have a thorough understanding on causes of plant resisting diseases in the natural environment, the causes could be related to diseases resistance in internal factors and external factors. The internal factors of diseases resistance are connected to the plant's biochemical metabolisms and structural features. For example, when the bacteria intrude the resistant plants, plants may produce the phytoalexins which can fight against pathogens or increase plant lignifications to weaken pathogens. The external factors refer to special environmental protection. For example, suppressive soils can protect crops and reduce diseases. Some soils are suppressive. The causes for their suppressive function probably include the soil pH value, microorganisms, and inorganic or organic antagonism. Reports show that soil conditioners were used to improve the soil disease-resistance capacity. The crop health is also closely related to their capacity of resisting diseases. The relationships between crop nutrients and disease resistance may be listed as follows:

- Disease resistance may decrease due to excessive or insufficient supply of nutrients.
- Especially, when nutrients are insufficient, crops are weak and more susceptible to the attack of pathogens.
- If nutrients are supplied sufficiently and proportionately, the crops can grow normally.
- Direct or indirect fertilization are conducive to the crops resistance to diseases or restraint of pathogens.

Therefore, the crop capacities of resistance to diseases are more or less related to nutrients. Fertilization can help the crops strengthen their resistance, but cannot easily inhibit diseases thoroughly, since some pathogens cannot be easily controlled.

7.12.3 Relationships between Nutrients and Diseases

The plants absorb nutrients from soil and meet the needs for the growth and metabolisms of plants. Varieties and contents of nutritive elements would affect the internal and external structures and constituents of plants, and may also influence the plant resistance capacities to pests and diseases. Since the types of pathogens and the crop varieties differ greatly, it is impossible to use one way to explain various diseases. The following are explanations of each element and their cases accordingly:

1. The situations where nitrogen affects occurrences of plant diseases are very complex. Both excessiveness and insufficiencies may influence the occurring or not easy occurring of diseases. Or there are different responses due to the different pathogens.

    (1) Excessive or insufficient nitrogen fertilizer would increase diseases

    Excessive or insufficient nitrogen fertilizer would increase harms to the geranium brought by the bacterial stem rot which caused by *Xanthomonas pelargonii*. The harms are the most severe, especially when there is a sufficiency of nitrogen.

    (2) Excessive nitrogen fertilizer would increase diseases

    Leaf blotch of spring barley is caused by *Rhynchosporium secaalis*. Absolutely parasitic rust diseases of wheat are caused by *Puccinia* spp., and powdery mildew which is caused by *Erysiphegraminis*.

    (3) Insufficient nitrogen fertilizer would increase diseases

    Arbitrarily parasitic leaf spots are caused by *Alternaria* spp. The wilt and rot which caused by *Fusarium oxysporum*. Tomato bacterial spots and wilts are caused by *Xanthomonas* spp.

    Application of ammonium sulfate among nitrogen fertilizer could make soil become more acidic and reduce the full rot of the wheat, while nitrate fertilizers would increase occurrences of this disease. Ammonium nitrogen (NH_4^+) does great harms to some *Fusarium* bacteria, while nitrite nitrogen (NO_2^-) is greatly harmful to the *Pythium* and *Phytophthora* bacteria.

2. Phosphorus is a necessary element of plant metabolisms and energy generations, which influences plants' health condition greatly. The plant which is short of phosphorus would easily be attacked by fungi. The possible cause may be an improper proportion of nitrogen to phosphorus.

3. Potassium is a related element to the osmotic regulation of plant cells, water relation and enzyme activity. Generally, diseases were occurred due to insufficiency of

potassium fertilizer, just like the absolutely parasitic and arbitrarily parasitic diseases of above nitrogen fertilizer; the rice stem rot which caused by *Helminthosporium sigmideum*; the oceanic Bermuda grass leaf spot which caused by *H. cynodontis*, etc; soft rot which caused by *Erwinia* spp. and *Xanthomonas* spp.; tomato bacterial wilt which caused by *Pseudomonas solanacearum*.

4. Calcium is a constituent of the plant cell wall and could fight against the intrusion of fungal pathogens. Insufficiencies of calcium and potassium would increase occurrences of diseases, such as the bean twin stem which caused by *Sclerotium* spp., tomato fusarium wilt which caused by *Fusarium oxysporum*, and soft rot which caused by *Erwinia carotovora*.

5. Accumulation of silicon would lead to materialization of the silicon on the epidermic cells. Insufficiency of silicon would easily cause diseases, for example, the rice blast which caused by *Piricularia oryze* Cav. Or it easily brings about pest intrusion.

6. There are various types of micronutrients, but less information about their resistances to diseases. The following are enumerations:

   (1) Insufficiency of zinc easily causes diseases, such as *Oidium* is harmful in *Hevea brazilinsis*.

   (2) Insufficiency of boron easily causes diseases, such as soft rot and powdery mildew of wheat.

   (3) Copper is widely used as fungicide and it is also the microbial element needed by plants. Insufficiency of copper leads to unobvious resistance to the white powder. A severe insufficiency would influence its capacity of resistance to diseases.

   (4) Plants lacking of manganese are easily invaded by fungi *Fomes annosus* and then become diseased.

7. The soil pH value affects the absorption and availability of plant nutrients, and living and reproductive performances of bacteria. Therefore, changes of the soil pH value affect the occurrences and degrees of the plant diseases greatly. The following are explanations of pH value influences and cases:

   (1) Acidification of the soil increases diseases: for example, clubroot which caused by *Plasmodiophora brassicae* of the cruciferae.

   (2) Further acidification of soil decreases diseases: for example, "take-all", root rot which caused by *Gaeumannomyces graminis* of wheat.

7.12.4 Application of the Soil Microorganisms and Biological Control

The ecological system of natural soil is relatively stable. Land cultivation and use of agricultural materials bring about changes in soil properties more or less. For different pathogens, properties of different soils (physical, chemical, and biological property) may cause suppressive soil or conducive soil of specific pathogens.

For soils in regions of subtropical and tropical climate and soils on the slope land, organic matter is decomposed quickly in soils. Together with influences from the intensive way of cultivation, soil problems have been brought up often in recent years. The causes for problems are continuous cropping, or excessive and improper use of agricultural chemical materials, bringing changes to the soil physical, chemical and biological properties and unbalances to the soil ecology, resulting in pests and diseases, and poor quality of products simultaneously. Soil is basic asset of human being and it is provided with vitality. Conservation of soil is responsibility of every generation. For example, if the beneficial bacteria in soil are destroyed, it takes a lot of time and money to recover them. It is an asset having been conserved for thousands and millions of years, so we need to strengthen our protection and put emphasis on it.

Table 7-10. Interrelationships between Plant Nutrients and Occurrences of Diseases

Element	Content	Names of the Diseases
Nitrogen	Excessiveness	Bacterial stem rot, leaf spot, rust disease, powdery mildew;
	Insufficiency	Leaf spot, wilt rot, spotted wilt;
	NH_4^+	Inhibition of *Fusarium*;
	NO_2^-	Inhibition of *Pythium* and *Phytophthora*.
Phosphorus	Insufficiency	Reduction of plant resistibility.
Potassium	Insufficiency	Absolute and arbitrary parasites, stem rot, leaf spot, and wilt.
Calcium	Insufficiency	*Fusarium* wilt, pod rot, soft rot, clubroot (low pH value).
Silicon	Insufficiency	Rice blast.
Zinc	Insufficiency	Virus disease.
Copper	Insufficiency	A severe insufficiency could lead to reduction of resistibility.
Manganese	Insufficiency	Some fungal diseases.
Boron	Insufficiency	Soft rot; powdery mildew.

Effects and benefits of the application of soil microorganisms are listed as follows:

Improving Soil Fertility:

There are many species of beneficial soil microorganisms. Different microorganisms play different roles, among which, the most important is to supply plants with nutrients, to improve the soil physical property and to increase the nutrient availability. Examples are listed below: nitrogen fixing microorganisms increase nitrogen source in the soil; decomposed microorganisms could decompose organic residuals and supply nutrients to the plants; phosphate-solubilizing bacteria could solubilize the ineffective phosphorus and transform them into phosphate, which plants could utilize; microorganisms would secrete polysaccharides to make the granular structure of the soil better and ameliorate the physical property of the soil, etc.

Assisting Plant Nutrient Absorption:

Plants absorb nutrients mainly through the root hair. The more root hairs mean the surface area of the absorption is larger and the absorption capacity is stronger. Mycorrhiza fungi are symbiotic with plant roots. The hypha stretches out of the root, whose function is the same as that of the root hair and could absorb more nutrients. The most obvious is the absorption of phosphorus. Experiments of the mycorrhiza have produced effects of increasing the crop production.

The existing forms of nutrients would affect the absorption efficiency. Not all kinds of forms of nutrients existing in the soil would be easily absorbed by plants. Some soil microorganisms would secrete organic acids and then form the structures of chelation, which contributes to the absorption or increase the availability of nutrients. For example, siderophore has effects of resisting the pathogens.

Strengthening Plant Capacities of Resistance to Diseases and Drought:

Ecological controls and balances in the soil make pathogens unable to reproduce massively. Therefore, microorganisms can be applied in the resistance, which can be called "biological control." Inoculation of the mycorrhiza fungi or rhizosphere-protective bacteria could occupy the root domain and reduce the pathogens intrusion of the rhizosphere, forming a biological defensive front. The mycorrhiza fungi are growing symbiotically on the root system. Just like the root hair, they contribute to the crop absorption of nutrients and water, and also achieve the goal of combating against the drought. The secretions of microorganisms increment in soil and the decrement of the freezing point has an effect of fighting against cold.

Saving Energy Resources and Reducing Production Cost:

Production of nitroge chemical fertilizer needs consumption of large amounts of energy from non-renewable resources. For regions where the energy is limited or there is a lack of energies, use of nitrogen fixing microorganisms could reduce the application of nitrogen fertilizers so as to reduce the agriculture consumption of energies. Phosphorus fertilizer comes from the processing of mineral phosphorite together with acids. After it is applied into the soil, it can be consumed immediately. Most of them are fixed into phosphates with low availability. Nitrogen fixing microorganisms and the mycorrhiza fungi can help sources of nutrients, particularly, nitrogen and phosphorus. Small amount of microorganisms and low cost of their production could save fertilizers and then reduce the production cost for farmers.

Reducing Environmental Pollution:

The trend of environmental protection is springing up and agricultural pollution cannot be ignored, either. Excessive application of chemical fertilizers would contaminate the stream, the reservoir and the water heads. Eutrophication in the water makes organisms to reproduce massively, thus influencing the balance of water organisms. Particularly, pollution of nitrogen and phosphorus needs more attention. Application of soil microorganisms (such as nitrogen fixing microorganisms and the mycorrhiza fungi, etc.) could reduce greatly the use of nitrogen and phosphorus fertilizers. Their pollution of environment could also be reduced to the minimum.

All in all, application of soil microorganisms is a new technology, which can reduce environmental pollution, improve qualities of crop products, and increase the production. Instead of studying the single functional product, multifunctional compound biofertilizer and soil conditioner are needed for further research studies. The marvelous biological world in the soil waits for our development and utilization.

Strategies for biological control are not merely application of above microorganisms. There are various other approaches as well. The following are brief explanations:

To Increase Indigenous Antagonism Microorganisms in Soil and Rhizosphere:
Organics, crop rotation, fertilization, and physical and chemical treatments could be used to change the ecological system of soil and rhizosphere, and to increase the beneficial microorganisms. Especially, the crop rotation system assists greatly in plant resistance to diseases. For example, rotation of legumes and other crops could help reduce the root rot. Addition of organic materials could help achieve restraint of pathogens through the multiplication of beneficial microorganisms.

To Add More Antagonism Microorganisms into Soil or Rhizosphere:

Physical and chemical treatments could be used to influence the ecology in soil, seeds and seedlings. Large amounts of antagonisms could be added and the preparation of restraining the pathogens, the disinfectant or plant protective cultivation can be used coordinately.

A great number of researches and applications have explored into the above approaches of biological control.

7.12.5 Guidelines of Management of Soil Fertility and Cultivation

Proper management of fertility and cultivation is beneficial to the crop production and can easily compensate the detrimental factors. First the limiting factors in soils should be improved and then agricultural materials could be applied, including inorganic fertilizers, organic fertilizers and biofertilizers. After long periods of maintenances, improvement of soil physical, chemical and biological properties could make the soil become the foundation of sustainable agriculture. Suggestions for management of soil fertility and cultivation are listed as follows:

- The relationship between plant resistance to disease and nutrients has been confirmed. Especially, when plant nutrients are inconsistent and insufficient, or when there are toxic substances, plants are susceptible to attacks of the pathogens. For example, if the soil is too acidic, low availability and obvious diseases would exist in many nutrients. Improvement of the soil acidity could increase its capacity of disease resistance. Application of alkaline materials (such as lime, slag, silicate materials and oyster shell powder, etc) not only improves the soil acidity, but also increases nutrients such as calcium, magnesium and silicon, etc. And the crop capacity of disease resistance could also be promoted.

- Good drainage of the soil physical property should be kept so as to promote the development upland crops' root system and to achieve efficient absorption of nutrients.

- Chemical fertilizer and organic easily-decomposed fertilizer containing large amounts of nitrogen. It had better not to be applied excessively so as to avoid problems, such as too much nitrogen, and the fixed and unbalanced nutrients in the soil.

- Micronutrients supply should be paid attention to and the insufficiency or excessiveness should be avoided as well. It is better to increase beneficial bacterial species to the soil, to apply properly biofertilizer, to avoid the thorough soil sterilization, and to strengthen the soil health and resistance.

In general, it is important to bring up the management of the soil fertility and cultivation and the crop nutrients in agricultural production. With the control of pests and diseases, it will reach the goal of twice the results with only half the efforts.

7.13 Guidelines of Soil Management for Plant Nursery

7.13.1 To Ensure Healthy Seedlings by Plant Germplasm Resources and Environment

In order to cultivate healthy seedlings, good plant germplasm resources and soil environment are necessary. The plant germplasm resources should be both internally and externally healthy. There should be good hereditary characteristics within plant resources. In other words, there should be good varieties. The exterior of plant sources should not carry pests and diseases, or carry factors which may easily bring about pests and diseases. Apart from that the seed source should be emphasized, the soil environment is a key factor to determine the health of seedlings and the success of cultivation. Healthy seedlings could be cultivated out of good plant germplasm resources in good soil conditions. Therefore, soil management becomes the fundamental body of plant nursery.

Correct concept for management of nursery soil is really significant, since improper fertilization management of the nursery would easily lead to soil degradation and acidification, resulting in problems for seedlings growth, abnormalities or diseases of roots. When seedlings grow poorly, fertilizers or agricultural chemicals could correctly used. It should be first known and studied what problems probably are. Since the prevention is more important than the cure, the soil conservation of the nursery is also worth paying attention to.

7.13.2 Common Problems in Soil Environment of the Nursery

In recent years, pesticides and chemical fertilizers (nitrogen, phosphorus and potassium) have been emphasized much in nursery management, but problems gradually occur to the soil in nursery after several years of cultivation. Pesticides and chemical fertilizers cannot solve all problems. The followings are problems of the common nursery soil environment. Specific remediation could be applied after problems are known. Problems of every nursery may be different. Therefore, the first step is to understand problems. The common problems are listed as follows:

Soil Acidification:

The soil acidification is a common problem with soils in the subtropical and tropical regions. Particularly, not only surface soil but also subsoil should also be paid attention to. Soil acidification is the most common problem in nurseries of greenhouse cultivation.

Low Organic Matter Content in the Soil:

Low organic matter content in soil is a common phenomenon for nursery soil which is lack of soil conservation. This problem is the most severe in nursery on slope lands. Soil organic matter content is less than 2%, would cause lots of soil problems.

High Electric Conductivities:

Excessive chemical fertilizer is often applied into the nursery on farmlands. For the whole year around, the salt accumulates and then leads to high conductivity with saltification and alkalization. After the soil has been dried, white powder or crystalline salts would appear if the problem is too severe.

Insufficiencies or Unbalances of Nutrients:

Lack of calcium and magnesium is the most common. If it is the soil of calcareous rock, lack of calcium would not happen. Insufficiencies or unbalances of nutrients are related with fertilization and soil acidification.

Degradation of Physical Property:

The loss of soil is the most severe in the degradation of physical property of the slope land. The tiny and fine soil has been lost. Hardening of coarse soil causes the death of seedlings cultivation. Other problems such as poor drainage and compact of the nursery soil on flat lands merit attention as well.

Pests and Diseases:

Due to continuous cropping, pests and diseases would appear in the soil, which is often used as the soil in the nursery. It would cause poor growth of root system or shoot parts. The most severe one would cause death.

Other Problems:

Apart from above problems, for example, accumulations of phytotoxic substances and pesticides, and other unknown causes would probably affect soil environment of nursery. Different problems would influence one another or the disadvantages are cumulative. All these problems related to the nursery management and these are worth our attention.

7.13.3 Guidelines of Soil Management for a Healthy Nursery

Generally, the three most important needs for crops with the prerequisites for the healthy soil are supplying sufficient nutrients, water and oxygen. And, good physical, chemical and biological properties are required. Soil in the nursery should be paid particular attention to since it provides a good cultivating seedbed for the seedlings. And healthy seedlings promise healthy plants and good harvests. The soil conservation management, which should be underlined, for the soil in the nursery is listed as follows.

Pay Attention to the Soil Aggregates, Water Conservation, and Drainage of Soil Physical Properties:

The sizes of seeds being sowed differ from each other due to differences in the crops. Especially, requirements of soil physical properties for the tiny seeds are higher. Big seeds (more than 2 mm) are easier to sow than tiny seeds. Tiny seeds cannot be sowed deep into the soil. Therefore, there are bad influences on the surface soil with too sticky, coarse, and hard crusts. The surface soil should be loosened; otherwise, the depth for the seeds would be inconsistent. The germinating percentage would be bound to be mostly affected. Particularly for those tiny seeds, a layer of sandy soil or river sand can be covered on the seedbed before the sowing so as to guarantee germination in order. This can ensure that those tiny seeds would not fall into deep layers. Since poor water holding capacity exist in sandy soil or river sand, a shade or a covering should be often used. Straws, residuals of grasses or the organic matter could serve as the covering.

Good water holding capacity and drainage should be existed in the soil that with the nursery. When seeds are germinating, water should be sufficient. But ordinary crops had better not be flooded in water for a long time. Particularly, upland crops among agronomic crops cannot stand too much water. Drainage should also be paid attention to tiny seeds of forest trees. The soil conditioner is a good choice for increasing the water holding capacity. Hardly decomposed peat, humic acid, or thoroughly decomposed bark compost had better serve as soil conditioners.

Pay Attention to pH Value, Electric Conductivity, Effective Nutrients and the Organic Matter' Content of Soil Chemical Property:

The seed germination is greatly influenced by soil chemical property. The pH value of the soil in the nursery had better be neutral, unless the crop favors the acidity. Calcareous materials, such as agricultural limes, dolomite dust and oyster shells powder, etc, could be applied into strongly acidic soils. Particularly, the strong acidity in bottom layer of soil should be improved.

The best improvement to the nursery soil with high electric conductivity is to plant aquatic crops. The salinity can be decreased through the way of steeping. High electric conductivity would not easily happen to the nursery on slope lands, unless improper irrigation or excessive fertilization causes it.

Insufficiencies or unbalances of effective nutrients in the nursery are very common. And it cannot be easily detected by our eyes. After soil analysis and plant analysis have been made, problems could be detected, and then remedies can be made against specific cases. Generally, the strong acidic soil is short of calcium, magnesium and boron, etc, so that these insufficient nutrients need to be supplemented.

Application of organic matter into the nursery is rather significant. The more thoroughly-decomposed organic matter is better. The more hardly-decomposed organic matter is better. Since the organic matter has many effects, more hardly-decomposed organic matters with small amount of salts are more beneficial to improve soil, germination of seeds and growth of seedlings.

Pay Attention to the Activity of Soil Biological Property and Symbiotic Relationships of Crops:

Microorganisms in the soil play essential roles in the recycling of nutrients. For the maintenance of soil microorganisms in the nursery, apart from application of organic fertilizer, beneficial microorganisms, such as nitrogen fixing microorganisms, decomposing bacteria, mycorrhizal fungi, phosphate solubilizing bacteria, and rhizosphere protective bacteria, etc., could be inoculated to effectively improve the soil activity and exert controls and balances the harmful microorganisms, such as pathogens and nematodes, etc. If seedlings are infected with bacteria, the growth of seedlings and the transplant seedlings would be greatly influenced.

From the soil conservation's perspective, even though safe use of pesticides is an effective treatment, the entire soil disinfection should not be done to soil under the condition of disease in the nursery. Disinfection would kill beneficial microorganisms in the soil. Therefore, disinfection should not be the first choice, if there is no special need; otherwise, the soil would be transformed into inactive soil. The fundamental solution is to change soil properties of the problematic nursery. If the soil is strongly acidic or acidic, calcareous materials and organic fertilizer could be used coordinately to change the soil properties; if the soil is strongly alkaline, the sulfur powder and organic fertilizer could be used coordinately. Application of large amounts of organic matter (not containing large amounts of salts) or green manure (such as *Sesbania*) could effectively change the soil properties of the problematic nurseries and improve the activities of beneficial microorganisms.

Attaching importance to the soil management of nursery is important work of the crop production. Good soil management promises healthy seedlings and ensures productions of high outputs and high qualities.

7.14 Management of Soil and Fertilizer in Production of Vegetables

7.14.1 Main Concepts of Soil and Fertilizer Management for Vegetables

Intensive production of vegetables absorbs a great deal of the soil nutrients and requires large amount fertilization for vegetables growth. As a result, there are more soil problems in vegetables fields than in fields of other crops. Whether the management of soil and fertilizer is good or bad determines the quality of vegetable production. Therefore, management of soil and fertilizer is the key to high-quality vegetables' production.

In different regions, due to differences in the climate, the soil, cropping system and management, the first thing needs to be considered for improving the production of vegetables' quality is whether there is existence of limiting factors in the soil. And then improvements should be made against specific cases. Soil conservations of the soil physical, chemical and biological properties should be viewed as important to the soil management of vegetables fields, which could contribute to the maintenance and conservation of the soil fertility. Increasing organic matter in the soil can effectively improve the fertilizer availability, reduce the fertilizing amount, and decrease the accumulation of salts and environmental pollution. This is also one of the effective ways to reduce problems for continuous cropping.

Fertilization aims at not only increasing the production, but also improving crop's quality. Only by applying proper amount of fertilizer, the quality of vegetables can be guarantee. Excessive fertilization causes harms instead, and even contaminates the environment and the water source. Application of the slow release fertilizer or the long-term effective fertilizer will be a significant direction for the agriculture in the future. Adaptation to the needs for the soil and crops, together with use of proper amount of nitrogen, phosphorus, potassium, calcium, magnesium, sulfur and micronutrients are conducive to improving the quality of the products and the nutrient of vegetables. Nutritive values of the food should be pursued more, instead of using external beauty of the product alone. Only by the establishment of correct concepts of soil conservation and fertilization can the vegetable quality be guaranteed.

7.14.2 Important Roles of Vegetables

The aim of soil and fertilizer management is to raise crops production and improve the quality of agricultural production. In recent years, the life quality has been improved and the requirements for the agricultural product quality also raise accordingly, including the quality of agriculture, gardening, farming and raw materials for food processing. Among them, to improve the vegetables' production quality is the focus and emphasis of our life. Although the quality of crop production is influenced by the variety of the crops, the climate, cultivation management, the soil, and the control of pests and diseases, in some regions of specified climates, the production quality of some specified varieties of crops is, to some degree, influenced even more by the management of the soil and fertilizer.

In different countries, the average intake of vegetables per person per year differs. But it is shown that vegetables play important roles for people's health. According to reports, almost half percent of human diseases are caused directly or indirectly by insufficient or improper nutrients. Particularly, the amounts of beneficial substances or active substances in the food should be paid attention to. Application of proper amount of fertilizer is beneficial to crops while excessive fertilization would bring about harmful effects to crops, to the environment and to our human. Therefore, it merits particular attention. Soil management and application of proper amount of fertilizers are really significant.

7.14.3 Qualities of Vegetables

The quality of vegetables is an important standard to determine the sale and the price. People have quite different requirement standards for the crop quality. Their judgments of the quality also differ due to different opinions. It can be generally classified into the commercial quality and the food quality:

The Commercial Quality:

The commercial quality refers to the quality of products in the market. In terms of the direct sales of agricultural products, qualities such as the external beauty of color and aroma, the character of being able to be stored for a long period of time, the taste and the sweetness are emphasized. In terms of food processing, the quality requirements for products, such as oil, sugar, starch and protein contents are different according to different processing needs. The price of the product is closely related to whether the commercial quality is good or bad.

The Food Quality:

The food quality refers to the nutrient value of the food, namely, the internal beauty of food. The consumers want to get healthy, nutritious and nontoxic agricultural products. Human and livestock animals need agricultural products with high nutritive values to gain health. Agricultural products of high nutrient values need healthy and unpolluted soil so as to guarantee healthy and safe food. The quality of food nutrient cannot be detected from their outward appearances, so it does not commonly influence the price. However, the food nutrient values need to be paid more attention to.

7.14.4 Concepts and Influencing Factors of Vegetable Quality

The quality of food nutrient cannot be accurately and fast measured so that some concepts could only be drawn from inferences and observances. Among them, the naturality of the food is the key for disputes. The natural school holds the concept that naturally produced food is good. Unnaturally produced or man-made goods are not. From this point of view, it can be inferred that the food produced with application of natural fertilizers such as compost and manure is good and valuable, while the food produced with application of man-made fertilizers is secondary. Opponents think that judgments and inferences of natural school of food are not entirety correct. The naturally produced food is not all good. There are bad and toxic wild foods as well.

The quality of crop products is influenced by a lot of factors, such as crop variety, soil, climate, cultivation management, and the control of pests and diseases, etc. The hereditary factors of crop variety would determine the characteristics of basic quality. Climatic factors, such as sunshine, temperature, etc., and factors related to soil water and supply of nutrients, etc. could assist hereditary potentials' performances or change the results.

The skill of fertilization would influence the nutrients efficiency to the plants and adjust the function between fertilizers and plants. Fertilization plays the central role of crop quality. Application of inorganic and organic fertilizers could improve the quality of crop products. However, improper fertilization may also cause harmful effects.

Recently, these questions are regarded to the quality: Could the use of traditional fertilization methods in the past or at present, especially the use of organic compost, help get high quality crops? Although application of a great amount of chemical fertilizers increases the production, would the nutritive value decrease? In order to yield high production, would the application of chemical fertilizer for increasing the production per unit area in the agriculture decrease the quality? All these questions are worth paying attention to. To answer these questions may require researches, experiments and inferences. Opinions are divided, so more thorough researches and experiments are needed. It is generally considered that the use of proper fertilization could improve the quality. Yet, improper or excessive fertilization would possibly bring about harmful effects.

7.14.5 Nutrient Supply and Quality of Vegetables

The quality of vegetables is influenced by the management of the soil and fertilizer. Good management makes the soil supply sufficient nutrients, water and oxygen so as to reach goals of high quality and production. The cost of production should be emphasized. The crop nutrients that provided by fertilizers would directly affect the quality of products. Chemical and organic fertilizers should be applied properly. Especially, when the soil could not supply sufficient nutrients, fertilization has the best effect (we often increase the production output and the quality). When the soil could supply proper amount of nutrients, increasing fertilizing amount sometimes could not improve the quality. Nevertheless, excessive fertilization would often lead to damages or decrease the quality, and even pollute the environment or the source of water. Consequently, excessive fertilizer should not be applied to the high cash crops blindly.

In different regions, the climate, the soil, the cropping systems and managements are different. To improve the production quality for vegetables, firstly, it requires an understanding to limiting factors of production in the soil. And then, improvement should be made for the specific case. Maintenances of the soil physical, chemical and biological properties should put emphasis on soil management of vegetable fields and this could contribute to the maintenance and conservation of the soil fertility. Promoting the soil organic matter could help improve the fertilizer efficiency, reduce the fertilizing amount, and decrease the accumulation of salts and environmental pollution. This is also one of effective ways to avoid continuous cropping of vegetables problems. To adapt the needs of soil and plant, together with application of proper amounts of nitrogen, phosphorus, potassium, calcium, magnesium, sulfur and micronutrients could be conducted to qualities of nutrient and vegetables products. The external beauty of the product could not be the only thing to be pursued, while the nutrient value of the food is needed more. Establishment of correct concepts of soil conservation and fertilization could guarantee the vegetables' quality. The relationships between different nutrients and the quality of vegetables are listed below:

The Relationship between Nitrogen Fertilizer and the Quality of Vegetables:

In the soil management of vegetables, supply of nitrogen fertilizers is very important, because nitrogen fertilizer in the soil is the main factor of controlling crops production. It has close relationships with the quality of production. The followings are explanations of the nitrogen fertilizer affects production quality:

The Content and Value of Protein:

Protein is an important nutrient in food. Nitrogen is a basic constituent of proteins. Protein is the polymer of amino acids and contains nitrogen. Consequently, application of proper amount of nitrogen fertilizer is conducive to the synthesis of amino acids and proteins. But excessive application of nitrogen fertilizers easily causes dilution and lowers the unit content instead. Moreover, surplus nitrogen would be accumulated easily in the roots, stems and leaves of vegetables.

When There is Excessive Application of Nitrogen, Common Accumulated Substances, which Contain Nitrogen, within Plants are Listed as Follows:

- Nitrate (NO_3^-) accumulates easily in the leaf of plants. Especially, when sunshine is little, the reduction amount of absorbed nitrate decreases and the accumulation is greater. In the food, there should not be too much nitrate-nitrogen. Generally, less than 50 ppm in plant would be accepted, so as to avoid potential dangers of the formation of nitrite or nitrosamines and NOCs. Besides, if the vegetables are stored in circumstances where there is lacking of oxygen, nitrate contained in the leaves would easily be reduced into the harmful nitrite nitrogen. After the vegetables have been taken, the iron in the haemachrome would be oxidized and thus influences the capacity blood to carry the oxygen.

- Generally, not too much nitroamines can be accumulated in the plant. But the potential dangers of these harmful substances should be paid attention to.

Application of Nitrogen Fertilizers Increase Brings to the Changes of Some Substances in Vegetables:

- The most proper application of nitrogen could increase contents of chlorophyll and carotene.

- Aplication of nitrate fertilizer easily leads to the accumulation of oxalic acid in vegetables, which is harmful. When nitrate fertilizer is applied into dry farming soil, most of ammonium nitrogen would be transformed into nitrate and then be absorbed by plants.

- The content of vitamin C would decrease. Particularly, when excessive nitrogen fertilizer has been applied, it would decrease remarkably.

In brief, application of nitrogen fertilizers' proper amount would increase the activity and germination rate of plant. However, excessive amount has many disadvantages mentioned above. Too much nitrogen fertilizers would also easily make the crops susceptible to diseases and would lead to bad phenomenon such as excessive crops growth, too many branches, poor flowering fruit trees and vegetables, and falling fruits and flowers.

The Relationship between Phosphorus Fertilizer and the Quality of Vegetables:

Phosphorus is an important substance to determine crops production. Its availability in the soil is main factor to determine the supply of phosphorus. Phosphorus takes important roles in plant metabolisms and quality control. The index of phosphorus' function on crop quality includes the content of phosphorus and the proportion of the phosphorus compound. Phosphorus also affects the contents of beneficial and harmful substances in plant. It influences crops health and the quality of the seeds. More supply of phosphorus would change the content of phosphorus and constituents of phosphorus compounds. Explanations are listed as follows:

Increasing Total Amount of Phosphorus in Plant:
When phosphorus fertilizer is applied into soil, the crop absorption of phosphorus is limited by the soil. Therefore, there is a limit to the increase in crops. And it is not easy to have a high amount of accumulation in plant.

The Change of the Proportion of Phosphorus-Containing Substances:
When there is no accumulation of phytine in green parts of plant, the increament happens in the form of inorganic phosphate. Phytine can be accumulated in grains while the inorganic phosphate accumulates in the leaf and stem. And there is only a slight phosphate increase in nucleic acids, while the amount of phospholipids remains unchanged.

Proper Phosphorus Fertilizer is Conducive to the Content of the Following Substances:
- Increasing the content of carbohydrates (sugar, starch).
- Increasing the content of crude protein in leaves.
- Increasing the content of vitamin B1.
- Decreasing the content of oxalic acid in leaves.

The Relationship between Potassium Fertilizer and the Quality of Vegetables:

Potassium fertilizer in soil affects the quality of crops in the way different from nitrogen and phosphorus fertilizers do. But the key for nitrogen and phosphorus is their contents in plant. In the food for animals and human, the balance between potassium and sodium is rather important. Excessive sodium in the food had better not be taken in too much, so that the importance of potassium cannot be ignored.

Absorption of much potassium by crops would lead to the reduction of calcium, magnesium and sodium absorption. Consequently, excessive application of potassium fertilizer would easily cause lacking of above elements. Potassium has great influences on adjustment of the cell turgor pressure and enzyme activities, and would thus affect the whole metabolisms and quality of vegetables.

Excessive application of potassium fertilizer would bring about the crop changes in the following aspects:

- It would strengthen photosynthesis and increase the content of carbohydrates, which shows that crops contain a great deal of carbohydrates, starch and crude fiber. Potassium chloride applied to crops that are susceptible to chlorine. The function of potassium lays in which, increases the content of carbohydrates; while the chlorine, with the opposite function, would decrease starch in the leaves.

- Use proper amount of potassium could reduce the content of oxalic acid.

- However, excessive application of the potassium fertilizer would increase oxalic acid instead. This is an unfavorable increment.

- The content of vitamins could be increased, such as the carotene which is precursor of vitamin A, vitamin B_1 and vitamin C. The amount of vitamin B_1 decreases due to excessive potassium fertilizer as well, which may have something to do with chloride ion in potassium fertilizer.

- The dilution caused by indirect carbohydrates increase leads to content of crude protein's reduction, but may also possibly increase the amount of protein instead, which is caused by application of potassium and reduction of other substances with nitrogen. Besides, the application of potassium reduces the content of chlorophyll as well.

- For the tuberous roots containing the starch (for example, potatoes), application of potassium could reduce the decomposition of starch. Then it would contribute to starch increment and help achieve the effect of improving the potato's quality.

The Relationship between Calcium, Magnesium, Sulfur and the Quality of Vegetables:

Apart from three elements (nitrogen, phosphorus and potassium), calcium, magnesium, and sulfur are elements most needed by crops. Calcium is an important constituent to adjust the liquid flow and is an essential component of the pectic substance on the plant cell wall. Increasing the calcium content would also reduce the concentration of the magnesium and potassium due to antagonism. And there is also a limit to the increment of calcium content. Lack of calcium would affect the quality of fruits and vegetables, and reduce value of the commodity. Especially, lack of calcium would lead to diseases.

Magnesium is an essential constituent of plant and a factor to adjust the enzyme activities, and thus exerting great influences on the quality. Application of proper amount of magnesium fertilizer helps increase the content of chlorophyll and carotene, and even increase the content of total carbohydrates.

Sulfur is the basic constituent of some essential amino acids. Lack of sulfur would reduce the quality of protein. Some plants such as the cruciferae and onions contain products of sulfur secondary metabolisms, such as mustard oil and onion oil, etc. Insufficiency of sulfur would affect synthesis of these substances containing sulfur. The sulfur is not only significant to the food quality, but also promotes the plant ability of disease resistance, which affects the quality indirectly.

The Relationship between Micronutrients and the Quality of Vegetables:

Among the conditions for the crops production, it should not lack of any nutrient element of the plant. Insufficiency of micronutrients would lead to problems in the plant metabolisms and then the quality of crop production would be certainly affected. Therefore, good quality of crops at least requires proper amount of micronutrients' supplies. Excessive supply of micronutrients would also bring about poisons and produce poor qualities. The following are explanations of the relationships between micronutrients and crop quality.

Iron in vegetables is an essential element for human, but there are great variations of iron content in the plants. It is mostly influenced by the soil reaction (namely, pH value) instead of the fertilization.

The content of manganese is a significant index in food quality. Nevertheless, only under circumstances of good mobility and availability of the soil manganese can be applied manganese, which achieves the effects of promoting the crop absorption of manganese. Application of proper amount of manganese helps increase the content of vitamins (such as vitamin C and carotene, etc.).

Application of copper could increase the copper's content in plants. Copper is important to the content and quality of protein. Lack of copper easily leads to flecks on fruits. Excessive copper would cause heavy metal hazards.

Application of zinc could increase the content of zinc in the plant. Excessive application would also increase the content of zinc in the plant up to several times as the original content. The content of zinc is an essential feature for the quality of crops as well.

Application of boron could increase the boron's content, so as to increase the production. Application of proper amount of boron could increase the content of sugar and improve the commercial value in terms of the quality of vegetables and fruits. Insufficiency of boron would easily cause flecks and wrinkles, and thus decrease the quality a great deal.

The content of molybdenum is also an important standard for the quality. Availability of molybdenum is connected to the soil characteristics. Molybdenum is a necessary constituent of nitrate reducing enzyme. The amino acids and proteins can be successfully synthesized with molybdenum. Therefore, it affects the quality a lot.

With regard to the relationships between the soil, the plants and our human, sometimes there are no symptoms in plants showing the insufficiency, but there are insufficient supplies of micronutrients in the soil and their content in the plants remains also low. This also affects human health. For example, from their green outward appearances, there seems no problems with plants. Nevertheless, if the content of molybdenum is relatively low and there are no symptoms of the plants for that, using of these foods that lacks molybdenum for a long period of time would do harm to the teeth of human beings.

7.14.6 Fertilization, Food Quality and Human Health

Human health depends on vegetables. Vegetables growth depends on the soil. Therefore, the linkages between the soil, the plants and human are close. Human health is affected by the food chain. In the end, it is related to the soil supply. There is a bearing of fertilization on the crop nutrient value since the crop constituents' quality or the effects from lacking of nutrient are connected to functions of the food. The followings are explanations:

Quality of Constituents in the Food:

Contents of valuable constituents in food cannot easily be valued in an integral way, since proportions of necessary substances, beneficial substances, and harmful substances in all kinds of food are different. The common valuable constituents are listed as follows:

Essential Amino Acids:
Contents of 8 or 9 essential amino acids are important components of the protein. Nitrogen fertilizers and crop varieties would have influences on the content of protein with amino acids.

Fatty Acids:
Three essential fatty acids are important components of the fat.

Vitamin:
There are about 15 kinds of important vitamins, including vitamin A, vitamin D, vitamin E, vitamin K, vitamin B_1, vitamin B_2, vitamin B_6, vitamin B_{12}, vitamin C and vitamin H, etc.

Mineral Substances:
About 20 kinds of important mineral substances are necessary for humanbodies.

Beneficial Substances:
Color, aroma, and taste of beneficial substances and antagonistic substances may be important constituents of food quality. The soil or the compost may contain some amounts of antibiotics. More researches are needed to be proved for disease resistance of plants after antibiotics being absorbed.

Effective Indexes of Food:

With regard to improvement of interior food quality after fertilization to plants, it may increase production output slightly (based on studies on mice) and it could increase the weight (based on studies on babies). It has more effects on increasing more blood cells and abilities of resisting diseases. At least, proper fertilization has not decreased the quality remarkably.

7.14.7 Conclusion

Correct concepts of fertilization should be established. Do not apply the fertilizer as much as possible, nor make crops grow as luxuriant as possible. The interior quality of crops should be valued. So, proper fertilization is the most significant. Drawbacks of excessive fertilization can only be detected if the production output decreases. Also, excessive fertilization often lowers the crops quality but it is not easily to be sensed. Particularly, the quality of vegetables should be given special priority to, so that the health of human can be guaranteed.

7.15 Conservation and Application of Irrigation Systems in Soil Problems of Facility Cultivation

7.15.1 Soil Problems in Facility Cultivation or Greenhouses

Facility cultivation aims at changing and adjusting the growing environment of crops, controlling the temperature, light, humidity, water or carbon dioxide according to the needs and setting up rain-proof, cold-defense, snow-defense and wind breaking buildings such as glasshouses, tunnels and greenhouses constructed out of different materials so as to achieve goals of protecting the crop production or adjusting the production periods. For the soil in the facility cultivation, the common point is that leaching of the soil by the rain or snow water has been reduced. Soil in the long periods of facility cultivation, under influences from control supervision, would result in problems such as soil acidification, salinization, unbalance of the nutrient, and problems for continuous cropping.

7.15.2 Guidelines for Soil Conservation of Facility Cultivation or Glasshouse

In order to slow down occurrences of soil problems of facility cultivation or glasshouse, guidelines for soil conservation are explained as follows:

To Apply Neutral Fertilizer and Fertilizer with Less Salt:

The most severe soil problem for facility cultivation is the acidification and alkalization. Neutral fertilizer and fertilizer with less salt such as the urea, rock phosphate powder, and potassium nitrate are better than ammonia sulfate, single superphosphate, and potassium sulfate since the later all contain sulfuric acid. Fertilizer with less salt is referred to the fertilizer containing low content of salts or nutrients. For example, livestock-dung organic fertilizer contains high content of salts. Excessive application of high-salt fertilizer could also easily lead to problems, such as acidification and alkalization.

To Adopt Fertilization of Proper Low-Concentration Nutrients:

To adopt fertilization of proper low-concentration nutrients is just like to use drip fertilization of the liquid fertilizer and proper small-amount fertiliziation. The fertilizer had better not be applied excessively at each time, so as to avoid problems such as long-term accumulation of salts.

To Use Good Water and Spraying Systems:

Since different regions have different water qualities of irrigation, it should be paid attention to whether the water quality needs to be firstly considered in the facility cultivation. Otherwise, water with too much calcium or irrigated iron and even polluted water would quickly accelerate occurrences of soil problems. Besides, washing or leaching from the top of plant to the bottom should be adopted in the irrigation system, so as to make less salt accumulate in the surface soil.

To Use Crop Rotation and Plant Crops with Strong Fertilizer Absorption:

Crop rotation means not to plant the same variety or the same type crop continuously in the same area. At the same time, we can plant crops with strong ability to absorb fertilizer once a year or once a couple years, such as corn, this could reduce the accumulation of surplus nutrients and salts in the soil.

To Adopt Organic Materials with Strong Absorption Capacity:

When salts accumulate in soil, and after the soil adsorption reaches saturation, soil problems would quickly appear. If materials with strong absorption ability to surplus nutrients, such as peat, charcoal, active carbon, it would thoroughly decompose wood and high-fiber materials could be applied, which would reduce the harms for accumulated salts.

7.15.3 The Irrigation Systems in the Facility Cultivation

According to the degree of proofing rain and different crops, the facility cultivation has different needs for irrigation systems. Particularly, plastic-cloth houses or tunnels need the irrigation systems. In general, types of facility irrigation systems include: furrow irrigation, spray irrigation, mist irrigation, drip irrigation and underground irrigation, etc. The choice of facility irrigation system is determined by the way of facility cultivation, the need of crops, and consider with climatic environment. The cost of different types of irrigation systems should be also taken into consideration.

Design of irrigation systems needs to pay attention to points of crop physiology, soil, water conservation and automatic control. Nevertheless, many common products in markets are very convenient to operate, and can be used coordinately with nutrients and agricultural chemicals. The following are explanations of advantages and disadvantages of different irrigation systems and their application guidelines:

Furrow Irrigation:

It is a simple way of irrigation for a traditional ridged field or a ditch. The water goes from the ditch or border method of irrigation into the field. The only disadvantage is that furrow irrigation takes a lot of water. For those fields with poor drainage soil, it would easily lead to long periods of accumulated water and long irrigation time.

Spray Irrigation:

According to different materials, it is classified into spray irrigation with sprinkler head or that with perforated pipe. The irrigation of sprinkler head is often used in intensive cultivation of flowers, vegetables and turfs, etc, while the irrigation of perforated pipe is often used in cultivation of large plants, such as melons and fruit trees. Under improper circumstances, spray irrigation easily gives rise to diseases and the falling flowers and fruits of plants. It takes more water than drip irrigation, but it is more convenient and labor saving than the drip irrigation in the installation, use and maintenance. Only poor water quality and large water consumption easily lead to cumulative colors and crust on the surface of leaves. The automatic control system can be used coordinately. In terms of time of spray irrigation and control of the water amount, pesticides and nutrient solutions could be sprayed in a coordinated way to achieve more purposes.

Drip Irrigation:

According to different dripping materials, it is classified into dripping head, permeation tube, and double-layered tube of drip irrigation, etc. The application can be determined in crop variety, their need of water and the convenience. The best advantage of drip irrigation is that it saves water and prevents the overgrowth of the weeds in the whole garden. The disadvantage is that it is time- and labor-consuming in the installation and maintenance. The small water outlet is easily blocked, and the landform or improper constructions affected by water pressure. The blockage is not easily cleaned out. Drip irrigation could also be used in coordination with the auto control system and aditional nutrient solutions to achieve more purposes.

Mist Irrigation:

Mist irrigation supplies water through spraying mist. In spraying mist materials, it could be divided into high-pressure sprinkler head (about 3.5 to 7 kg/cm^2), medium-pressure sprinkler head (about 1.5 to 3.4 kg/cm^2, fan humidifier, water-cushion fan, etc.). It is mainly used in the cuttage reproduction of the crops, with which increases the humidity in the facility. However, in the high-temperature summer, for example, diseases may easily appears in the poor environment. Mist irrigation can be used coordinately with all kinds of automatic control systems.

Underground Irrigation:

It is an irrigation system in the under surface or below the medium. It can directly or indirectly supply the water or nutrient solutions to the root. Underground irrigation could often maintain moisture evenly in soil or medium. But it also easily blocks the water outlet and may be infected with diseases reciprocally. Moreover, it is hard to be repaired or replaced.

7.16 Guidelines for Fertilization and Management in Golf Course Turf

7.16.1 Proper Management could Reduce the Occurrences of Soil Problems

During the use of any piece of land, under improper management, soil problems would raise. Management of turf in golf course is no exception, and proper management should be emphasized. Improper fertilization management has harmful effects on turf and could result in more pests and environmental problems as well. Correct fertilization management could reduce soil problems to minimum.

The soil of a golf course includes sandy soil and original soil. The service area, the green and the lane are all covered with sand and grasses grow there. The sandy soil area has weak capacity of fertilizer and water conservation, so that its management requires special attention.

The fertilizers include easily dissolved fertilizer and hardly dissolved one. The easily dissolved fertilizer gets lost easily, while hardly dissolved fertilizer supplies the nutrient slowly so it gets lost hardly. Therefore, a choice of a proper fertilizer can reduce the loss problems. The plant absorbs the fertilizer rather slow in terms of the speed and the amount. Generally, weak capacity of utilizing fertilizers exist in the plants. The best soil has only 50% of recovery rate of nitrogen and potassium fertilizers being applied. This rate is even lower, if the soil fertilizer conservation capacity is weak or when there is much rain. The recovery rate of phosphorus fertilizer is about 20%. The phosphorus is easily fixed by the soil, so the leaching loss is less. Therefore, correct fertilization could reduce the leaching losses.

The play area has its necessity of covering sand to the surface of soil for growing grasses. The texture of sandy soil is rather thick and it is well elastic. The grasses would not be stamped to death easily and has good resistance to compression, and good drainage. They could recover themselves easily after irrigation or rain. Moreover, they could leach easily for secretions of plants to reduce autointoxication and keep its green and glossy characters in seasons. However, they have weak in fertilizer and water conservation capacities, and may easily bring about losses of agricultural materials, such as fertilizer and agricultural chemicals, etc. Their good aeralation may incur pests and diseases. Therefore, proper fertilization and strengthening of soil fertilizer and water conservation capacities should be attached great importance to reduce losses of agricultural materials. The cultivation of more trees in areas that are out of bound could reduce the occurrences of problems effectively, maintain the health of turf, take account of environmental maintenance at the same time, and then reduce burdens to the environment.

The theme of this article is to understand the turf environment and guidelines for fertilization management. Next is to report how to improve golf course's fertilizer conservation capacity and water holding capacity so as to achieve the objectives of saving the fertilizer and the water.

7.16.2 Basic Understandings of Turf Growing Environment

The growth of turf is influenced by climate, soil, cultivation management of grasses, and control of pests and diseases. It is very important to have an understanding of whole environmental conditions so as to achieve four standards of turf in a supplementary way: (1) the turf is glossy and green; (2) elastic, resistant to compression; (3) easy to maintain; and (4) not susceptible to pests and diseases.

In other words, it is visually verdant; it is elastic if steps on it; it springs the ball moderately; it possesses great resistance to compression; it is easy to maintain and it is not susceptible to pests and diseases.

To know whether the turf has achieved the above four standards, we can observe the following aspects of turf: the leaf color, the number of leaves, the density and thickness of leaves, the hardness of leaves, the number of tillers, the pests and diseases of plants, etc. All of these have direct or indirect close relationships with the growing environment of turf. The following are three turf environments to be known, which are the main factors influencing the growth of turf.

The Climatic Environment:

We should first have an understanding of variances of the golf course in region climatic environment and seasonal changes, such as rainfall, distribution of seasons, monsoon, sunshine and temperature, etc. Climate is the main factor to influence the turf growth and its management method. Under conditions of rainy and dry seasons, the way for turf management differs. Unchanged way of management would easily cause problems. For example, in terms of fertilizer types, too much ammonium nitrogen fertilizer (NH_4^+) had better not be applied in rainy seasons; otherwise, it may lead to ammonium poisoning and grass withering after raining. In rainy seasons, easily decomposed fertilizers had better not be applied excessively since fertilizers may get lost. Before the dry seasons approach, too much of nitrogen fertilizers had better not be applied; otherwise, it may easily lead to wide leaves, and reduction of drought resistance capacity. Phosphorus-potassium fertilizer had better be co-applied to increase the capacity of drought resistance. In dry seasons, the spraying irrigation system should be used coordinately so as to reach the best efficiency. Particularly, control of soil water holding capacity and water saving value should be attached importance to. In addition, the opportunity of control for pests and diseases should be paid attention to, since the climate affects their severity.

The Soil Environment:

The soil management of golf course needs to be specially understood so as to reach the goal of double the results with half efforts. To understand the soil of golf course is to understand soil parent materials, soil chemistry, and physical properties. For example, there are great differences in the characteristics of red soil, mud-rock soil, sandy soil and clay soil. Their managements are not the same. Soil management of red soil is quite different from that of mud-rock soil. As a result, soil managements in different golf courses are more or less different. Among all soil characteristics, soil pH value, electric conductivity, fertilizer conservation capacity, water holding capacity, the drainage and permeability, organic matter and nutrient contents and proportions, etc. need to pay particular attention to. For example, the turf that lacks of nutrient or excessive nutrient would result in the diseases increment. Particularly, calcium and phosphorus are the most important. Healthy soil promises healthy turf. A healthy turf could reduce occurrences of pests and diseases; that is to say, it could reduce the use of pesticides.

Adaptation to Grass Species:

The choice of grass species is significant. The chosen characteristics and adaptation of grass species should be understood before the grass is planted. For example, in subtropical-tropical Taiwan, adaptation to high temperature and much rainfall should be taken into consideration. Special conditions of golf course, such as drought resisting, resistance to monsoon, salinity, strong acidity of soil (red earth), or strong alkaline (mudstone soil), resistance to pests and diseases should also be paid attention to. A wrong choice of grass species would waste a lot of efforts, care and fund to make a remedy.

7.16.3 Fertilization Management of Turf: Increase Nutrient Availability and Reduce Fertilizer Loss

Methods for fertilization management in golf course differ from each other according to climate and soil of different regions, while managements for sandy soils are similar. Guidelines of fertilization are different, if there are great differences in climates. The green and the fairway are key areas. Fertilization and spraying in those areas should be relatively improved and should be thus emphasized more.

The goal of fertilization is to keep the grasses glossy and green, elastic, resistant to compression, easy to maintain, and not susceptible to pests and diseases. Guidelines for turf fertilization to sandy soil and rough soil area are different. Fertilizer conservation capacity should be paid attention to in sandy turf soil of golf course, while fertilizer efficiency should be emphasized on the fairway soil area. Therefore, slow release fertilizer could often be applied in the sandy turf soil, while the choice of fertilizers and the improvement of soil should be paid attention to in the fairway soil area. The following are explanations of the guidelines for fertilization to turf soil.

Application of Slow-Released Fertilizers:

Slow-Released fertilizer is hardly dissolved fertilizer that made of chemical fertilizer (easily dissolved) through special methods, such as outer coating, sulfur wrapping and chemosynthesis. It has long-term effects and would not be easily dissolved and lost in rainy seasons and during the spraying water. Nitrogen fertilizer is the easiest to gets lost among all the fertilizers. Among all nitrogen fertilizers, nitrate gets lost most easily. Rock phosphate powder has long-term effects among phosphorus fertilizers. The mixture of chemical fertilizer with peat could reduce the loss.

The Fertilizer should be Applied According to Different Seasons:

Fertilization needs to take temperature and rainfall of different seasons into consideration. When temperature is rather low or too high, the turf grows quite slowly, so the amount and the frequency for applying fertilizer could be decreased. Especially in the region where there is much rain in the winter, the urea and ammonium N had better not be applied excessively. Low temperature and much rain could lead to the poisoning of the ammonium and the nitrite nitrogen. When the temperature is high, the grasses grow rather quickly. At that time, nitrogen fertilizers had better not be applied excessively; otherwise, it may easily bring about pests and diseases. Particularly, in the region where there is much rain, it causes fertilizer loss more easily. Nitrogen had better not be applied excessively before drought so as to avoid resulting in big leaves and too many stomata, which leads to great water evaporation and weakening of drought resistance capacity. Phosphorus-potassium fertilizers should be applied together in the fertilization.

Foliar Application:

The foliar application is conducted through dissolving soluble fertilizers into water and then applying them with spray irrigation. The fertilizer used in foliar application has low concentration, but there is a great loss, so that it had better not be applied excessively, but a small amount each time and more time application again. Foliar application has fast fertilizer efficiency and can be used to supply nutrient insufficiency. However, it is only a temporary solution. The basic solution should focus on improving the long-term fertility of soil.

Organic Fertilizer Application:

Organic fertilizer has many effects for soil. Besides supplying nutrients in a slow release way, it can increase water and fertilizer conservation capacity of sandy soil, accelerate the detoxification of pesticides, and promote the health of turf. Therefore, organic fertilizer should be emphasized and put to good use. In the golf course, organic fertilizer can be applied in sand mixture at the new built golf courses or applied when digging a hole or covering golf course with the sand. Since there are different organic fertilizers, they can be classified into easily decomposed materials and hardly decomposed ones. Easily decomposed organic fertilizers such as the compost applied for supplying nutrients, while those hardly decomposed ones such as peat applied for improving water and fertilizer conservation capacities of soil. Therefore, organic fertilizers used in the new built golf courses should be mainly hardly decomposed organic fertilizers. In order to improve soil properties of golf course, animal dung organic fertilizer had better not be applied. Stinking organic fertilizers and those which are succetibable to pests and diseases had better not be applied in golf course. Easily decomposed organic fertilizers had better not be applied excessively, since they may lead to great loss of nutrient in raining seasons, or sometimes bringing about too much reproduction of fungi and causing a lump of poor drainage area and turf withering.

Fertilization to Special Soil:

There are grasses and trees in those rough areas where sand has not been covered. Hence, appropriate management is also needed. There are different fertilization guidelines for the special soil, such as the red-yellow soil, mud-rock soil and sandy soil, etc. The soil texture of the red-yellow soil is acidic and has low availability of fertilizer. Calcareous materials can be used for the improvement, so that the fertilizer would not be wasted during the fertilization. The mud-rock soil is strongly alkaline, and has unstable soil layers, so that acidic materials can be used in the improvement. The sandy soil has weak fertilizer conservation capacity, so hardly decomposed organic peat can be added.

Application of Microbial Fertilizers:

The turf has dense root systems. To maintain the whole year round of healthy growth requires keeping the health of rhizosphere. And the cleanness, nutrient adjustment, availability of nutrients and fighting against germs around the rhizosphere all need the coordination of microorganisms. These microorganisms include mycorrhizal fungi, phosphate-solubilizing bacteria, nitrogen fixing microorganisms, rhizo-proctective bacteria, etc. Therefore, in order to maintain the rhizosphere of turf's health, the care of the microorganisms is really significant so as to make the turf easily to be managed and maintained, and to reduce the occurrences of pests and diseases and other problems. Application of microbial fertilizers during water spraying is really a great approach for turf management.

7.16.4 Diagnosis of the Nutrient Problems

When grasses lack of nutrients, symptoms would appear just like other crops. Some can be easily diagnosed by our eyes while others cannot. Hence, plant analysis and soil analysis can be used to detect the causes for the problem. The symptoms for the turf plants that lack of nutrients have something to do with nutrients movements in plant. The following are brief explanations of the functions and symptoms for deficiencies of elements.

Nitrogen:

Nitrogen is one of the three major nutrients required for plant growth. Nitrogen is a very important constituent of amino acids in the plant, such as proteins (including enzyme), nucleic acids and chlorophylls, etc. Nitrogen can easily be moved in the plant. Lack of nitrogen would make the old leaves easily get yellowing for older leaves and severe deficiency would make plant wither. Excessive nitrogen would lead to excessive growth of grasses, long and thin stem, and bigger and thinner blade. Thus, it is easily bringing about diseases and decreasing its tolerant ability to drought.

Phosphorus:

Phosphorus is one of the three major nutrients required for plant growth. It is the constituent of substances for plants conserving and transferring energy, and nucleic acids, etc. It can easily be moved in plant. Lack of phosphorus would often lead to symptoms turning a reddish-purple appearing in old leaves or stems, and old leaves are easily withering. Excessive phosphorus would cause grass leaves and stems smaller.

Potassium:

Potassium is also one of the three major nutrients required for plant growth. It is an important constituent to adjust osmotic pressure and enzymes, and can be moved easily within the plant. Lack of potassium would first lead to scorching from leaf tip margins and interveinal chlorosis which begins at the base of old leaves. Excessive potassium would cause grass leaves and stems smaller.

Sulfur:

Sulfur is one of the constituents of plant amino acids and can be moved easily within the plant. Lack of sulfur would often cause new leaves yellow first, sometimes this phenomenon followed by older leaves. Nowadays, many fertilizers contain sulfur, so that deficiencies of sulfur are rare. Excessive sulfur would lead to acidification of the soil.

Magnesium:

Magnesium is one of the constituents in plant chlorophyll. It is a co-factor of the photosynthesis, nitrogen transformation and many enzyme activities, so that it is extremely essential. Lack of magnesium would cause the vein of leaves green, but leaf intervein takes on a strip of yellow color, which appears on the old leaves first.

Calcium:

Calcium is the structural substance on plant cell wall and cannot be moved easily within the plant. Lack of calcium happens to the new leaves first, making the shoot tip become deformed and shrink, and leading to problems such as growth stunted or rotten easily after raining. Severe deficiencies would lead to the withering of the stem's top.

Boron:

Lack of boron would easily result in deformation of the growing shoot tip. Severe deficiencies would lead to terminal buds die, witches' brooms form. Symptoms are the same as those when lacking of calcium.

Microelements such as Iron, Manganese, Molybdenum, Zinc, etc.:

All these microelements cannot be removed easily from the plant. Lack of them would lead to symptoms, such as the growing shoot tip on the new leaves are becoming yellow (lack of iron) or becoming yellow-white (lack of zinc), etc.

7.17 Managements for Fertilizer Conservation, Water Conservation and Water Saving in the Golf Course

7.17.1 To Strength the Fertilizer and Water Holding Capacity of Turf Soil is the Fundamental Method of Soil Management

In order to make the turf in the golf course glossy and green, elastic, resistant to compression and not susceptible to pests and diseases, fertilization, pest and disease control and irrigation are necessary, just like the crop field management. Therefore, reduction of the fertilizers loss and agricultural chemicals is as important as the crop field management. Good turf and turf management could also contribute to environmental protection. The golf course includes sandy and rough soil area. The sandy soil in the green and fairway area has the weakest capacity of the water and fertilizer conservation. Water holding capacity is to be primarily strengthened in the soil of rough area.

The cause for the fertilizer loss is that fertilizers are easily dissolved in water and then gets lost. The sandy soil has the weaker fertilizer conservation capacity than other soil, so it is necessary to improve the fertilizer conservation capacity of sandy soil, so as to reduce the loss of fertilizers. Sandy soil in golf course has good drainage and this also leads to the weak water holding capacity. During dry seasons, the sandy soil needs water spraying to increase water content in the soil. Then water flows fast and the dissolved fertilizers can be moved down easily. The results for increasing water conservation of the sandy soil are saving water and reducing the fertilizer loss. In order to improve the soil and water holding capacity in the rough soil area, more trees could be planted. If trees could grow densely to a forest, they can help achieve water conservation and ecological protection. The turf management of the golf course needs to examine fertilization, spraying and pest and disease control in an overall way. All of these have close relationships with turf water and fertilizer conservation capacity. Good management can keep the turf healthy.

7.17.2 Key Methods for the Fertilizer Conservation of Capacity Improvement

Guidelines for the Improvement:

To increase the soil fertilizer conservation capacity is mainly to increase soil adsorbing capacity of fertilizers. In other words, it needs to conserve anions and cations dissolved from fertilizers for increasing the capacity of soil positive and negative monovalent ions or their ion exchange capacity. Generally, the soil surface is dominated by the negative charge, so that it mainly conducts the adsorption or exchange of cations. It means that the soil possesses the cation exchange capacity (CEC). The soil is weak to adsorb anions of fertilizers. Soil organic matters often play the role to adsorb anions. Therefore, soil cation exchange capacity and its organic matter content are the indicators for fertilizer conservation capacity. The sandy soil has low cation exchange capacity and less organic matter content, so materials with strong cation exchange capacity or hardly decomposed organic materials need to be added.

Materials with high fertilizer conservation capacity also usually possess strong cation exchange capacity. They have surface characteristics of porosity, great surface area and containing anions, such as peat and other special minerals, etc. Hardly decomposed materials, for example, peat, have not only great surface area and contain lots of cations and anions. They also have great capacity of holding ions and substances. Therefore, materials with conserving cations and anions of fertilizers are all effective. Materials of strong fertilizer conservation capacity could also reduce the pesticides movement (herbicides, insecticides, fungicides, rodenticides, pediculicides, and biocides).

Methods for the Improvement and Notes:
- Materials with high cation exchange capacity could be added and mixed into soil layer of sandy area in the under construction golf courses. For those to built golf courses, the above materials can be mixed with the sand when the sand is covered or when the sand is used to fill up the burrow. Or they can be used for sand mixture in renovation of fairway.

- Easily decomposed materials such as animal dung, meals and grasses had better not be used to increase fertilizer conservation capacity, but can be used to supply nutrient since they are decomposed very quickly. Animal dung compost should not be applied excessively, so as to avoid bad odor which affects the air quality.

- Materials that are added had better not be strongly acidic or alkaline. The materials could be used after their pH values have been adjusted; otherwise, the turf growth would be inhibited.

7.17.3 Guidelines and Methods for the Water Holding Capacity Improvement

Guidelines for the Improvement:

To increase the soil water holding capacity is to increase the soil capacity of conserving water and reducing the leakage of water. Water holding capacity refers to the capacity to hold water. The sand in the sandy area of the golf course has low water holding capacity but high permeability, so that it drains away the water very quickly. The drainage and the water percolation capacity under the sandy soil layer of the fairway soil determine the characteristics of the fairway soil layer. And, it is determined based on the degree of compaction, water permeability, porosity and texture of the soil profile.

Materials with high water holding capacity possess good water conservation and water absorption abilities, and have high porosity, expansion and large surface area, such as layer and expansive clay mineral, peat and water-absorbing polymers. Water-absorbing polymers can be added into the sandy area of the fairway which has low capacities of water conservation and absorption. Water holding capacity and soil property of the rough soil area are related to plants on the ground. Woody plants had better be planted, since they can give function of carrying the water into deep soil layer.

Methods for the Improvement and Notes:

- Materials of good water conservation ability can be added and mixed into soil layer of sandy area in the under construction golf courses. For those built golf courses, these materials can be mixed with the sand when the sand is covered or when the sand is used to fill up the burrow. Or, they can be used for sand mixture in renovation of fairway.

- Woody plants had better be planted in the rough soil area along the two sides of the fairway. It is better that plants can grow into a forest. However, woody plants such as evergreen trees with foliole are appropriate compared to the deciduous trees with heaps of fallen leaves, which may easily result in problems, such as blocking of the water drainage. Trees have the best effects on the soil capacities of water retention and conservation.

- Organic matters are materials to conserve the water. However, easily decomposed materials have short durability, so hardly decomposed materials are needed, such as peat. The woody peat is the best choice.

- Added materials had better not be strongly acidic or alkaline. The materials could be used after their pH values have been adjusted.

7.17.4 Integral Function of Fertilization and Water Spraying Together with Fertilizer Conservation, Water Conservation, and Water Saving

Of all the fertilizers, slow release fertilizer has the best effect of fertilizer conservation. Water needs to be sprayed onto the fairway except in raining day. Application of slow release fertilizer together with fertilizer conservation materials can effectively control fertilizer's loss. If chemical fertilizer is applied, it may get lost or run into the bottom layer when it rains or during water spraying. The fertilizer conservation materials alone cannot produce good effects. It's better not used the chemical fertilizer excessively in sandy areas, since the sand has low ability of fertilizer conservation and easily causes the dissolved fertilizer to be lost or run into the bottom when it rains or during water spraying. Besides, it is an important concept not to apply excessive fertilizer. Efficient fertilization coordinated with good ability of fertilizer conservation of the soil which can yield best result.

Water spraying is a special method of irrigation which is needed for the golf course in special circumstances. The aim of water spraying is to maintain the growth of grasses, so the spraying time and its amount are especially significant. It is well known that water spraying yields the best effect at night time, since there is no sun and low temperature, less evaporation. Therefore, water application can be more effectively infiltrated into the soil. The standard for the appropriate water spraying amount is that the water does not overflow, but how to save water should be thought carefully and to be put into practice. The following are explanations for the guidelines of water saving:

Do not Apply Excessive Fertilizer:

If too much fertilizer has been applied into soil, soil water content for permanent wilting point of plant would be higher, and then the need for spraying water amount would be greater; otherwise, the plant would easily reach the permanent wilting point during dry seasons, due to high evaporations and then withers. Therefore, proper amount of fertilization could save water.

Promote the Ability of Drought Grass Tolerance:

Different varieties of grasses have different abilities of drought tolerance. Some varieties have rather strong abilities of drought tolerance, while others have weaker. In order to promote turf ability of drought tolerance, the starting point can be the evaporation reduction from shoot of plant and water absorbing capacity improvement of root system. The following are explanations of the management for balancing the water:

CH7 Guidelines of Improvement of Fertilization and Soil Management

Proper Fertilizer Quantities:

In order to reduce evaporation from shoot of grass, the leaf area and the number of stomata should be reduced. In terms of management, do not apply excessive nitrogen fertilizers. If nitrogen fertilizer is sufficient, the leaf area of grasses would be relatively greater, the evaporation and the number of stomata would also increase relatively, and the ability of drought tolerance would decrease. Do not apply excessive nitrogen fertilizer, especially when dry seasons are approaching.

Plant Training:

Before dry season comes or two and three weeks before dry seasons, water spraying amount to turf should be reduced to the degree that just to make the leaf roll slightly at noon (do not make it excessive!). The turf going through the water deficient training could help promote the extension of root system and adjust its growing so as to attain drought tolerance and to save water.

Beneficial Microorganisms' Application:

Symbiotic beneficial microorganisms of turf root system, for example, the length of mycorrhiza fungal hypha, sometimes reaches several tens cm, can increase the distance and area of water absorption, contribute to the function of drought tolerance. Beneficial microorganisms around the rhizosphere assist the nutrient availability, absorption and decomposition of agro-chemicals and other toxic substances, thus making the plants grow strong and increase the ability of drought tolerance.

The Design of Water Spray Irrigation System should Correspond to the Slope and the Flow Direction of Water:

Design of common water spray irrigation system should only take the spray area into consideration, but fails to consider flow direction of the spraying water. This may lead to too much water in part of areas. Therefore, how to control the spraying time should be based on a practical understanding of the conditions in the local area.

Grow Tree Plants along both Sides of the Fairway:

The amount of water needed for the fairway is related to day evaporation, sunlight and wind speed. In terms of the microclimate, the amount of water could save a lot of water by the obstruction of tree plants. A forest on the slope side of the fairway may increase water permeability and conservation, thus saving water for the fairway. The concept of planting many trees along both sides of the fairway should be emphasized. More trees can be served for multiple purposes, so why not?

Application of the Leaf Humectants:

In dry seasons and when the water is insufficient, the last line of defense is to spray nontoxic and harmless waxy water-protective materials to the grass leaves for reducing the plant evaporation and thus saving water.

7.18 Guidelines of Rational Fertilization to Fruit Trees

7.18.1 Preface

In order to seek strategies of rational fertilization, scholars and experts in the field of soil fertilization have recently launched several tasks, such as various kinds of fertilization handbooks, and monographs on the soil and fertilizer. They advocate proper fertilization periods, amounts of fertilizers, types of fertilizers and fertilization methods. Recommendation of the most proper fertilization amount is based on analysis and diagnosis of crops, so as to achieve optimum production and quality. In rational application of the fertilizer, environmental maintenance should still be given consideration to. Fertilization to fruit trees is not an exception. For orchards of large areas, rational fertilization has its important values. For crop lands at districts of water source, rational fertilization should be paid more attention to.

The least amount of fertilizer to achieve rational fertilization with best fertilization effects requires not only proper fertilizer at proper fertilizing period by proper fertilizing methods. How to promote growth, absorption and health of crop root system are important guidelines. Therefore, coordination of beneficial materials in fertilization is important, such as beneficial microorganisms and humic acid, etc.

7.18.2 Understand Characteristics of Crops before Fertilization

Fruit trees are perennial crops. A plant would go through processes of the seedling stage, vegetative stage, reproductive stage, and mature stage, during which the number of years and the productivity have close relationships with the local climate, crop variety, soil environment, cultivation management, and pests and diseases control. A healthy and strong seedling stage lays the foundation for fruit tree's later productivity since a fine root system and shoot would absorb more nutrients from soil and conserve more organic materials. These nutrient supplies the ability of production. Every year during seasonal changes, they are absorbed, synthesized and conserved and then produced. This process will be repeated again and again by years.

Therefore, fruit trees keep absorbing nutrients of fertilizers in the whole year round. But in order to achieve the best growth and production, in terms of supplement of fertilizer nutrients, adjustments should be made according to seasons. In fertilization at the vegetative, mature green and red fruit stages, correct fertilizer should be formulated. With regard to the fruit tree size and its production, the total amount of fertilization should also be taken into consideration.

According to deciduousness of fruit trees, they can be divided into two types: deciduous ones and evergreen ones. With regard to fertilization management, emphasis and guidelines are also not the same. Deciduous fruit trees include plum, peach, pear, apple, grape, persimmon and chestnut, etc. Evergreen fruit trees include orange, banana, lichee, longan, loquat, guava, lotus seedpod, mango, pawpaw, carambola, cherimoya, pineapple, Indian jujube, etc. Even though some varieties possess different abilities of absorbing and utilizing the fertilizer, understand beforehand helps rational fertilization. During deciduous periods, nutrient absorption of deciduous fruit tree reaches the lowest, and it is also the same in adjustment of production period and heavy cutting of the whole plant. Evergreen fruit trees do not defoliate leaves. Their growth and absorption of nutrient are influenced by seasons. Therefore, guidelines are different when the fertilizer is applied.

The above is from the point of growth and their physiological characteristics of fruit trees. The following conditions should be taken into consideration in order to achieve rational fertilization:

- Types of fruit trees: deciduous or evergreen.
- Age of the plant: age of growth.
- Flowering characteristics: maturity of branches and located part on the fruit tree.
- Characteristics of growth: growing potential and degree of excessive growth.
- Ability of nutrient absorption: distribution of root system and absorption ability.
- Seasonality: climatic factor.
- Soil factors: characteristics of soil and irrigation.
- Field management: pruning and trimming.
- Adjustment of production period: change of the production period.

7.18.3 Understand the Characteristics of Fertilizers

In the whole, rational fertilization to any fruit tree should be based on the local recommendations for crop fertilization, which published by the department or specialists of agriculture. It explains the recommended amount of three elements (NPK), fertilization period, distribution rates, fertilization methods or other element application such as lime and micronutrients. On account of the subtle differences in soil, variety, climate and cultivation management of different regions, the highest skill of fertilization is to adjust and utilize the least amount to achieve the most effective production and qualities.

Rational fertilization applied to different fruit trees, with proper amount and appropriate fertilizer, with proper methods and at proper times. Besides knowing the specific characteristics of the crop, it is necessary to pay attention to and understand the types of fertilizer, including inorganic fertilizer, organic fertilizer and biofertilizer.

Attention should also be paid to how to use the advantages of different fertilizers and how to make the best combination. How to reduce the fertilizer amount and its losses in different fertilizers is also an important strategy. The following are brief introductions of the characteristics and application of three types of fertilizers.

Chemical Fertilizers:

Chemical fertilizers are made of chemical synthesis or transformation. Its ingredients have certain amount of content and the nutrient supplement for plants, such as nitrogen, phosphorus, potassium, magnesium, calcium, sulfur, boron, copper, zinc, manganese, iron, molybdenum, etc. Chemical fertilizers can be produced to single fertilizers and compound fertilizers. Chemical fertilizers are often different from natural substances. For example, rock phosphate is a natural material and after acid treatment, for example, treated with sulfuric acid; it would make the chemical fertilizers, called single superphosphate. Chemical fertilizers can be quick-soluble fertilizers or slow release fertilizers. Dissolution properties include the soluble and the insoluble. The soluble fertilizer could be applied to soil and also serve for foliar application. The following points should be paid attention to when chemical fertilizer is applied:

The Guidelines for Applying Chemical Fertilizer are Reduced to Given Proper Amount of Proper Type of Fertilizer at Proper Times by Proper Methods to Proper Parts according to Crops Needs:

For example, application of ammonium sulfate over a long period of time and application of single superphosphate which has not been through after-ripening and is of poor quality could lead to soil acidification. Another example is that too much nitrogen fertilizer applied to fruit trees may lead to excessive growth, no or less flowering, and premature drop of flowers and fruits. Excessive or improper use of chemical fertilizer is not only directly influences production and quality of crops, but also indirectly influences human health from foods. So it should be treated cautiously. With regard to the crop quality in precision agriculture, the soil plays an extremely important role. Researches of the crop quality's management and soil nutrient cannot be ignored.

The Concept of Chemical Fertilizer's Balanced Application is Rather Significant:

The crops have a proportion of absorbing soil cation nutrients such as potassium, sodium, magnesium, calcium and part of micronutrients. For example, potassium fertilizer could increase the content of potassium in the leaves and reduce the contents of other cations. If the soil lacks of magnesium, and then potassium fertilizer is applied, the crops would lack more in magnesium. This is called "antagonistic action of ion"; therefore, the concept of competition in nutrient absorption and that of balanced fertilization are absolutely necessary.

Organic Fertilizer:

Since the ancient times, human beings have known to apply organic fertilizer. People in the past applied compost or animal dungs, while nowadays there are a great number of types of organic fertilizers due to great differences in the sources of materials and degrees in decomposition of organic fertilizer. Generally, the sources of organic materials are two types: one is organic fertilizer in farms, including barnyard manure, compost, green manure and waste organic residues, etc.; the other is commercial organic fertilizer, sources of which include residues and waste of plants, animals, and microorganisms, peat, or extracted and condensed organic fertilizers. Commercial products are produced according to specific standards and thus have fixed ingredients. The speed of decomposition also differs. Green manure, animal dung, meals, compost and residues of animals and plants generally decompose quickly, whereas organic materials containing much lignin and humus decompose rather slowly, such as peat and bark compost, etc.

Generally speaking, when organic fertilizer is applied to soil that contains insufficient organic matter content (less than 2%), hardly decomposed organic materials, such as peat should be added, should increase soil organic matter content year by year. Nevertheless, easily decomposed organic materials mainly supply decomposed nutrient, but there is less organic matter which can accumulate year by year. Therefore, it is indeed necessary to combine hardly decomposed and easily decomposed organic materials when organic fertilizer is applied. It can not only supply long-term nutrients, but also achieve the effect of soil conservation, which is really a choice to meet multiple purposes.

Biofertilizer:

Biofertilizer takes advantage of active organisms to achieve effects of increasing soil nutrient and nutrient availability, protecting root, and promoting nutrient absorption. Therefore, organisms are often microorganisms, referred to microbial fertilizers or microbial inoculants. Biofertilizers are new scientific and technological fertilizers in modern times, and there are a lot of varieties of microorganisms. They are distinguished from each other mainly according to microbial functions. For example, application of nitrogen fixing microorganisms can help increase the sources of nitrogen in the soil. Nitrogen fixing microorganisms are divided into symbiotic, associative and non-symbiotic ones; mycorrhiza fungi mainly assist the crops to absorb phosphorus fertilizer; phosphate-solubilizing bacteria have lots of varieties and can dissolve ineffective phosphorus into effective phosphorus fertilizer; microbial fertilizers have less impacts on the environment, so they would not break the original ecological system, due to the limited nutrients and antagonistic systems of soil. Consequently, when beneficial microorganisms are applied, no pollution will cause to the soil environment.

The basic concept for the application of microbial fertilizer is to have full contact to crop root or seed. Only by that way, microbial inoculants can be given full play to. Besides, organic decomposers need to work in contact with organic matter. When microbial fertilizer is mixed with organic matter, such as humic acid, carbohydrates and inorganic nutrients, it can bring its integral effect into play in a better way.

7.18.4 Understand why the Fertilizer Efficiency Decreases

In modern agriculture, fertilization is an indispensable method to increase production and improve the quality of agricultural products. In order to balance and supply sufficient nutrients to crops, the needs of the crop and the soil should be taken into consideration in fertilization. The same amount of the same fertilizer is applied into soils in different regions may have different abilities to supply nutrients, since adsorption, fixing and fertilizer conservation capacities of soils somehow differ from different fertilizers. Therefore, improvement of physical, chemical and biological properties of the soil would directly or indirectly improve the soil utilization of nutrients in the fertilization.

The fertilizer applied into soil would not be absorbed quickly by plant. The plant cannot recover all applied fertilizers, either. The soil characteristics play an extremely important role in the fertilizer efficiency. The soil first adsorbs and conserves the fertilizer, then transforms it, and gradually supplies them for plants to absorb. However, the fertilizer efficiency often decreases after the fertilizer has been applied into the soil. The following phenomena may decrease the efficiency of the fertilizer:

Fixation to the Precipitation of Insoluble Materials:
They are mainly phosphorus, calcium, and micronutrients.

Loss of Fertilizers:
They are mainly nitrogen and potassium.

Volatilization of Fertilizers:
The most severe is the denitrification of nitrogen.

Immobilization of Fertilizers by Soil Organism:
The most distinct is nitrogen and micronutrients.

Insufficient or too much Water in Soil:
It leads to fertilizers difficultly being absorbed or injury to root system.

Improper Application of Fertilizers:

Different fertilizers have been improperly mixed so that reactions make the fertilizer difficult to be absorbed.

Absorption Problems in Crop Root System:

The fertilizer efficiency decreases due to the pressure coming from soil environmental stresses or pests and diseases of crops.

In order to reduce the loss and waste of fertilizers, the fertilizer efficiency must be improved, so as to raise the income for agricultural investment. To use the least fertilizer amount to achieve the highest production and the best quality is the goal of precised agriculture and the improvement of soil fertility.

7.18.5 Guidelines in Rational Fertilization

Adapting to the Requirement and Character of Crops:

Different crops respond differently to the use of fertilizer. Different varieties of crops respond to fertilizers are different as well. In order to achieve the effect of rational fertilization, the requirement of crops should be taken into consideration. Some crops or some varieties require more fertilizers, so more fertilizers can be applied in order to increase the production or improve crop quality. However, if some crops are given too much fertilizer, it would be just a waste. The concept of "limited fertilization" should be established. Guidelines are listed as follows:

Understanding the Character of Crops:

Different varieties of crops have their own fertilizer recommendation rates and fertilization methods, or their extension handbooks. Fertilization and cultivation after an understanding from the character of crop varieties could help achieve double the results with only half the efforts. Otherwise, when there is fertilizer injury, it would be too late.

Adapting to the Growth Period or the Age of Crops:

During the seasonal changes, the crops require different nutrient for its growth and reproduction. Generally speaking, for perennial fruit trees, the size of the tree determines the amount of fertilizer. Small plants should not be given excessive chemical fertilizer, which not only has poor fertilizer efficiency, and may also lead to fertilizer injury. In this circumstance, organic matter and biofertilizer should be added to the chemical fertilizer so as to mitigate the effect. Foliar fertilization can also be supplied periodically.

Other Means of Management:

The fertilization effects for crops are often reduced due to pests and diseases. Therefore, elimination of the occurrences of pests and diseases is also a non-ignorable means of improving the fertilizer efficiency.

Reducing the Loss and Volatilization of Fertilizers:

When there is a heavy rainfall, the loss of fertilizer cannot be ignored. Some fertilizers such as the urea and potassium chloride are easily dissolved in the water and then lost; some fertilizers are in the forms of small particles and fine powder, which are easily carried away in runoff. Losses of fertilizers happen on sandy soil with poor fertilizer conservation capacity, on slope fields and on mountain slopes, and thus more preventions are needed there; otherwise, it is not only one individual's waste, it may also lead to pollution to the reservoir, stream and water source. Therefore, it should be dealt with care. The loss of fertilizer includes not only losses along with the running water but also losses through volatilization in the air. The most distinct loss is the volatilization of nitrogen fertilizer. Nitrogen volatilization losses are in two ways: volatilization of ammonia and denitrification. In alkaline soils, urea and ammonium sulfate volatilize easily in the form of ammonia gas. After urea is applied into the soil, it would be affected by urease enzyme, and NH_4^+ would then be released, which may also take the form of NH_3 in the alkaline soil. And NH_4^+ easily volatilizes and gets lost at high temperatures or when it is dry. NH_4^+ is transformed into NO_3^- through nitrifying bacteria, and then they volatilize not so easily into the gas. However, the NO_3^- nitrogen gets lost easily in runoff or denitrifies into nitrogen gas or nitrous oxide (N_2O) in anaerobic environment and then leads to loss of nitrogen. The guidelines of management are listed as follows:

The Fertilizer is Applied through Plough under the Soil or Deep Root Feeding:
When the labor power is possible, fertilizer should be applied deep into the soil. Deep application of fertilizer in orchards can achieve the best result since it is most effective to deep-rooted crops.

Do not Use a Large Amount of Fertilizer Apply to Soil Surface during the Raining Season or the Rain Tends to be very Heavy.

Orchard Grass Cultivation:
In the orchard, there is grasses cover on the soil surface of fruit trees. On soil surface, grasses would absorb and compete for part of fertilizers, but it is more advantageous to those long-term perennial crops. Particularly, orchards on slop fields and mountain lands need more grass cultivation under the tree. The grasses used in the orchard need to be selected. The best grasses are those which cannot grow higher and bigger, or grow poorly during the winter. The grasses could be cut off and covered on the soil surface, which can increase sources of organic matter and reduce the occurrences of fertilizers losses and the surface soil. Long-term soil and water conservation is extremely important to the orchard. Grass cultivation not only contributes to soil and fertilizer conservation, but also helps water conservation. Grasses loosen the soil, making the water go easily deep into the soil. And this really answers multiple purposes in orchard grass cultivation.

To Increase Soil Organic Matter Content:

The soil organic matter content has great capacity of fertilizer and water conservation. After chemical fertilizer has been applied into the soil, the soil fertilizer conservation capacity is closely related to soil texture and the amount of soil organic matter. The more the organic matter is, the better the fertilizer conservation capacity is. The sandy soil has coarse texture and good aeralation, which helps decompose the organic matter faster. Therefore, the ordinary sandy soil has less organic matter. Leaching leads to fertilizer losses. Fertilization to the sandy soil had better be done more time and again, but each time just a small amount of fertilizers.

To Reduce the Volatilization of Ammonia:

The highly alkaline agricultural materials should not be mixed with urea and ammonium fertilizer or have direct contact with them; otherwise, it may lead to volatilization of ammonia. Although grasses on the surface can also absorb some of ammonia, the open and wide surface increases the volatilization of ammonia.

Use of Coarse-Grain or Slow/Controlled Release Fertilizers Could Reduce the Losses in Runoff.

Paying Attention to Soil pH Value:

After the fertilizer has been applied into the soil, the soil pH would affect the fertilizers availability. Both macronutrients and the micronutrients are influenced by soil pH. For example, the availability of phosphorus would decrease, the fertilizer efficiency would be poor and crops could hardly absorb and utilize the fertilizer if the pH value is higher than 7.5 or lower than 5.5. The guidelines of management are listed below:

Application of Neutralizers:

- Calcareous and alkaline materials such as limes for agricultural use, limestone, magnesia lime, dolomite and oyster shell powder can be applied to acidic soils. The amount of limes is determined by the soil acidity and lime materials, and neutralization of the soil should be done gradually, step by step every year.

- Acidic materials such as sulfur powder, gypsum and dilute the sulfuric acid can be applied into the alkaline soil. The neutralizer cannot be mixed and applied with most chemical fertilizers; otherwise, it reduces greatly the nutrient availability of fertilizers, but it can be mixed and applied with organic fertilizer. Use of neutralizer can be the fundamental measure to improve the fertilizer efficiency, but use of excessive neutralizer may change acidity into the alkalinity or the vice versa. The best way to apply neutralizers can combine with organic fertilizers.

Application of Organic Fertilizers:

Application of organic fertilizer into the acidic or the alkaline soil contributes to nutrient availability and increase of fertilizer efficiency.

Foliar Application through Urgent Supply of Nutrients:

If the soil reaction (acidity/alkalinity) has not been timely improved, spraying fertilizers to leaves can act as a temporary remedy of insufficient nutrients, but it is not a fundamental way to solve this problem. Priority should still be given to the soil pH value improvement. Foliar fertilization to leaves is quite effective for the urgent supply nutrients to those short-term crops. Fertilization to the leaves often supplies micronutrients and nitrogen which crops are in deficiency. It should be noticed that the amount should not be excessive so that it could achieve the highest production and quality.

Improvement of Fertilization Management:

The purpose of fertilization is to supply sufficient and balanced nutrients for crops. The nutrients for plants absorption from the soil include a great number of macronutrients and micronutrients, such as nitrogen, phosphorus, potassium, calcium, magnesium, sulfur, iron, boron, silicon, zinc, copper, manganese, molybdenum, chlorine, sodium, and cobalt, etc. Except the inorganic nutrient, plants would also directly absorb or utilize some organic substances from soil, such as organic acid, plant hormones, which have smaller molecules, and even humic acid and fulvic acid. They have great importance for the health protection and the crops quality. The guidelines of management are listed as follows:

Making Soil Diagnosis and Plant Analysis:

The primary thing is to know about the nature of the cultivated soil, at least to know pH value of the soil, and to further know about the physical, chemical and biological properties of the soil. The plant analysis is an important indicator in the problem soil, and following the correct fertilizers should be applied according to the diagnosis result. Blind fertilization may bring double injuries to both soil and plant.

Understanding the Nature and Effects of Fertilizers:

The effects of nutrients differ in terms of crops need, and so does the needed amount. Lack of one of needed nutrients could lead to other nutrient efficiency decrement. Therefore, an integral concept should be established in fertilization. Complementary effects can be achieved when the major elements have good fertilization efficiencies and the soil could supply sufficient micronutrients. For example, when there is poor growth and metabolisms of the crop caused by lack of a micronutrient, it makes no differences by using large amount of other applied fertilizers.

Paying Attention to the Mixture, Transportation and Storage of Fertilizers:

It is necessary to prevent the fertilizer from absorbing moisture and caking. Fertilizers with different pH values should also be avoided being mixed and stored together. The mixed fertilizers should be used up quickly, so as not to affect its fertilizer efficiency. During transportation and storage, it should be avoid that the fertilizer should not be exposed to the sun and rain, or stored under high temperature and high humidity.

Applying Proper Amount of Organic Fertilizers:

The application of organic fertilizer can bring lots of benefits. The biggest benefit is that it can increase the nutrients efficiency of fertilizer. Nutrients supplied from the soil are influenced by organic matter in the soil. The humic acid in organic matter can not only improve soil physical and chemical properties, but also promote the crop roots growth and soil microorganisms.

Using Beneficial Microorganisms or Biofertilizer:

Application of microorganisms could promote the nutrient absorption of roots, improve the nutrient availability, protect the root system, and promote its growth, etc. It is an important method used in rational fertilization.

Other Means of Soil Management and Fertilization Skills:

Soil physical improvement and chemical properties is beneficial to the fertilizer efficiency. For example, deep plough, irrigation and drainage are all helpful. The water contributes to the nutrient absorption, promotes the movement and diffusion of nutrients in the soil, and supplies sufficient nutrient for plants. Therefore, sufficient supply of water is rather important to the fertilizer efficiency.

7.19 Soil Medium and Fertilization Management in Rose Cultivation

Rose is a kind of perennial and evergreen woody plant, so it has its own characteristics for the soil, medium and fertilization management. There are many varieties of roses. According to plant types, there are erect shrubs, climbing or trailing with stems. There are also high roses and short ones. In terms of the flowering characteristics and its growing season, different types also differ from each other. Therefore, in fertilization management, attention should be paid to its woody, flowering and growing characteristics. In excellent soil or medium, if coordinated with proper fertilization management, the rose cultivation would become rather easier. If the rose is cultivated in poor soil, it is necessary to strengthen the improvement of the soil, so as to reduce the occurrences of pests and diseases for plants and then cultivate the roses with high proficiency.

7.19.1 Soil Adaptability of Rose

The most suitable soil for the growth of rose is the soil layer between 35 and 40 cm and requires such following factors: (1) the texture of loam; (2) pH value between 6 and 6.5; (3) good drainage; and (4) rich amount of organic matter (> 2%). Therefore, if the soil is in poor conditions, it should be improved for yielding double result with half the effort in the cultivation.

7.19.2 Soil Improvement during the Cultivation

The soil layer for the rose cultivation should be rather deep, so the depth for the improvement should better be at least 34 to 40 cm. The soil on the seedling bed can be fully mixed with 70% soil together with 10% compost (such as cow dung compost and peat), 10% bone meals and 10% residues (such as soybean meal or other meal). After being fully mixed, it can be used on the soil. After the organic material is added on the soil, the seedling bed should be made by ploughed and turned over, so as to fully mix them, and after more than 10 days, it can be used for rose cultivation. In the soil improvement in the seedling bed, some chemical fertilizer and microbial fertilizer can also be applied.

7.19.3 Approaches to Improve the Poor Soil Properties

Soil improvement requires prior understanding and analysis of the cultivated land. Afterwards, problems can be detected and it can later be determined whether the land needs improvement or not. The following are explanations of the common approaches to improve the poor soil, so as to make the soil suitable for the cultivation of roses.

Improvement of the Soil Texture:

If the soil is clay, coarse sand and organic matter (such as peat and compost) can be added and mixed to loosen the soil. Conversely, if the soil is of the sandy type, fine soil and organic matter (such as peat and compost) can be added and mixed to increase the soil water and fertilizer conservation capacity.

Improvement of the Acidic and the Alkaline Soil:

If the soil pH value is lower than 6, organic matter and lime materials (such as magnesium lime, oyster shell powder and slag) can be mixed to make the improvement. Conversely, if the soil pH value is higher than 7.5, organic matter and acidic material (such as sulfur powder and peat) can be mixed to the soil for the improvement. The added amount of neutral materials need to be tested by the pH meter and then correctly applied.

Improvement of the Incompletely Drained Soil:

The incompletely drained soil can be improved through setting up a higher plot of land with furrow drainage together with exposed pipes. Underground pipes can also be used to drain off water.

Improvement of the Soil with Low Organic Matter Content:

If the soil organic matter content is less than 2%, which are not easily decomposed materials (such as peat, bark compost, and cow dung compost) can be applied to make the improvement.

Soil Improvement for Continuous Cropping:

When the soil quality in the garden declines, the soil should be kept improving for 3 to 5 months, so as to be not inhibited the growth for continuous cropping. The approach of improvement is to mix microbial fertilizers with organic fertilizers (such as compost and bone meal). Then exchange the surface soil with the subsoil or plough and turn over the soil to make the improvement. Lime materials can be applied together with organic fertilizers in the improvement of the acidic soil. During the improvement period, the soil should be dried at least once.

7.19.4 The Medium and its Features for Potted Rose

Rose is a perennial flower. Since the soil and the medium in a pot are limited, it is necessary to pay attention to water conservation, fertilizer conservation and loose of the soil property. The soil as the materials of the mixing of potted media proves the longest cultivation. If only the single peat moss is used, the rose may wither up due to everyday negligence in watering or maintenance. In mixing of potted media, 70% loam soil, 10% peat moss, 15% cow-dung compost (do not add compost containing too much chemical fertilizer), and 5% dreg and bone meal are appropriate. The standard for an excellent cultivation medium is in good physical, chemical and biological property, as listed below:

Proper Physical Property:

The cultivation medium needs proper physical property, including porosity, texture, structure, and loose. The physical property influences the medium water holding capacity, drain ability, adhesive ability of rose roots and rooting rate. Water holding capacity further influences the frequency of irrigation. Poor drainage would influence the rooting and the growth of plant root system. Adhesive ability would influence the adhesion of the medium to the root, when the seedlings are taken out from pots. It may also influence the survival ratio of field cultivation. Therefore, good physical property of the medium requires proper water holding capacity, good drain ability and aeralation ability, high rooting rate, and normal growth of the seedling. To take out the seedling completely and easily (medium do not be loosened) from pots is also required. It is related to whether the rose root system grows well or not.

Proper Chemical Property:

The cultivation medium needs proper chemical property, including pH value, electrical conductivity, cation exchange capacity, nutrients and other chemicals. The chemical property influences fertilizer conservation capacity, osmotic pressure, and nutrient-supply capacity of the cultivation medium. It thus can further influence the growth rate, height, solidity, and branch type of rose plants, etc. In the cultivation of nursery, differences in the amount of nutrient supply and the supply period could have decisive effects on the transplanting success and the adaptive ability of seedlings. Excessive growth of seedlings reflects unhealthy transplant. Therefore, the chemical property of a good medium is required for producing healthy, properly-growing and well-adaptive plants. Sandy loam or their medium can be generally adopted as the cutting medium. The mediums containing too much residuals or non-decomposed organic matter are not preferred since they may decrease the survival ratio of cuttings greatly.

Good Biological Property:

The cultivation medium needs sound biological property, including beneficial bacteria, such as rhizosphere-protective bacteria, mycorrhiza fungi, phosphate solubilizing bacteria, nitrogen fixing microorganisms or detoxifying bacteria, etc. The biological property of the medium influences the plant rooting rate, nutrient absorption, growth, transplanting success, and pest and disease resistance, etc. Good microorganisms in the medium would decrease pests and diseases in seedlings. If there are pathogens, pest eggs or induce factors of disease in the medium, this medium is a poor mediums. Healthy microorganisms in the medium may play a protective guard for plants. Therefore, good and healthy seedlings should not only have strong outward appearance. More importantly, they need to be inoculated with beneficial bacteria into the rhizosphere and the endophytic part. This has close relationships with the transplanting success and the growth after transplanting.

7.19.5 Concepts and Methods of Fertilization

The plant nutrients are mainly nitrogen, phosphorus, and potassium; next are calcium, magnesium, sulfur and some other microelements, such as boron, zinc, manganese, copper and molybdenum, etc. Nitrogen is an essential component of the plant chlorophylls, proteins (including enzymes), and nucleic acids. It facilitates the fast growth of leaf and stem, and is also a nutrient easily lost from the soil. However, too much nitrogen would lead to excessive lateral buds growth around the rose plant, which decreases the quality of flower. Phosphorus fertilizer is an essential part of nucleic acids, ATP, and metabolic reactions. It is an important element which assists flowering, not easily lost from the soil but easily fixed. It is less efficiently absorbed by the plant. Potasium fertilizer possesses special functions of regulating plant osmotic pressure and assisting special functions of nitrogen and phosphorus. Other secondary and micro elements are nutrients which cannot be neglected in the plant and need to be paid attention to.

The ways of fertilization to the rose are listed as follows:

Soil Fertilization:

Apply the solid or liquid fertilizer to the soil surface. The fertilizer can be ploughed into the soil or permeate into the soil after irrigation.

Foliar Application:

Dissolve the fertilizer in the water and spray it onto the leaves of plant. The leaves of rose are waxy. When foliar application is conducted, the surfactant (to lower the surface tension of a liquid) should particularly be added to the nutrient solution.

Mulch Fertilization:

Mulch the organic materialss, such as compost, leaf residues, and peat onto the soil surface. After spraying and watering onto mulching materials, plant nutrients can be washed into the soil for plant absorption.

7.19.6 Types and Methods of Fertilization

There are various types of fertilizers. Generally, fertilizers could be classified into three types: chemical fertilizer, organic fertilizer and biofertilizer. Each of the three fertilizers has their own characteristics and they can be used in a coordinated way so as to achieve complementary effects. Fertilizer forms include mainly solid fertilizer and liquid fertilizer. Solid fertilizer cannot be spread immediately in the soil while liquid fertilizer could be spread and then they move promptly in the soil and approach to the rhizosphere. The liquid fertilizer had better not have high concentration; otherwise, it may lead to fertilizer injury.

Chemical Fertilizer:

Types:

Chemical fertilizers include straight (single) fertilizers and compound fertilizers. Straight fertilizers contain only one primary nutrient element while compound fertilizers are two or more primary nutrient elements which were made from mixing straight fertilizers or by chemical syntheses of other raw materials. The advantage of straight fertilizer is that it leaves easily for the adjustment of nutrient ratio in a typical fertilizer, while compound fertilizer has a fixed proportion of nutrients. However, it is sometimes inconvenient to get a right nutrient ratio for a typical crop. The proportion of nitrogen, phosphorus and potassium is signified by N-P_2O_5-K_2O. High-concentrated nitrogen and phosphorus are often applied in rose field, for example, 16-16-8 or 8-8-4. Microelements are often included in foliar fertilization as supplement for the insufficient soil absorption.

Amount of Fertilizers:

Fertilizer amount is determined according to the amount of effective nutrients in the soil. Generally, if the proportion 16-16-8 is employed, the fertilizer could be applied three to five times a year, or it can be used many times but each times only with a little proportion. However, before rainy seasons, the fertilizer had better not include high nitrogen concentration.

Fertilization Period:

With the florescence as the rule, fertilizer should be spread six weeks before and it can thus exercise a positive effect on the quality of flowers. Accordingly, with regard to different regions, varieties and greenhouses, the florescence in the whole year should be recorded for the further planning of the fertilization period.

Application Methods:

Generally, irrigation should be conducted one day before chemical fertilizer is applied, so as to avoid fertilizer injury for the root. The surface soil should be softly pushed aside, and then apply the fertilizer with a distance of > 10 cm to the stem of rose. After that, spray some water or water them.

Organic Fertilizer:

Types:

There are a great number of organic fertilizers. They mainly come from residues and wastes of animals, plants and microorganisms, including dung, dreg, animal powder (fish meal and bone meal), compost, and peat. There are liquid organic fertilizers as well, such as fish meal, humic acid, seaweed and amino acids, etc. They work very quickly. Granular organic fertilizer now has compounded with a variety of raw materials. It is convenient to use and has its own characteristics.

Amount of Fertilizers:

The fertilizer amount is determined according to the amount of the original organic matter in the soil. This fertilizer can be applied once a year. The liquid organic fertilizer could be applied once every one to two months.

Fertilization Period:

Organic fertilizer ought to be applied at least eight weeks before the year budding period or be applied after the florescence.

CH7 Guidelines of Improvement of Fertilization and Soil Management   277

Application Methods:
Organic fertilizer can be applied into the soil or put on the surface soil as mulch. To apply the fertilizer into the soil requires at least 5 to 10 cm digging into the soil. After the organic fertilizer is applied, the plant should be earthed up. The organic matter covering the surface soil should not include the dung which is prone to fly.

Biofertilizer:

Types:
Biofertilizers are mainly active microorganisms and there are a lot of types. Biofertilizers applied in rose cultivation include nitrogen fixing, phosphate-solubilizing bacteria, and mycorrhizal fungi, etc, which not only help plants absorb nutrients and keep them healthy, but also reduce the use of chemical fertilizers, thus diminishing soil acidification, soil degradation and soil salinization. This really achieves many things at one stroke.

Amount of Fertilizers:
The fertilizer amount of biofertilizer should cover at least 20% of the rose root in order to produce significant effect. Therefore, when irrigation is conducted, the fertilizer amount should be 8 ~ 20 L of liquid or 2 ~ 4 kg of powder per 0.1 ha. The amount differs based on the number of bacteria and the age of rose plants.

Fertilization Period:
Biofertilizer could be applied at any period, but the best application period is 4 ~ 6 weeks before the florescence. It could be applied 3 to 6 times per year. Mycorrhizal fungi are symbiotic microorganisms. The application should be inoculated at the seedling stage in order to get best effect.

Application Methods:
It is a common to dilute the original liquid or powder of biofertilizer. The dilution rate is determined according to each specification of each product. Generally, the liquid is diluted at 200 to 1,000 multiples and the powder is diluted at 1,000 to 3,000 multiples. The diluted liquid could be directly applied into the rhizosphere. After the rain, liquid with higher concentration and then less irrigation can be conducted. Biofertilizer is viable microorganism and in order to bring its best effect, humic acid and other organic liquid fertilizers are often used together with biofertilizer. To make biofertilizer go deep into the rhizosphere of plant, the biofertilizer should be applied with less water in the soil or several days after the irrigation.

7.19.7 Summary

Rose is kind of perennial crops. No matter it is growing on the land or in the pot, it is necessary to improve the soil or the medium, since it is especially essential for the soil and provide the medium with adequate nutrients, water and oxygen all the year round. Furthermore, sound physical, chemical and biological properties of the soil and the medium could help provide a healthy environment. Management of soil fertility should be conducted in coordination with chemical fertilizer, organic fertilizer and microbial fertilizer. Only by bringing the integral function into play in an unbiased way can the high productions and fine qualities of rose cultivation be achieved.

7.20 Cover Grass Cultivation and Management of Soil Fertility

7.20.1 Functions of Cover Grass Cultivation

There is a wide definition of grass. Referring to Gramineae family, usually herbaceous plants are with narrow leaves. Or, other refers to all the herbages except crops. Vegetation is the ground cover provided by plants in one specific area, particularly those grasses, shrub or tree growing on the soil surface. Weed is generally known as herbs growing in an undesirable place. However, since weeds contribute to both crop growth and soil and water conservation on the slope lands, their existence is also considered to be beneficial. Therefore, in terms of agriculture, grasses could be classified into beneficial ones and harmful ones according to different places, sites, main crops and types of grasses.

Grasses could be used for cover, mulch and being ploughed into soil for agricultural production. Especially, grasses could play important roles in the orchard of perennial crops. Fruit tree in the orchard have a wide space between each plant, and grasses grow just between those spaces of tree. This is called cover grass cultivation, which is thought to be conducive to soil fertility management. Grasses benefit soil fertility greatly, such as improving physical, chemical and biological properties of the soil. Physical property Improvement includes increasing more rain water to be permeated into the soil, reducing floods, enhancing the water-retaining capacity, improving the aggregated structure of soil, preventing soil surface washout and superficial landslide, slow down drastic changes of ground temperature, reducing the acute growth and absorption of crops, and preventing the reduction of organic matter content in soil due to high ground temperature. Improvement of soil chemical property mainly includes increasing organic matter in soil and then influencing the physical, chemical and biological properties of soil. In rainy seasons, loss of fertility is severe. Grasses may absorb many nutrients. After mowing, they decompose and release many nutrients for the crops to use. This phenomenon could be considered as nutrient storage in grass plants, which contributes to long-term fertilizer recovery rate increment for crops.

In crop fertilization management, chemical fertilizer is always chosen by farmers. Holes of grass root system help the fertilizer go into the soil layer, thus reducing the fertilizer loss through runoff water. Grass management could be considered to be green manure application. If coordinated with crops and season management, and with the reduction of use of herbicides, for the crops and conservation of soil, it can bring the effect of cover grass cultivation better into play, which can answer multiple purposes.

7.20.2 Effects of Grasses on Soil Physical Property

The effects of grasses on soil physical property are in three aspects: vegetation cover, vegetation mulch and ploughed into the soil.

Effects of Vegetation Cover:

The effects of vegetation cover mainly from extension of the grasses root system and action of microorganisms around the rhizosphere, which can raise soil porosity and stability of rhizosphere particles. Therefore, vegetation cover tends to increase water permeability of soil on slope fields notably, thus decreasing the surface runoff volume on slope lands and then achieving the purpose of the water and soil conservation. Grasses are commonly growing in the orchard. Since the physical property obstructs sunshine, grasses could lower the soil temperature and reduce a great deal of soil organic matter decomposition. At low temperatures, grasses could reduce heat radiation, adjust soil temperature and thus reduce the temperature difference between day and night, so as to diminish summer damage and winter damage for crops. Water conservation and aeralation indirectly influence the availability and utilizability of soil nutrients and thus need to be coordinated with soil fertility maintenance.

Cultivation of grasses on the slope land could increase porosity and water permeability of soil physical property. When spreading fertilizer, they can help the fertilizer go deeper into the soil. On the one hand, they can help conserve water and soil, reduce the fertilizer loss and maintain the soil fertility. They can give full play to the use of fertilizers and then facilitate the growth and production of crops.

The Effect of Vegetation Mulch:

Mulching is to cut off grasses and then to put them on the soil surface. Its effect on soil physical property is mainly to reduce the loss of surface soil and water. Improvement of soil water-retention capacity can help preserve nutrients in soil and increase their availability. Vegetation mulch can help increase the amount of organic matter in soil, promoting the structure and stability of soil particles. Stubble mulch possesses good ability in absorbing and retaining water, which could raise the utility value of water and soil conservation of agriculture on steep slopes. Therefore, vegetation mulch effect on the physical property of soil could finally achieve the purpose of improving soil fertility.

The Effect of Plowing Grasses into the Soil:

Grasses could be easily ploughed in paddy fields, while it is inappropriate on slope lands since plowing grasses may lead to loss of water and soil. The effect of plowing grasses into the soil on the physical property of soil is mainly to loosen the soil, to improve porosity of soil and structure of particles, and then to have sound effects on promoting soil water which retains capacity.

7.20.3 Effects of Grasses on Soil Chemical Property

The effects of grasses on soil chemical property are in the three aspects: vegetation cover, vegetation mulch and plowing grasses into the soil.

The Effect of Vegetation Cover:

The effect of grass cover has on soil chemical property is mainly the activity of grass residuals and the root system. Although at the early stage of grass growth, grass cover may compete for and consume soil fertility of the main crops. In particularly for those perennial crops, they could improve the soil fertility, when grasses are aging, and the residuals are decomposed by microorganisms. The grass could accumulate a great deal of residuals, over a long period of time, outstandingly increasing the soil organic matter. If the grass belongs to the bean or pea family, such as hairy vetch and subterranean clover, it can help increase total nitrogen of soil. The whole-year grass cover can increase the organic matter in soil and improve the function which influences soil fertility, its physical and chemical property, and biological property. Among these, there are direct and indirect influences on the soil fertility, including improvement in the storage capacity, exchange capacity and availability of nutrients. The improvement is particular in the availability of phosphorus increment and zinc fertilizer. Organic matter in soil may form a complex with phosphorus and metal ions of microelements, and thus improves the availability of phosphorus and metal ions of microelements.

The Effect of Vegetation Mulch:

The effects of grass mulch on soil chemical property come mainly from effects of matters brought by leaching, infiltration, and decomposition of residuals. Organic matter content in orchards on the slope fields and their strong acidity are limiting factors for crop growth. Grass mulch could increase organic matter and provide lots of decomposed and released nutrients. It is often used in farmlands and orchards on the slope fields. For example, after Bahiagrass is mulched, leaching and decomposition of organic matter could increase sources of nutrients and total nitrogen in the soil.

The Effect of Plowing Grasses into the Soil:

When the grass has been brought into the soil through plowing, it can decompose and release nutrients faster than those grasses growing on the soil surface. It can influence soil chemical property directly as well, including pH value and Eh value. At the early stage of decomposition, the grass would immobilize the nutrients, in particular nitrogen. This is called N-immobilization. After grasses are decomposed by microorganisms, the mineralization can be occurred. At the later stage of decomposition, the mineralization may increase nutrients and their availability, and the effect of increasing organic matter in soil. Its effect is similar to that of green manure. If it can be utilized together with grasses in the fallow period of land, it can decrease use of herbicides. For soil and environmental protection, it can serve for multiple purposes.

7.20.4 Effects of Grasses on Soil Biological Property

The effects that grasses have on soil biological property are mainly from grasses roots and the role of residual decomposition. Since soil microorganisms could get carbon and energy of the secretion and residual from plant roots, soil microorganisms are more active in rhizosphere than in non-rhizosphere. Where there are grasses, there are earthworms. This can bring the grass residuals into the lower part of soil. This could help increase organic matter content of soil.

Phase transition of soil microorganisms is influenced by soil from plowing land. After those fresh grasses get into the soil, they help increase the action of soil microorganisms, leading to the result that 80% of the organic residuals decomposed in the first year due to mineralization. This may release a lot of nutrients for crops, thus improving fertility of soil. Further researches should be made for investigating the effects of grasses on the phase transition of soil microorganisms.

Chapter 8

Soil Organic Matter and Organic Fertilizers

8.1 Introduction to Soil Organic Matter

8.1.1 Soil Organic Matter Content is a Standard for Good Agricultural Soils

A good agricultural soil consists not only the three essential conditions for fertile soil -- nutrients, water and oxygen -- that can fully provide crops need, but also contains favorable soil physical, chemical, and biological properties, which makes farmland easier to be managed and reduce plant diseases and pests so as to spare human efforts and keep farmland productive. For example, soil with high water-holding capacity does not need frequent irrigation and can resist long drought period, while soils with high nutrient holding capacity does not require constant fertilizer application and does not tend to suffer from fertilizer injury. Good buffering protects soil from easy acidification. Strong water-draining strength avoids water accumulation and increases water infiltration. Disease-inhibiting capability prevents plant diseases and insect pests, and reduces corresponding loss and pesticide application. There are many methods for improving soil to a good agricultural soil. Among all, increasing soil organic matter content is an important one. To achieve this, the most available and direct way is to apply organic fertilizers. Soil organic matter, with many advantages, is the characteristic of a good agricultural soil. Healthy crops need healthy soils, and good soil requires sufficient organic matters.

The components of soil include inorganic matter, organic matter and microorganism. Soil organic matter refers to the organic substance after decomposition in soils and it contains two kinds of substances: Humic and nonhumic substances (Figure 8-1). Humic substances are just a part of soil organic matter. The nonhumic substances refer to the residues of animals, plants and microorganisms and the decomposed compounds, which are in a free state and not polymerized in soils, such as carbohydrates, amino acids, fats, proteins, nucleic acids, etc. Residual mass should not be regarded as humic substances, because humic substances are high molecular weight compounds that are composited and polymerized from decomposed matters. Due to the distinction between polymerized matters and the consuming time for forming humic substances, the characteristics of the polymerized matters are also different. Generally, humic substances can be divided into three types of polymers according to their solubility characteristics during extraction: Humic acid, fulvic acid and humin. Humic acid is a kind of acid in humic substances that is soluble in alkaline solution but not soluble in strong acidic solution (pH 1). Fulvic acid is soluble in both alkaline, and acidic solution and humin cannot solubilize in either acidic or alkaline solution.

Narrowly speaking, soil organic matter does not include residues, litters, and soil organisms. However, when measuring the content of soil organic matter, it is usually estimated in accordance with organic carbon amount in the soil. By this method, the measured soil organic matter refers to a soil organic matter in a broad sense, including residues and microorganisms. Sampling of soil often contains a great deal of broken roots and plant residues; hence, a seemingly high content of soil organic matter is often mistaken as the soil organic matter.

For example, organic fertilizers such as compost and animal manure cannot be regarded as humic substances and humic acid, because it contains many components of nonhumic substances that are easy to be decomposed in soil. Peat soil has high

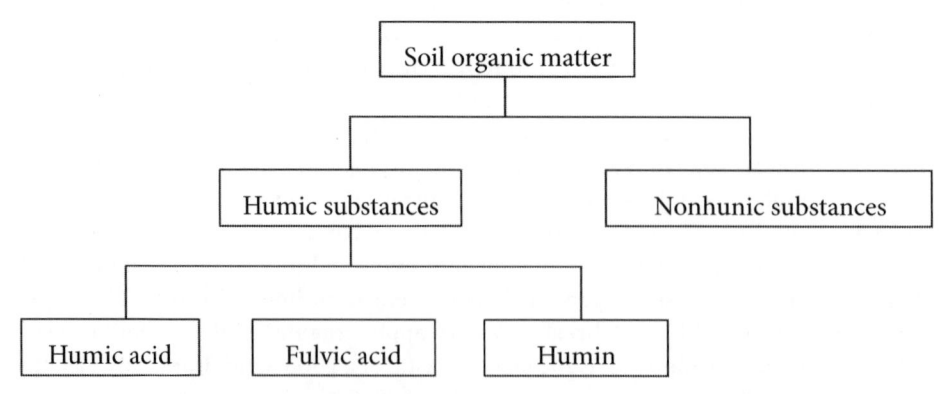

Figure 8-1. Major Soil Organic Matter Components.

amounts of humic substances and other inorganic matters. The common fibrous wood materials are, in fact, not peat or humic substances. Various effects are caused by different organic matters and needed to be distinguished.

8.1.2 The Best Feature of Soil -- Humic Acid

In recent years, the application of humic acid has been developing very rapidly and gained worldwide attention as an active ingredient in the field of organic fertilizer. However, humic acid is different from soil organic matter, peat, and humic substances, but people often get confused with these terms.

The Origin of Humic Acid and its Characteristics:

Humic acid exists in soil and coal mine layers, and also in seawater and fresh water sediments. Normally, there are high contents of humic acid in peat, and in mine layers with > 80% humic acid according to a report. Many agricultural soils have a low content of organic matter (about 2%) with a relatively small amount of humic acid (approximately lower than 0.5%).

Humic acid consists of various phenolic polymers with nitrogen element. There are many ways to to form soil humic acid, mainly through the polymerization of decomposed residue products via microorganism metabolic enzymes reaction or soil catalysis. Besides main phenolic compounds, humic acid also consist of amino acids, purines, sugar acids, amino sugars, pentoses, hexoses, aldoses, methyl carbohydrates, fatty acids, etc. Natural humic acid has weathered through thousands of years of modification and combination to form into complexes with amorphous structures and is of higher molecular weight than fulvic acid. Humic acid has a relatively dark color with hydrophilic and acidic characteristics, yet with less carboxyl and hydroxyl functional groups compared to fulvic acid.

The surface of humic acid molecules is full of oxygen-containing groups, which can react, chelate or adsorb metallic and nonmetallic substances. In terms of metallic elements, various metal ions, metallic oxides, hydroxides and metal organic compounds can react with humic acid. Soil humic acid also contains microelements. Humic acid can fix excessive iron, aluminum, manganese, zinc and other elements in the soil. Furthermore, phosphorus fertilizers are easily fixed by iron, aluminum, and manganese. With these ions or oxides fixed by humic acid, the system can improve the availability of phosphorus fertilizers.

Humic acid can also react or adsorb organic compounds. For example, humic acid can easily adsorb many pesticides and change their function and mobility. Humic acid can also adsorb other organic matters, such as urea, amino acids, proteins, fatty acids, and carbohydrates; therefore, humic acid extracted from natural products contains various kinds of organic matter complexes.

Another significant and valuable characteristic of humic acid is its ability to resist decomposition by soil microorganisms. This special characteristic enables humic acid to remain stable for a long period of time in soil and serves as a long-term organic matter.

Functions of Humic Acid:

The functions of humic acid are beneficial to crops and especially to the soil listed as follows:

- Decrease soil compaction.
- Improve water mobility in soil.
- Improve the availability of P, Fe and Ca.
- Increase microorganism community in soil.
- Increase ion exchange and nutrient holding capability of soil.
- Increase soil buffer capability.

The importance of humic acid's function to crops is clearly mentioned above. Humic acid is the essence and play a very significant role in agriculture and the soil. It provides much convenience for cultivation management, which is necessary for agriculture and the soil. Over million years of improvement, the soil humic substances increment in the topsoil is most valuable treasure. The soil that lacks humic substances or humic acid demands more efforts and costs in soil and crops care in the cultivation management.

Proofs and experiments have shown that plants can absorb humic acid and its effects have revealed on plant physiology and metabolisms. Due to the distinct origin and extraction techniques of humic acid, there are differences in their structures, leading to a direct different physiological effect on plants.

Many studies have discussed the direct effects of humic substances on crop metabolisms. These metabolic processes include respiration and synthesis of nucleic acids and proteins. The influence of humic acid on plants is illustrated in Figure 8-2 with the following specifications:

- Affect on osmotic pressure of root cell membrane and ion carriers which results in the rapid and selective entry of necessary elements.
- Improve respiration and metabolic processes in crops, so as to increase ATP synthesis.
- Increase chlorophyll content and photosynthesis, so as to increase production of ATP, amino acids, carbohydrates and proteins.
- Improve nucleic acid synthesis, including DNA and RNA.

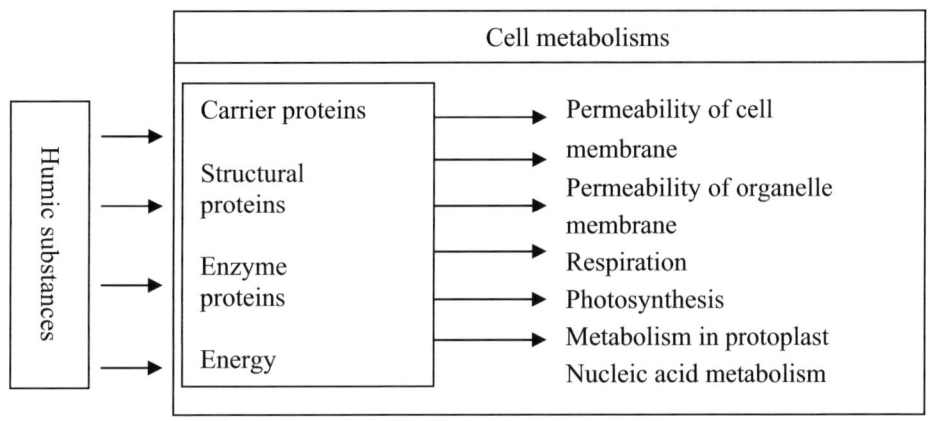

```
┌─────────────────────────────────────────────────────────┐
│                    Cell metabolisms                     │
│  ┌──────────────┐                                       │
│  │ Carrier      │ ──→  Permeability of cell             │
│  │ proteins     │      membrane                         │
│  │              │ ──→                                   │
│  │ Structural   │      Permeability of organelle        │
│  │ proteins     │ ──→  membrane                         │
│  │              │                                       │
│  │ Enzyme       │ ──→  Respiration                      │
│  │ proteins     │ ──→  Photosynthesis                   │
│  │              │ ──→  Metabolism in protoplast         │
│  │ Energy       │      Nucleic acid metabolism          │
│  └──────────────┘                                       │
└─────────────────────────────────────────────────────────┘
```

Figure 8-2. An Illustration of Humic Substances' Effects on Plant Cell Metabolisms.

- Affect protein synthesis and increase the production of enzyme, ion carriers, structural proteins, etc.
- Increase or inhibit enzyme activity based on different types and origins of enzymes.
- Show similar function as plant hormones.

Maintenance of Soil Humic Acid:

Agricultural topsoil contains certain levels of humic acid that has been accumulated for a very long period. Tens of millions years can only lead to a little amount of humic acid accumulation. Although humic acid cannot be easily digested by microorganisms, it can still be utilized and decomposed. Therefore, the content of soil humic acid or humic substances demands careful maintenance. Otherwise, inappropriate soil management will accelerate the decomposition and decrease the humic acid or humic substance contents in the soil.

To maintain, preserve or increase humic acid content in soils, the following methods should be noted:

To Apply Organic Fertilizers or Humic Acid:

During the decomposing process, a portion of the organic fertilizer undergoes mineralization into inorganic nutrients, which is necessary for crops. Another portion of the organic compounds are polymerized into high-molecular weight molecules, which are then modified and combined into humic acid or other compositions of humic substances. Apart from ordinary organic fertilizers, commercial humic acid or peat can also be applied.

To Plant Green Manures:

Green manures can provide an environment for the mass proliferation of microorganisms and offer abundant combined or polymerized products after digestion for the synthesis of humic substances. Among green manures, legume crops are available in increasing the nitrogen source and promoting the formation of humic acid or humic substances.

To Adopt Rotation Cropping System:

Lowland and upland rotation cropping should be adopted for farmland cultivation. Upland farming reduces the organic matter content, while wet field lowland cropping is beneficial to the preservation of organic matters. Therefore, wet field/upland rotation cropping system assists the production and maintenance of soil humic acid.

To Cover the Soil:

The decomposition rate of soil organic matters is strongly influenced by the temperature. Topsoil covering with cover crops has many advantages, such as decreasing soil organic matter decomposition rate and providing favorable factors for humic substances formation.

Attentions for Humic Acid Application:

There are many functions in soil humic and it is convenient for application without limitation. To achieve the greatest availability, the application of humic acid should be paid close attention as follows:

To Combine Humic Acid with Chemical Fertilizers:

The function of humic acid is to improve the soil and enhance plant growth. However, the supply of nutrients such as nitrogen, phosphorus, and potassium from the product of humic acid is low and limited, unless inorganic chemical fertilizers have already been added in. If we need to supply large amounts of nitrogen, phosphorus, and potassium for plants, one cannot completely rely on the application of humic acid. During crop growth period, great amounts of nitrogen, phosphorus and potassium are necessary, humic acid should be applied together with fast-effective chemical fertilizers to make crops flourish. Some products of humic acid were mixed with microelements, which provide multiple functions.

Notice the pH Value when Applying:

Humic acid is an acid matter in nature. Some products of humic acid are adjusted to basic, and thus provide similar effects as lime. When foliage spraying is applied, high acid or basic solution may show adverse effects on the leaves, and thus the spraying solution for leaves should be as neutral as possible.

Do not Mix Humic Acid with Pesticides that can be Adsorbed:
Many pesticides can interact or adsorb humic acid and change the effect of the pesticide. Therefore, in order not to reduce the effect of the pesticide, humic acid should not be mixed with pesticides in application.

To Apply for Upland Cropping:
Fruit tree and upland crops have higher demand for organic matter in soil than wet field crops, while soil organic matter is less available to aquatic crops. Humic acid has a better availability on non-wet fields, especially on slope land orchard soils, of which the organic matter is deficient.

Insoluble Humic Acid are not Suitable for Spraying:
Insoluble humic acid or humic substances are not applicable to foliage spraying because they form a dust-like cover on leaves that prevent crops from absorbing solar energy, and inhibit crop growth.

8.2 The Characteristics and Applications of Peat

8.2.1 Many Agricultural Soils Require Organic Matter

In hot and rainy subtropical and tropical regions, soil organic matters decompose rapidly and their contents in farmland are generally low. According to a previous survey, many farmland soils contain lower than 2% of organic matters. Besides, after years of farming, slope lands have a lower content of organic matter, particularly for the subsoil (about 15 cm or more beneath soil surface) in orchards with 1% of organic matters or lower. The soil organic matter content is an important indicator of soil fertility. Soil deficient in organic matter will result in weak preserving capability of water and fertilizer, reduce soil buffering ability, and cause soils to be damaged by harmful substances (such as pesticides, chemical fertilizers, waste water, etc.). These disadvantages will affect crop growth and quality in their natural environment. In order to improve the physical, chemical and biological properties of soil, application of organic fertilizer is the most efficient way.

8.2.2 The Formation and Characteristics of Peat

Peat is the most important and universally applied organic material in the pot culture. It has been used in vegetable and plant production as early as the 18th century. There are many studies and researches on peat.

The formation of peat is relevant to that of coal and petroleum; they are all formed from plant material under anaerobic or low oxygen conditions for several thousand years. The procedures are listed as follows: (1) plant material → humic acid and humin (peat) → ignite → bitumite → anthracite → graphite; (2) plant material →

sediment → humic acid, fulvic acid, humin, kerogen, petroleum. Peat is with low pH value. Owing to the differences in decomposing degrees and plant origins, there are various kinds of peat, which can be generally classified into sphagnum peat (peat moss), swamp peat and wood peat. In Table 8-1, high-maturity peat has a larger content of carbon, nitrogen and sulfur than sphagnum peat, but with a lower content of oxygen. If compared with ignite and bitumite of high carbonization degree, the distinctions are more obvious.

The composition of common peat consists of carbon (~ 50%), nitrogen (~ 1%) and mineral contents (~ 3.7%; Table 8-2). Peat consists of high contents of humic acid, fulvic acid and humin. Sphagnum herbs, bryophytes or woody plants can form peats of different physical and chemical properties. The content of humic acid in peat produced from common mosses is only about 5%, whereas the content in typical peat from woody plants is about 5 ~ 25%, or in some cases higher.

The most important characteristics of peat as a suitable potting medium is its high water holding capability, good aeration, low density (fine looseness) and high cation exchange capability (good potential for fertility preservation). Peat cannot be easily decomposed and it possesses abundant plant nutrients and carbon materials (see Table 8-2). These qualities are most applicable for soil conservation and the improvement of agriculture's quality and growth, horticulture, and livestock farming.

8.2.3 The Slow Decomposition of Peat is the Best in Organic Fertilizers

There are substantially differences in many kinds of organic fertilizers regarding to the decomposition rate. Common organic fertilizers cannot persist in soil for long time, as most organic fertilizers would be decomposed and exhausted within a year. Only a small fraction would remain in the soil. Therefore, large amounts of organic fertilizers should be applied each year. In contrast, peat is different from those easily decomposed common organic fertilizers. Peat is formed from ancient grass and

Table 8-1. Elemental Analysis of Various Peats and Coals

Types	Carbon	Hydrogen	Oxygen	Nitrogen	Sulfur
			%		
Sphagnum Peat	48 ~ 53	5.0 ~ 6.1	40 ~ 46	0.5 ~ 1.0	0.1 ~ 0.2
Mature Swamp Peat	56 ~ 58	5.5 ~ 6.1	34 ~ 49	0.8 ~ 1.2	0.1 ~ 0.3
High-Maturity Peat	59 ~ 63	5.1 ~ 6.1	31 ~ 34	1.0 ~ 2.7	0.2 ~ 0.5
Ignite	70 ~ 80	4.7 ~ 5.0	20 ~ 25	1.0 ~ 1.5	0.6 ~ 0.8
Bitumite	80 ~ 85	5.0 ~ 6.0	9 ~ 12	1.1 ~ 1.4	1.0 ~ 2.0

wood over millions of years, and thus peats will not decompose easily (properties as in Table 8-2). Applying peat is like depositing money. A large portion of peat can be reserved in soil for improving the physical, chemical and biological properties of the soil in a long effective period of time. Peat can be regarded as one of the best solid organic fertilizers. On the other hand, peat moss is easier to be decomposed than peat soil with a half-life period (time for 50% decomposition rate) of about 1 ~ 5 years in accordance with different soil conditions.

8.2.4 Applications of Peat and its Advantages in Agriculture

The application of peat is very broad. Peat is usually used as a soil conditioner, as fuel, and for the extraction of humic acid used in agricultural and engineering purposes. Advantages of peat are specified by following features in its agricultural application:

- Peat can resist decomposition by microorganisms. The best among organic fertilizers and a long-term soil conditioner.
- Cation exchange capability is the best indicator for nutrient holding capability. Peat has a high cation exchange capability of about 121 ~ 224 cmol(+)/kg soil (at pH 7). Therefore, it can be used as a slow-released organic fertilizer for nitrogen and phosphorus.

Table 8-2. Composition and Properties of Common Peat

Properties	Average	Range
Carbon Content (%)	50.69	46.7 ~ 54.9
Carbon Nitrogen Ratio	45.64	15.4 ~ 99.2
Mineral (%)	3.70	0.3 ~ 8.9
Calcium (mg/g)	6.92	1.0 ~ 15.0
Total Sugar Amount (mg/g)	35.50	13.8 ~ 81.3
pH Value	3.76	3.0 ~ 5.2
Cation Exchange Capability (cmol(+)/kg)	148.50	121.0 ~ 224.0
Density (g/mm^3)	0.11	0.064 ~ 0.195
Water Holding Capability (%)	616.10	366.0 ~ 937.0
Initial-Period Decomposition Ratio (daily decomposed carbon amount, mg/kg)	124.80	59.8 ~ 319.2
Later-period Decomposition Rate (unit as above, mg/kg)	55.10	22.3 ~ 207.8

- The loose property of peat can improve soil aeration and water permeability. Hence, it shows great water holding capability.
- Peat is rich in natural nutrients, including various macro and microelements (Fe, Mn, Co, Zn, etc.) which are necessary for plants, and organic matters (humic acid, fulvic acid, humin, amino acids, carbohydrates, etc.). In addition to plant nutrients, it may contain "healthy material." However, further study is required. Healthy plants have relatively stronger power to prevent diseases and pests.

8.2.5 The Effects of Peat

Peat is an organic matter that is not easy to be decomposed. It has the following effects for improving the physical and chemical properties of soils:

- Organic matter can improve the physical property of soils, and soil aggregation structure to obtain good agricultural soils.
- Increase water holding capability of soils.
- Increase nutrient holding capability of soils, and adsorb or exchange plant nutrients, which will improve slow-released effect of fertilizers.
- Release nutrients required for plants after decomposition.
- Chelate micro nutrients and assist the solubility and availability of plant nutrients.
- Improve soil buffering capability to mitigate soil acidic and basic reactions.
- Provide favorable condition for microorganism activity in soils so as to strengthen microorganism power to resist pathogens.
- Enhance the decomposition of man-made (e.g., pesticides, etc.) or natural toxic substances.
- Black is the color that is beneficial to heat absorption and for early spring farming.

Peat has a high content of humic acid, which have been shown to affect cell metabolic processes directly, including respiration and synthesis of nucleic acids and proteins. These effects are listed as follows:

- The effects on osmotic pressure of root cell membrane and ion carrier proteins result in rapid and selective entry of necessary elements to roots.
- Improve respiration and metabolic processes in crops, so as to increase ATP synthesis.
- Increase chlorophyll content and photosynthesis, so as to increase production of ATP, amino acids, carbohydrates and proteins.
- Improve nucleic acid synthesis, including DNA and RNA.
- Affect protein synthesis; increase the production of enzymes, ion carriers, structural proteins, etc.

- Increase or inhibit enzyme activity according to different types and origins of enzymes.
- Show similar function as plant hormones.

8.2.6 Fundamentals and Attentions for Peat Application

Peat is a high-humified organic fertilizer. Application of peat is convenient with ordinary organic fertilizers. The application guidelines and attentions are listed as follows:

Application Rate:
The amount needs to be estimated based on the soil organic matter content. Normally, we can apply 150 ~ 300 kg peat for 0.1 hectare of soil, and reduce to 50% if the soil organic matter content is higher than 4%.

Combined Application with Other Fertilizers:
The valuable quality of peat is that it does not easily decomposed, and thus it acts as a good soil conditioner. In practice, it is necessary to combine peat with other fertilizers to assure that the initial period will be supplied with sufficient nutrients. For example, 1 ton of peat can be mixed with 50 kg of liquid fertilizer (N:P:K = 4.5:9:9) or with 2.25 kg of N, 4.5 kg of P and 4.5 kg of K.

Application Timing:
Generally, the main period of peat application is during the basal fertilization. Other periods can also be applied without any adverse effects.

Applicable Crops and Applying Methods (Suitable for all Kinds of Crops):
- Fruit Trees:
  Deep placement of peat can show best performance. Peat can be applied in ditches and is best when mixing with soil.

- Vegetables, Nurseries and Flower Pants:
  Apply peat as a basal fertilizer during land cultivation. After applying in the field, we can plow and loosen the soil to bring about better performance.

- Cereal Crops:
  Apply peat in the field before plowing. Mix it while plowing is the best.

- Tea Tree:
  Apply peat between rows of tea trees and plough it under soil surface or use it as a cover.

- Orchid:
  Mix peat with tree fern, perlite or small stones to make the potting medium for cattleya. Small stones are needed at the bottom for better draining.

- Turf Grass on Golf Courses:
  Before planting turf grass, mixing peat with rock phosphate powder and sand layers will show best performance. If grass has already been planted, it can also be buried in holes or applied on the surface during turf maintenance period.

- Others:
  The application method for crops includes tobacco, special crops and medicinal plants, which is the same for vegetables and fruits. Apply peat mixed with basal fertilizers will show best performance.

The exploited peat is originally acidic or strong-acidic. Therefore, when it is applied for ordinary crops that cannot stand strong acid, the pH value of peat needs to be adjusted to about 6.5. Generally, commercial products have already been adjusted. If the acidic peat is used for alkaline soil, there is no need to adjust pH value, and can be directly applied after exploitation.

8.3 Application Guidelines for Organic Fertilizers

8.3.1 Reasons why Organic Matter can be Used as a Fertilizer and a Soil Conditioner

It can be Decomposed and Provide Nutrients for Plants:

The residues of plants, animals and microorganisms, animal wastes, and even nontoxic organic wastes can all be used as organic fertilizers. If certain organic matter cannot be consumed by human, it can be produced as animal feed or used as a fertilizer as long as it does not contain heavy metals or toxic substances. Organic matter contains carbon, hydrogen, oxygen, nitrogen, sulfur, phosphorus, potassium, calcium, magnesium and microelements. All high-molecular weight organic matter can be decomposed by soil microorganisms into low-molecular weight organic compounds, which are later broken down into CO_2 and plant nutrients N (NH_4^+, NO_3^-, etc.), P, S, and various microelements. The metabolic process of organic matter transforms into inorganic matter after being decomposed by microorganisms. This procedure is called "mineralization". Therefore, organic matters can be used as fertilizers for plant nutrients. Certainly, many free decomposed organic compounds with low-molecular weight can be absorbed and utilized by plants.

Effects of Decomposed and Released Low-Molecular Weight Organic Compounds:

As stated above, organic matters release many low-molecular weight organic molecules after the mineralization process, which have the effects of improving the physical, chemical and biological properties of the soil. For example, polysaccharides and carbohydrates play important roles in the formation of soil aggregates, and thus it can improve soil water draining and aeration properties.

Properties of Humic Substances after Decomposition and Re-Polymerizing:

After organic matters are decomposed into small molecules, a portion is decomposed by microorganism and mineralized into inorganic matters, while another portion will be polymerized into high-molecular weight humic substances, such as fulvic acid, humic acid, humin, etc. The polymerized high-molecular weight substance that is formed over a long period of time can resist degradation by microorganisms and remain as long-term stable organic matters in soil. Such humic substances are the most valuable "assets" in soil, and can contribute significantly to the physical, chemical and biological properties of soil.

8.3.2 The Source and Application Guidelines of Organic Fertilizers

Various organic fertilizers are distinctive and can be separated based on the source materials and the rates of decomposition. Therefore, it is necessary to be aware of different organic fertilizers' characteristics to assist us in selecting the right organic fertilizers for application. This section provides the specifications of organic fertilizers based on the source of organic materials.

Generally, the source of organic materials can be classed into two categories: Farmland organic fertilizer and commercial organic fertilizer. The composition of farmland organic fertilizers is versatile, while commercial organic fertilizers is mixed and produced in accordance with certain standard and each of these have their respective features. Both types of organic fertilizers can show positive effects and improve soil quality if we follow the specified guidelines as follows:

Farmland Organic Fertilizer:

This type of organic fertilizers includes animal manure, plant residues, compost, green manure, organic waste matter, etc.

Animal Manure:
- Source:
  Animal feces, including solid and liquid wastes, are common materials for organic fertilizer, such as feces of poultry, cattle, pigs, and sheep. Large amounts of wastes from massive poultry farms require appropriate handling or leading to environmental pollution. Types of wastes and their nutrient contents differ in various animals. Besides, the feed type and digestion ability of each animal are also important factors that determine their values as organic fertilizers. Normally, chicken feces have a high content of N, and pig feces have higher content of P than cattle feces. For animal wastes, the solid and the liquid can be disposed separately, or can be mixed with absorbing materials (plants or adsorption materials) as animal manures. Heavy metals (copper, zinc) and antibiotics are often added into the feeds for pigs and

chicken in farming companies, and thus these wastes require special attention for disposal and application with appropriate methods.

Application Guidelines:
- Pay Attention to the Application Rate:
  Animal manure contains a relatively higher salt and nitrogen content. Applying fresh urine without fermentation can easily result in fertilizer injury and excessive nitrogen. Long-term application of solid and liquid feces should be limited.

- Suitable as Basal Fertilizer:
  Animal manure with a high content of nitrogen is suitable as a basal fertilizer. When used for fruit trees, it is not applicable during the flowering period as excessive nitrogen or its degraded products will cause flower and fruit drop or poor blossom. Early application should be noted.

- Applied in Combination with Other Fertilizers or Plant Residues:
  After application, animal manure can be rapidly degraded and, lose 2% ~ 30% of its nitrogen. The loss will be even greater under high temperature and strong wind. Single superphosphate can be applied to stabilize ammonium to reduce the loss of nitrogen. Application method with balanced nutrients needs to be valued. Application of organic fertilizers also need to pay special attention to the nutritional balance. Plant residues are the best adsorption materials for poultry and livestock wastes, as they are also convenient and beneficial to organic fertilizer application.

- Pay Attention to Hygiene during Application:
  The application of animal waste organic fertilizers should avoid spreading on top of the soil surface, or would result in bad odor and provide a propagation environment for pests. This disadvantage can be reduced after certain management, but still it would be better to cover it with soil or plough into the soil. Especially for vegetable farmlands, special attention to hygiene and safety when applying organic fertilizers should be emphasized for avoiding the propagation of parasites and pathogens.

Rice Straw Residues and Compost:
- Source:
  Crop residues can be used as animal manure, compost or upland organic fertilizers. Although the nutrient contents of crop residues are not high, their application can improve soil structure, produce organic acids during decomposition, enhance the availability of soil nutrients, and is particularly important in providing microelements.

Compost is the product of the animal's stacking, plant wastes and the garbage after partial decomposition. Because there are a wide variety of material sources, the following basics can be referred to application and production of high-quality compost:

- The purpose of applying compost is to increase soil organic matter, promote crop growth, maintain soil fertility, and improve the physical, chemical and biological properties of soils.
- The decomposition of organic matters rely on microorganisms, and thus the environment for producing compost should also suit microorganism activity, such as enough moisture and aeration. Stacking and turning are required procedures along the processing, and sufficient nitrogen is also necessary for microorganism reproduction. In order to get a good start, the use of matured compost or microorganism inoculants at the initial stage and this can accelerate the composting process. During decomposition, the compost can produce high temperature (up to 70°C) that could help kill pathogens. Therefore, it is useful to cover the compost for increasing the temperature.
- For compost materials with high C/N ratio, adding minerals and fertilizers, especially nitrogen (usually around 0.1% ~ 1%), can effectively accelerate the compost maturation. In addition, lime (usually around 0.3% ~ 0.5%) can be added to reduce the acidity of the compost and is beneficial to the quality of the compost.
- Soil animals are also considered to show positive effects in accelerating the decomposition of organic matters.

Application Guidelines for Residues and Composts:
- Crop straw residues (such as rice straw, corn straw, etc.) should be spread or mixed into soil as early as possible before sowing, which can prevent nutrients (nitrogen, microelements, etc.) from being absorbed and fixed by microorganisms. Otherwise, seedlings would be influenced by the residues and cause nutrient deficiency and yellowing of crops. Therefore, immature crop straw residues are necessary to be applied in combination with chemical fertilizers.
- Farmland with poor draining ability should avoid applying large amounts of crop residue or immature compost. Under poor soil draining ability and anaerobic conditions, organic matters are easy to degrade into harmful substances that will damage the crop root.
- Straws with pathogens or weed seeds are not suitable for direct application to prevent the propagation of pathogens and weeds. Otherwise, the loss outweighs the gain.

Green Manure:
- Source:
  Green manure indicates the direct use of the whole green plant or roots as a fertilizer. There are many crops that can be used as green manure, among which are legume crops, including sesbania, sun hemp, yokohama bean, Chinese milk vetch, alfalfa, clover, Chinese trumpet creeper, soybean, as well as non-legume crops, such as rape, turnip, etc. Legumes can establish symbiotic relationship with rhizobia and fix nitrogen from the air, and thus increase the nitrogen source in the soil. Green manure crops, regardless of leguminous or non-leguminous crops, have the effect of preserving soil nutrients. Green manure crops can absorb soil nutrients and preserve them temporarily to prevent loss. Afterwards, the green manure will be decomposed and release the nutrients; meanwhile, it can improve the availability of nitrogen and microelements.

In intensive agriculture farmlands where crops are closely planted and the ease of chemical fertilizer application, the utilization of green manure is often neglected. However, in respect of long-term soil conservation benefits, green manure should be cultivated at least once every two or three years, or once a year if possible. Farmland that does not apply much organic fertilizers requires green manure plantation particularly.

Green Manure Application Guidelines
- Select optimal green manure crops in accordance with the season: Green manure in winter includes Chinese milk vetch, rape flower, green husk beans, and alfalfa, etc. Green manure in spring and summer includes sesbania and sun hemp, etc.
- Coordinate green manure with soil characteristics and the next-rotation crop: Leguminous green manure is most suitable for soil that is short of nitrogen fertilizer and organic matters, especially in dry fields. Conversely, the application of leguminous green manure in wet field with excessive nitrogen should consider the soil condition because legumes may supply content with high degrees of nitrogen fertilizer and this leads to non-productive tillers of rice. Therefore, the cultivation of leguminous green manures in wet field should coincide with the soil condition to reduce the nitrogen content. Otherwise, this would easily result in excessive nitrogen or rice blast issues.
- Green manure should be applied cautiously on farmland with poor draining capability. Cultivation of non-legume plants, such as rape and turnip, are more suitable for acquiring green manure after the flowering of green manures or until green manures is aged. This is because younger plants of green manures are prone to decompose after plowing into the soil, and they may be toxic or compete with the crops for consuming nutrients.

Commercial Organic Fertilizer:

Commercial organic and organic/mineral fertilizers are very similar. The varieties are specified in Table 8-3.

- Peat Organic Fertilizer:
  Peat has been adopted as a commercial organic fertilizer for a long time. Peat is the result of long-term deposition and transformation of ancient organisms. It degrades quite slow in soil and is the most effective material for long-term soil organic matters increment. The maturity and transformation degree differs in distinct production areas. Peat contains large contents of the soil

Table 8-3. Categories of Commercial Organic Fertilizers

		Categories	Source of Organic Matter	Approximate Components
1	Organic Fertilizer	Organic matter fertilizer	Peat, organic residues	30 ~ 50% organic matter
		Organic nitrogen fertilizer	Feathers, keratin meal feces	5 ~ 14% N
		Organic phosphorus fertilizer	Bone meal	13% P
		Organic nitrogen Phosphorus fertilizer	Mixed keratin feathers and bone meal	4 ~ 10% N 2 ~ 10% P
		Organic nitrogen Phosphate potassium fertilizer	Bird dung, fish scraps, animal residue wastes	4 ~ 6% N 4 ~ 5% P; 2% K
2	Organic/ Mineral Fertilizer	Organic/mineral mixed fertilizer	Animal and plant residues, urban wastes and sludge	25 ~ 40% organic matter
		Peat	Organic matter above 35%	1 ~ 3% N, P, K
		Peat mixed fertilizer		1 ~ 3% N, P, K
		Organic/mineral nitrogen fertilizer	Lignin	15 ~ 19% N
		Organic/mineral nitrogen phosphorus fertilizer	Animal and plant waste residues	6% N; 3 ~ 4% P
		Organic/mineral nitrogen phosphate potassium fertilizer		4 ~ 14% N, P; 3 ~ 14% K

essence -- humic acid, fulvic acid and humin. Peat is light and acidic and with high organic matter content and ash content at around 1 ~ 2%. Commercial peat products are often pre-neutralized and supplied with nutrients, so they are considered as stable and persistent soil conditioners. Peat moss with shorter formation time is also a common organic fertilizer commodity.

- Humic Acid:
  Solid peat has a high content of humic acid. In recent years, this sort of humic acid fertilizer has become a good organic fertilizer. Humic acid is an acid phenolic polymer that dissolves in alkaline but not in acid. Because humic acid precipitates in acidic condition, commercial humic acid products are derived by extracting with alkaline solution to liquid or powder form after being dried. The alkaline powder product can dissolve in water. Humic acid is hard to degrade and has a stable structure. It can be used as a long-term effective organic fertilizer and a great soil conditioner. Generally, products without chemical fertilizer additives cannot be regarded as a source for abundant nutrients. Humic acid thus should be mixed with chemical fertilizers in application, so it could be applied as a product that supplies nutrients. Liquid humic acid has many effects on the soil improvement. Deep fertilization in soil is beneficial and can effectively increase the organic matter content in deep soil.

- Organic Fertilizer from Animal Wastes:
  This class includes animal feces (chicken, pig, sheep, etc.) and discarded residues (fish meal, bone meal, feathers, fur, etc.). The composition of animal feces (Table 8-4) is greatly influenced by the feed and additives, which in turn affect the quality of the organic fertilizer. For example, the addition

Table 8-4. Main Nutrients in Organic Matters

Rich in Nitrogen	Animal materials: fish meal, crab shell, chicken and pig feces Plant materials: soybean meal
Rich in Phosphorus	Animal materials: bone meal, fish meal
	Plant materials: cotton seed flour, husk ash, chicken and pig feces
Rich in Potassium	Animal materials: sheep and cattle feces
	Plant materials: plant ash, cotton seed flour, seaweed, tobacco stem
Rich in Calcium	Animal materials: bone meal, crab shell powder, oyster shell powder, cattle, chicken, pig and sheep wastes
	Plant materials: plant ash, coal ash, tobacco stem

of excessive rice hull can affect the composition of chicken feces. Special attention should be paid to quality when purchasing. Both fish scrap and feathers contain high nitrogen content. The latter also contains high protein content, and can be adopted as nitrogen fertilizers. In some commercial products, proteins are degraded into amino acids to produce amino acid nitrogen fertilizers. Bone meal is high in calcium phosphate content and serves as a slow-released organic phosphorus fertilizer (Table 8-4).

- Organic Fertilizer from Plant Residues or Wastes:
  This class includes those that are degraded, half-degraded, and un-degraded. Mature compost is a common degraded organic fertilizer. The quality of mature compost can be based on the source material, the amount of nutrients, and the degree of maturity. Generally, maturity is determined by the C/N ratio of the compost. The lower the C/N ratio, the higher the maturity degree is. However, the C/N ratio can be decreased by the addition of nitrogen fertilizers in the immature materials, such that, a low C/N ratio does not necessarily indicate a high maturity. Therefore, the maturity needs to be determined by the appearance of the compost. Tree bark and plant residue composts are both good organic fertilizers.

Common immature organic fertilizers include soybean cake, various seed extract residues, fiber wood scraps, etc. Soybean cake and seed extract residues are high in nitrogen content, while hard shell residues are persistent organic matters that can be used as long-term soil conditioners.

- Extracted or Concentrated Organic Fertilizer:
  Among commercial organic fertilizers that are produced from animals or plant extracts appearing as liquid or concentrated powder could contain various organic substances, nutrients (including microelements), enzymes, plant hormones, and antibiotics. Common integrated nutritional agents can be extracts of seaweed or fish meals.

Among the above commercial organic fertilizers, some are pure organic fertilizers while some are added with fast-released chemicals or supplements for special purposes in order to satisfy the full demand for the decomposition process or the crops.

8.4 Function, Quality and Selection of Organic Fertilizers

8.4.1 The Functions of Organic Fertilizers

There are a great variety of organic fertilizers with different characteristics for distinctive purposes for crop growth. Generally, we can class organic fertilizers into decomposition level: easy or hard. The main effect of easy degradable organic fertilizers is to provide plant nutrients, while the hard degradable organic fertilizers retain in soil and improve soil physical and chemical properties.

The following illustrates the functions of organic fertilizers from the soil physical, chemical and microorganism aspects:

Improvements in Soil Physical Properties:

Improve Soil Structure:
Organic particles or fragments can loosen the soil. Organic matters can increase microbial products, and thus lead to soil aggregation and a better soil structure.

Increase Soil Water Holding Capability:
Organic fertilizers can directly assist water retaining or indirectly increase water holding capability by improving soil structure.

Develop Soil Aeration:
Organic fertilizer can increase oxygen supply for roots and allow easy diffusion of CO_2 from roots.

Increase Soil Temperature:
The dark color of organic matters can absorb much heat. Moreover, soil structure improvement can indirectly assist the soil water draining ability in the rainy season, which is favorable to increase the soil temperature.

Improvements in Soil Chemical Properties:

Increase Stored Nutrients in Soil:
Exchangeable forms existed on the surface of organic matters to increase stored nutrients in soil, which is especially important for soil with less clay content.

- Supply Nutrients and Energy for Plants after Decomposition:
  During decomposition, organic matters will release inorganic nutrients (such as N, P, K, Ca, Mg, S, microelements, etc.), organic nutrients (such as amino acids, carbohydrates, etc.), and CO_2 for photosynthesis. Decomposed products promote the mobility and availability of stored inorganic nutrients.

Available nutrients will be directly released from the decomposition of organic matters. Alternately, the decomposed products will result in organic acidification, or reduce the redox potential for improving the reduced elements' availability and mobility, such as Fe, Mn, or indirect to P and Mo.

- Nutrient Immobilization:
  At the early stage of the decomposition process of organic matters, microorganisms bloom and absorb a great deal of nutrients when C/N ratio is

greater than 30. This is called "the nutrient immobilization" and often results in the competition of nutrients between crops and microorganisms.

- Organic Fertilizer Contains Active Substances for Plant Growth:
  During the decomposition process, massive organic compounds are released, including growth stimulators, inhibitors and possibly substances similar to antibiotics, which can directly affect crops growth or soil microorganisms. These organic compounds differ among different organic fertilizers.

Improvements in Soil Biological Properties:
- Organic fertilizers provide nutrients and energy for soil organisms. Various organic matters consist of different components and will influence their decomposition rate. The decomposition of lignin is slow as few microorganisms utilize lignin.

- Humic substances with low content of nitrogen can improve the activity of soil nitrogen-fixing microorganisms that fix nitrogen into nitrogen compounds that can be utilized by organisms.

- Increase in soil beneficial microorganisms can antagonize pathogens in soil.

8.4.2 The Quality of Organic Fertilizer

The specific purpose of applying organic fertilizers affects our judgment on their quality, as different criteria should be adopted based on our needs. As detailed in the previous section, organic fertilizers posses numerous functions and effects of each organic fertilizer contributes differently. In other words, the main effect of some organic fertilizers is to provide nutrients, while others function as long-term effect of soil conditioners is to improve soil physical and chemical properties. Therefore, the quality of the organic fertilizer should be determined based on its purpose.

Generally, the quality of organic fertilizers can be specified in accordance with the following two main application purposes:

Application Purpose as Nutrient Provider:

Provide Optimal Nutrients:
There are differences in the quantity and quality of nutrients provided by various organic fertilizers. For example, chicken feces compost and soybean cake can provide much nitrogen, while sawdust is less, unless nitrogen fertilizers are added in. Bone meal can provide much Ca and P. Usually, the more nutrients are provided in a certain quantity, the better; yet, organic fertilizers should not be over-used, or they could result in counter effects.

High Maturity:

The maturity of organic fertilizers refers to their decomposition degree. Fresh organic matters (before composting) when applying to soil will result in the blooming of decomposing microorganisms, which in turn absorb a great deal of nutrients from soil and compete with crops. On the other hand, application of matured organic fertilizer to the soil can immediately provide nutrients to crops and will not result in nutrient competition. Generally, the maturity degree is related to the fermentation and decomposition time during the maturing process. The longer the time, the more mature the organic fertilizer is. C/N ratio is usually used as the indicator for the maturity degree. Lower C/N ratio indicates higher content of N. The measured C/N ratio will become lower if nitrogen fertilizers are added into immature organic fertilizers. After application, the N added immature fertilizer will not result in the competition for N between microorganisms and crops.

Fast Release of Nutrients:

Different forms of organic fertilizers lead to differences in the speed and capability of release nutrients. Usually, organic fertilizers with high content of lignin or bones degrade relatively slow. Thus, the released nutrients are also slow, and they would be insufficient in nutrient supply if massive nutrients are in urgent need from the plant. Besides, the longer effective period in nutrient supply is the better.

Application Purpose as Soil Conditioner:

Features for Long-Term Effects:

The application of organic fertilizers will improve the soil physical and chemical properties to an extent. Long-term effect of the applied organic fertilizer is beneficial to soil physical and chemical properties, and will also improve soil conservation. The long-term effect of organic fertilizers to improve soil properties is closely related to the functional groups in the organic materials.

Features for High Stability:

As mentioned above, the long-term effect of organic fertilizers for improving soil properties is closely related to the functional groups in the organic materials, while the stability is closely related to the structure and the form of the organic material. The more stable the structure and the form is, the less it will be degraded. In fact, the features of long-term effect and high stability cannot be clear in practice separately. Usually, organic materials with long-term effect are also highly stable, such as humic acid and peat. Peat contains humic acid, fulvic acid and humin, etc. Humin is very stable and dissolves in neither acidic nor alkaline. Humic acid dissolves in alkaline yet precipitates in acidic. Fulvic acid dissolves in both acidic and alkaline. They are different in the structure and functional groups..

The required qualities of organic fertilizers are different based on various application purposes. In addition to the above qualities, the following general qualities of organic fertilizers also need to be considered:

No Pathogens and Pests:
The occurrence of pathogens and pests can be brought along in harmful organic fertilizers.

No Toxic Substances:
Toxic substances include organic and inorganic matters, such as phytotoxic substances, pesticides, heavy metals, etc.

No Obvious Odor:
Foul smell is also a kind of pollution to the environment and human.

Neither too Acidic nor too Alkaline:
Organic fertilizers that are too acidic or too alkaline are harmful to plants and soil. It can be improved by mixing with neutralizing materials.

Largest Amount of Dry Matters for a Certain Price:
If sold for the same price by unit weight, those with high content of water or oil (such as bean) are not favorable to buyers for organic fertilizers.

8.4.3 Guidelines of Right Organic Fertilizers and Application Selection

Different organic fertilizers have diverse forms, decomposition rates, maturity degrees and prices. Prices should be considered not only in ordinary economic view, but also from the viewpoint of investment reward. Limited capital can afford a larger amount of organic fertilizer with lower price, so low price is an important factor when massive fertilizer application is required. However, under normal circumstances, commodity with higher quality is more expensive. Therefore, qualities, effects and prices should be considered inclusively in the selection of organic fertilizers.

Different organic fertilizers should be applied to meet different requirements. The following section specifies how to select right organic fertilizers for various application purposes:

Selection Based on the Soil's Demands:

If the content of organic matters in soil is quite low (< 1%), the amount of organic fertilizers application comes in a priority that improves the overall qualities of soil organic matter content. It is better to fertilize and affect deep soil. Therefore, organic fertilizers with the largest amount should be chosen if capital is limited. In this case, the organic matter's supplement in most soil can be achieved based on the organic fertilizer quantity. If the content of organic matter in soil is high (> 3%), massive organic fertilizer application is not necessary, unless for gravel soil that needs more organic fertilizer. Soil organic matters can be increased in barren soil areas with a great quantity of green manure plants (such as sesbania, sunn hemp, yokohama bean, etc.).

Selection Based on the Crops' Demands:

Different categories of crops have various features in their growing, blooming and fruiting periods. Reasonable cooperation of organic fertilizers with different characteristics can bring out the effects of organic fertilizers. Otherwise, they may cause damage. For example, applying a great deal of immature organic fertilizer to rice fields will make rice seedlings flourish. Yet, increasing useless tillers during tillering stage, it may decrease the overall rice production due to the excessive supply of nitrogen. Organic fertilizers with high N content should not be applied too much for fruit trees. Massive nitrogen supply is inappropriate after medium fruit stage; otherwise, the fruit quality will drop.

Before planting short-term crops, organic fertilizers with high maturity degree are needed in order to avoid the disadvantages in early decomposition period. In terms of perennial crop, organic fertilizers with different maturity degrees can be applied in land early after harvest. Slopeland orchard can apply immature organic fertilizers as coverage and apply mature organic fertilizers after fruits starts to grow. Immature organic fertilizers is better to combine with chemical fertilizers (N, P, K, Mg fertilizers, etc.) or lime materials. in order to meet the crops' demands.

Selection Based on the Geography of Environment's Demands:

Generally, organic fertilizers are applied in ditches, and, particularly, on slope lands. It also needs to be covered with soil, so as not to be washed away by rain. If only surface spread method is adopted, the forms of chosen organic fertilizers should be of large sizes or particles. Organic fertilizers in the form of powder can be easily washed away by rain, and should be applied into the soil. Deep application and liquid spread/irrigation have good performance in applying organic and chemical fertilizers on slope lands. For example, liquid organic humic acid is very effective in improving the overall soil properties.

All in all, it is important to select high-quality, economic and effective organic fertilizers and apply them with the most appropriate method with the most suitable quantity, in coordinate with beneficial season and crops requirements. It is the basic and essential application for fertilizer application to consider the optimal time, applicable place, appropriate method and suitable amount.

8.5 The Application of Amino Acid Fertilizers

Amino acids are nitrogenous organic acids with the generic formula:

$$\begin{array}{c} NH_2 \\ | \\ R-C-COOH \\ | \\ H \end{array}$$

In this formula, R group represents organic compounds that consist of carbon, hydrogen, oxygen, nitrogen and/or sulfur. Twenty common amino acids are naturally incorporated into polypeptides of a vast variety of proteins and enzymes. Proteins exist in cells of living organisms and microorganisms and would consist of about half of dry weight. In general, the amount of animal protein content is higher than those of plant. For example, fish and meat have higher protein contents, and beans have higher protein contents than other plants. Each protein is formed by a chain of amino acids linked together through peptide bonds, and these natural amino acids mainly consist in protein polymers.

Amino acids have different R groups, which include nonpolar, hydroxyl (-OH), carboxyl (-COOH), amine ($-NH_2$) or sulfur (-SH or -S-). Among them, there are linear and ring-shaped amino acids.

Amino acids have the function to protonate or deprotonate (H^+): ($RCOOH \leftrightarrows RCOO^- + H^+$).

When amino acid is in its anionic state, they can form ionic bonding with cations (such as potassium, sodium, calcium, magnesium, copper, and zinc ions) to form ionic compounds (such as $RCOO^- K^+$). Amino acids have one or two amine groups ($-NH_2$). These amine groups will become cationic ($-NH_3^+$) in acidic conditions and will form ionic bonds with negative ions or will be absorbed by the electronegative on soil surfaces.

8.5.1 The Source and Function of Amino Acid Fertilizers

Source:

Commercial amino acid products can be derived from residues of animals, plants or microorganisms. These residues contain proteins that can produce plenty amino acids of different structures after hydrolysis. On the other hand, amino acid products for animal feed mainly consist of amino acids and can be used as fertilizers. Due to the extent of hydrolysis, the products may differ in the decomposition degree. Pure amino acids or amino acids with ammonium fertilizers can be obtained.

Function:

Fertilizer Functions:

Amino acids contain nitrogen and thus amino acid fertilizers are mainly applied as nitrogen fertilizers, which are very effective for the shoots' growth and crops' leaves. Because the growth impact of amino acid fertilizers is obvious even with our eyes, it is easily accepted by farmers as good fertilizers. A small portion of amino acids contain organic sulfur, and thus it can provide sulfur nutrients and increase the content of sulfur in soil.

The Function of Organic Matters:

Amino acids are organic acids of organic matters. Amino acids can be degraded by soil microorganisms and polymerize into humic substances. The formation of small-molecular weight organic compounds from degradation of amino acids can be absorbed by plants, activate the activity of soil microorganisms, and also improve soil physical, chemical and biological properties. Therefore, according to different requirements of crops and nutritional balance, amino acids could exert distinct reactions in the soil.

8.5.2 Guidelines and Attentions for the Application of Amino Acid Fertilizers

Apply with Optimal Amount:

Amino acid fertilizers are mainly applied as nitrogen fertilizers. Crops that could not endure excessive nitrogen should apply amino acid fertilizers with suitable amount. The effects of nitrogen fertilizers are mainly shown on foliage growth and branching, but excessive application of amino acids or high proportion of nitrogen fertilizers compared with other fertilizers can easily result in problems, such as overgrowth, flower drop, fruit drop and low quality fruits.

Apply on Appropriate Land:

Amount of application should be adjusted based on different soil conditions. If the soil has been applied with much nitrogen fertilizer (such as urea, ammonium sulfate, ammonium nitrate, potassium nitrate, etc.), amino acid fertilizers or other chemical nitrogen fertilizers should be reduced. Therefore, the application of amino acid fertilizer on soils with high contents of nitrogen or organic matters should reduce to a suitable amount and be paid special attention to.

Apply at Favorable Time:

The developmental stages of perennial crops and seasonal differences are factors to consider when applying amino acid fertilizers. Amino acid fertilizers are organic matters. Thus, massive application is not appropriate in rainy season, or it would result in the eutrophication of organic matters and excess of nitrogen. Amino acid fertilizers should not be applied in excess during the young fruit and the medium-size fruit stages of perennial crops, or it may cause overgrowth, fruit drop and quality decrease.

Apply with Appropriate Method:

Amino acid fertilizers are organic nitrogen fertilizers. Liquid amino acid fertilizers should be diluted before use, and should be applied to the place where there are many roots. On the other hand, solid amino acid fertilizers can be spread in the farmland with water or after irrigation to allow solid amino acid fertilizer powder to enter the soil and perform its functions.

8.6 Applications of Green Manure and its Effects on Soil Conservation

8.6.1 Green Manure is the Moisture Controller of Soil Conservation

Nurturing and maintenance of soil nutrients are important tasks in agricultural. Continuously cropping of various crops on the same soil will exhaust massive soil nutrients after harvesting. One should not underrates the value of soil and regard it as just a medium or stone powder for crop growth. We should view soil as a mother that nurtures her offspring, and thus it is important that we take good care of our soil in order to ensure the health of descendant. Soil with intensive production is like a mother that gives birth frequently, and eventually become weak and barren. Regards to soil management, fertilization is a must for agricultural cultivation and microbial fertilizers. Green manure application is a kind of organic fertilizer application.

Green manure crop is grown for a specific time of cropping, and then plowed under and incorporated into the soil when plants are still green or shortly after maturity. These green manure plants include legumes or non-legume plants. Common leguminous green manure plants include sesbania, sunn hemp, yokohama bean, soybean, Chinese milk vetch, Egyptian clover, alfalfa, Chinese trumpet creeper, pea, broad bean, lupine, etc. Common non-leguminous green manure plants include rape, turnip, etc. Green manure application is greatly beneficial to soil and is considered as the moisture controller for the maintenance of soil.

8.6.2 The Effects of Green Manure

Green manure has a number of effects as it influences the physical, chemical and biological properties of the soil. Its main effects are explained as follows:

Increase in the Source of Soil Organic Matters and Microbial Activity:

Green manure is a fresh organic matter source that is plowed into the soil and utilized and degraded by microorganisms. The little portion that is difficult to degrade will accumulate in soil, while the degraded matters will polymerize into humic substances and increase the soil organic matter source. Mature green manure crops have more lignin that is difficult to degrade, and thus they are favorable as they not only increase soil organic matter source, but also is beneficial to microbial activity of soil. Application of chemical fertilizers without the use of organic matters will exhaust massive amounts of soil organic matters.

Increase Soil Nutrients and Reduce the Application of Nitrogen Fertilizers:

Green manure will degrade and release its nutrients to provide the next rotation crop after it is plowed in the soil. Leguminous green manure can undergo nitrogen fixation and can provide massive nitrogen after decomposition. Any green manure after decomposition can release nitrogen, phosphorus, potassium nutrients and other micro and trace elements. Green manure, in particular, can degrade rather fast and provide nutrients efficiently.

Increase Soil Nutrient Availability:

Green manure can produce organic chelating agents or organic acids after decomposition and transform insoluble forms to soluble nutrients. This is especially available in promoting P and Zn. Moreover, green manure can greatly improve the availability of P in alkaline soils as its decomposition could release CO_2.

Reduce Washing and Erosion of Soil:

There is usually a time frame between two sessions of cropping. During this period, exposed soil is easily to be washed away by rain and result in soil erosion. Therefore, green manure has the effect of protecting the soil, and reducing the washing and loss of soil.

Reduce Plant Diseases, Pests and Weeds:

Rotation of green manure crops in farmlands can reduce the population and occurrence of pathogens, pests and weeds. Further studies are required with regards to the possibility of green manures in antagonizing pathogens or pests, and technologies to accelerate the humification of green manures.

8.6.3 The Application Method, Amount and Selection of Green Manures

There are various application methods for green manures based on their varieties, crop age, soil characteristics, production volume, and the types of next crops (Table

8-5). Generally, green manures are directly plowed in the soil in application and the soil should be kept moist after plowing. In paddy fields, green manure should be applied one to two weeks before planting rice seedlings. Green manure can also be used as a cover, especially for slopeland farming to reduce soil loss and increase water retention. Covering slope land with green manure does not require plowing, as it can reduce the washing and loss of the topsoil.

In order to improve soil properties, green manure can be combined with lime material in application. The addition of 2 ~ 3 kg of lime material in 100 kg fresh green manure can accelerate the decomposition rate of green manure. Green manure should not be applied more than the common amount: 10 ~ 15 tons per hectare. The more green manures we plow in the soil, the longer it requires for the initiation of the planting of the next crop. Many green manures, including legume or non-legume crops, are plowed in the soil at their flowering stage as basal fertilizers.

The selection of green manure crops is dependent on the next crop and the soil characteristics. There are large and miniature types of green manures. Green manure crops of the large type, such as sun hemp and sesbania, can grow into tall and woody plants with high contents of lignin; while the miniature types, such as Chinese milk vetch and clover, are short. Because plant stature is related to the length of the growth period, younger plants have lower contents of lignin and are degraded rather fast. On the other hand, mature and old plants with higher contents of lignin degrade relatively slower, and can have prolonged effect on the improvement of soil properties, and also are beneficial to the perennial crops growth.

There are legumes and non-legume green manure crops. Because legumes can form symbiotic relationship with nitrogen-fixing rhizobia, the nitrogen contents in these plants are higher and thus are beneficial to crops in need of nitrogen fertilizers. For example, cultivating leguminous green manures before planting gramineous upland crops, such as corn and sorghum, will enhance their growth. For farmland with poor drainage capability, non-legume green manures (such as rape) can be planted in the cool season to increase the growth and production of the next crop.

In addition, legume green manure (such as sesbania) is applicable for soil that is prone to disease, and will reduce the occurrence of diseases in non-legume crops, such as solanaceae crops (the potato and eggplant family). The inhibition of diseases may be related to soil characteristics and pathogen types. Green manure cropping system merits further study in order to reduce the amount of pesticide application.

312 Soil and Fertilizer

Table 8-5. Briefings on Green Manure Crops

Varieties	Family	Seeding Time	Character	Seeding Quantity (kg/ha)	Sward Production (ton/ha)	Plant Nutrients (%) N	Plant Nutrients (%) P_2O_5	Plant Nutrients (%) K	Application Time to Soils (Days After Seeding)
Sesbania	Legume	Spring, summer	Not resistant to cold, not suitable for early spring	20~30	25~35	0.47	0.05	0.40	50~70
Sunnhemp	Legume	Spring, summer	Prefer high heat and moisture	25~30	20~30	0.37	0.08	0.14	50~60
Soybean	Legume	Spring, summer, autumn		50~70	15~30	0.71	0.08	0.46	70~90
Yokohama Bean	Legume	Spring, summer	Not resistant to cold and humid conditions	20~40	20~30	0.15	0.11	0.39	60~70
Rape	Non-legume	Autumn, spring (15~20°C)	Resistant salts, humidity, cold and drought	6~9	20~35	0.21	0.02	0.28	Full blooming stage
Egyptian Clovet	Legume	Autumn, winter (cold)	Not resistant to frost and heat, resistant to shade and salts	10~15	20~30	0.52	0.13	0.36	Flowering period
Chinese Trumpet Creeper	Legume	Autumn, winter		10~20	30~50	0.56	0.13	0.43	Flowering period
Chinese milk vetch	Legume	Autumn, winter	Prefer humid condition, acid-resistant	10~25	10~34	0.46	0.09	0.37	Flowering period

Table 8-5. Briefings on Green Manure Crops (Continue)

Varieties	Family	Seeding Time	Character	Seeding Quantity (kg/ha)	Sward Production (ton/ha)	Plant Nutrients (%)			Application Time to Soils (Days After Seeding)
Lupin	Legume	Aautumn, winter	Not suitable for summer	25~30	15~20	0.40	0.007	0.39	Flowering period
Buckwheat	Non-legume	Autumn, winter		50~60	6~17	0.4	0.15	0.33	Flowering period
Cabbage	Non-legume	Autumn, winter		8~12	7.7~15	0.24	0.09	0.52	Flowering period

8.6.4 Attentions for Green Manure Application

Select Appropriate Varieties of Green Manure:

It is important that the various green manures adapt to the climate. Green manures that are suitable in summer may not be so in winter. For example, Chinese milk vetch and clover should be cultivated in the cool season (fall or winter), while summer is too hot for the growth. Besides, there are also distinctions between different varieties. Therefore, a suitable planting season and soil pH should be recognized before planting green manure.

Legume Green Manures may Require Inoculation of Rhizobial Species:

The rhizobia population in some soils may be low and result in low nodulation in legume green manures, and thus rhizobial inoculation is required. This is particularly necessary for lands that had never cultivated legume green manures.

Apply Green Manures into Soil at Suitable Period:

Some green manures (such as sunny hemp, sesbania, etc.) are easy to be plowed in soil with machine during flowering stage when it is still young and soft. However, if the green manure crop have matured and aged into woody plants, it is more difficult to plow in soil and it requires heavy machineries.

Selection of Green Manure under Special Situations:

Excessive application of leguminous green manures in paddy fields may have adverse effects, such as excess nitrogen will result in over tillers. Therefore, green manures should not only be applied with suitable amount, but also require early plowing to decompose the green manure and reduce the disadvantages caused by green manure. This is particularly true for soils with poor drainage capability where excessive legume green manures should be avoided. Consequently, suitable amount in green manure application should be paid closely attention to.

8.6.5 The Necessity of Green Manure Application -- Planting Green Manures is Better than Fallow

Many people neglect the application of green manures, simply because they think there is no time for planting green manures within intensive production areas. This is like a man who is too busy without rest. His loss will eventually outweigh the gain after certain years. If an intensive cropping farmland can supply the soil with organic fertilizers, green manures will not be a necessity. However, the supplements to the soil are usually only chemical fertilizers. Therefore, the soil under such management will suffer, the soil will decline, and the demand for more chemical fertilizers will result in a vicious cycle that gravely destroys the soil fertility.

As mentioned above, green manure has various effects and it is also beneficial to the physical, chemical and biological properties of the soil. Fallow is not as beneficial as planting green manure, because weeds may spread during the fallow period and increase weed seeds and pests in the farm. Good green manure crops are usually with lower chance of pest infestation annually. Consequently, the application of green manures can reduce excessive application of herbicides and pesticides. In addition to effects of soil disease reduction rate, the application of green manures, in fact, shows many positive purposes. The soil with low organic matter contents especially requires green manure. Several green manures used for rotation is also needed. The application of green manures still requires further study and popularizing. We should protect our soil to guarantee the long-term productivity of the soil.

8.7 The Application and Recycling for Crop Residues

8.7.1 The Significance of Recycling for Crop Residues

Crop residues or straws are residual organic matters after harvest, which contains abundant organic matter (about 90%) and inorganic ash content (about 10%). The organic matter mainly includes cellulose, lignin, protein and carbohydrate, while ash content mainly includes inorganic K, Si, Ca, Mg and microelements. The organic matter is produced through photosynthesis via absorption of CO_2 and subsequent synthesis into high-carbon elements. Incineration of the organic residues and straws would return the CO_2 back to the air. In recent years, global warming and dramatic climate changes are serious threats. The main causes to CO_2 emission should be the focus of the whole world and restrained by some protocols.

There are many uses of residues and straws, such as recycling as compost, producing feeds, providing the heat, producing paper and other extraction purposes. Residues and straws cannot provide many nutrients as a feed. Even if fermentation is adopted, the low efficiency of fiber decomposition will prevent fermentation from taking quick effects. Using residues and straws to produce paper have many disadvantages, such as serious wastewater pollution issues, poor quality, collection difficulty, and mere provision after harvesting seasons. When used as a fuel, residues and straws are of low in heat value per unit volume and will release CO_2 into the air rapidly. If residues and straws are used for carbohydrate extraction, the high cost and low efficiency is yet another problem. All the above purposes will produce a great deal of wastes and pollution to the environment. Therefore, recycling of crop residues and straws to the soil is the best choice amongst all application methods, as what is from the soil is to be returned to the soil. By this method, massive nutrients will be recycled and will not be wasted. They can also reduce the occurrence of plant diseases and pest infestations. In short, recycling of crops residues and straws back to the soil is the best choice.

There are various kinds of crop residues and straws, including those from wheat, corn, rice, cotton, etc. Recycling of residues and straws is a good way to decrease the application of chemical fertilizers, improve soil characteristics, and lay the foundation for a stable grain production. As such, crop quality, crop health and safety will be increased along with farmer's income and allow the sustainable agriculture development.

8.7.2 The Purposes and Strategies for the Recycling of Crop Residues and Straws

In terms of recycling crop residues and straws, the most economical method is to transform crop residues and straws into organic fertilizers. There are many benefits for recycling, including: (1) improve soil organic matters and ensure sustainable, and stable production of high quality crops; (2) reduce soil degradation issues and acidification caused by long-term application of chemical fertilizers; and (3) reduce plant diseases, pests, and pesticide applications. Soil organic matters should be effectively improved, or the soil will suffer from continuous decline which will reduce the efficiency of chemical fertilizers and the application amount in turn. Consequently, serious plant diseases and pests will possibly occur and in turn increase the application of pesticides and end up in a vicious circle.

Soil is mother of all crops, but nutrients within it are limited. Application of crop residues and straws for other purposes will accelerate the decline of the soil. Incineration of crop residues and straws will release all organic C and N into the air and left with ash that only contains mainly potassium. In addition, releasing CO_2 will also result in fume issues. Therefore, efficient recycling of crop residues and straws is an important strategy and direction to soil improvement.

8.7.3 Methods to Recycle Crop Residues and Straws

Recycling of crop residues and straws mainly includes direct application, conventional composting and fast composting.

- Direct application indicates to remain the residues and straws on the field after harvesting, or shred into pieces with machines and spread or plow in the soil.
- Conventional composting is to first cut straws into pieces and then compost them for a long period (2 ~ 4 months), so that straws can be fermented and become compost. Straws can also be composted with added materials, such as poultry and livestock wastes and microorganisms.
- Fast treatment method (composting-free) is to rapidly complete treatment with physical, chemical and microorganism technology in a reactor.

The advantages and disadvantages of the above recycling methods are specified in Table 8-6.

Table 8-6. The Advantages and Disadvantages of Different Recycling Methods for Crop Residues and Straws

Method	Advantages	Disadvantages
Direct Application	It is convenient for harvesting machinery to be equipped with a cutting device.	1. Recycling crop residues and straws. Lacking of time to mature in the soil will decrease crop production or cause allelopathic effects. 2. After harvesting, the residues and straws need to be plowed into the soil. The soil moisture should be optimal to allow degradation and maturization to take into effect. A period of 1 ~ 2 months is needed before planting of the next crop. 3. Long-term direct application of crop residues in dry fields will result in soil diseases and continuous cropping obstacles, which will greatly influence grain production. 4. The effect of increasing soil organic matter is slow. Fresh straws recycled in soil are easily degraded by microorganisms, leading to slow formation of humic substances.
Conventional Composting	Matured compost can be produced with enough time and space.	1. It is difficult to control the quality of conventional compost when the nutrient contents are low. 2. Composting takes 2 ~ 3 months, during which turning and piling requires much time and labor. If machinery devices are used, the cost will be high. 3. Large space and house for composting is required, and they requires investment on machineries and facilities 4. Composting in the open area will cause wastewater pollution; composting with shelter takes big space and costs higher. 5. Composting can easily produce odor and provide the propagation environment for pests, which is even worse when wastes of poultry and livestock or meals are added.

Table 8-6. The Advantages and Disadvantages of Different Recycling Methods for Crop Residues and Straws (Continue)

Method	Advantages	Disadvantages
Fast Composting	1. Saves big space as compare to conventional composting. 2. No second wastewater, odor, pest propagation and pollution to the environment. 3. The quality of compost product is stable. It is easy to adjust the nutrient contents to adapt to different crops.	1. It also requires investment on machineries and facilities. 2. Costs of microbial reagents for every batch of production, similarly to the addition of microorganisms in conventional composting method.

8.8 The Application of Organic Wastes

8.8.1 The Necessity for the Development and Application of Organic Wastes

Agricultural production is essential for human survival. However, long-term utilization of farmland often leads to the soil fertility decline. One of the best approaches to improve soil fertility is to increase the content of soil organic matter, which is a significant indicator of soil fertility. The use of organic wastes is a chief way to maintain the content of soil organic matter, a necessity for sustainable development of agriculture and an indispensible task in modern agricultural management.

The recycling of agricultural organic wastes has many effects. It can directly increase the soil organic matter content, provide nutrients for plant growth, improve the availability of soil nutrients, and it can also indirectly develop the physical, chemical and biological properties of the soil. There is a large variety of agricultural organic wastes with quite different compositions. They can be mainly categorized into two kinds based on the C/N value, wastes from high C substances and wastes from high N substances, with distinct purposes. The characteristics of organic wastes include physical, chemical and biological properties, which affect their application methods. Generally, applying organic wastes is for recycling wastes. Besides, organic wastes contain abundant organic matter and nutrients needed by plants. Properties such as C/N ratio, pH value, EC value and water content are important indicators of the characteristics of the organic matter. Due to the different compositions of organic wastes, their application technologies vary according to different characteristics.

There are two common ways to deal with organic wastes before applying them to soil: one is composting before application and the other is direct application without composting. Composting takes advantage of the maturing process where organic wastes are degraded by microorganisms to reach the purpose of stabilizing the compost before application. On the other hand, direct application of organic wastes without composting achieves its application purpose by making use of soil microorganisms to degrade the wastes. Examples of such application include rice straws application to wet field and meal application to orchard. It takes some time in the soil for direct application method to show its effects. According to different application rates, plants can utilize a large quantity of released nutrients only after 2 ~ 4 weeks. Therefore, direct application of rice straws should be conducted before rice planting, and application of meal to orchard should be conducted a few months before blooming of deciduous plants or after picking fruits. In order to achieve the goal of waste recycling, the choice between the two application methods should be based on the characteristics of the organic waste and the growth conditions of the crop. These two methods have their respective advantages and disadvantages and the stabilizing processes are completed under different conditions. The stabilizing process in conventional composting is conducted in composting areas, while the direct application method without composting is conducted in the soil after application. Anyhow, their common goal is to reach stabilization and provide nutrients for crop growth.

8.9 The Effects, Production and Changes of Compost

8.9.1 The Effects of Compost

There are various sources of compost. Various organic substances have different characteristics and thus show distinct effects on crop growth. Generally, organic matters can be categorized into easy or difficult degradation. The main purpose of the easy degradable composts are to provide nutrients for plants, while the difficult degradable composts can persist in soil for longer time and thus can be utilized to improve physical and chemical properties of the soil.

The following specifies the effects of compost application to soil from the physical, chemical and microorganism aspects of the soil:

Improvement in Soil Physical Properties:

Improve Soil Structure:
Compost particles or fragments can loosen the soil; increase microorganism products and thus leading to soil aggregation and improve soil structure.

Soil and Fertilizer

Increase Water Holding Capability of Soil:
Compost can directly improve water retaining or indirectly increase water holding capability by improving soil structure.

Improve Soil Aeration Capability:
Compost can increase the soil oxygen concentration surrounding the roots and allow CO_2 to diffuse readily from roots.

Increase Soil Temperature:
The dark color of organic matters can absorb heat. Moreover, the improvement in soil structure can indirectly improve water draining ability of soil in spring and favors the increase of soil temperature.

Improvements in Soil Chemical Properties:

Increase Nutrients Storage in Soil:
Exchangeable or stored forms of surface organic matters are especially important for soil with less clay fractions.

Supply Nutrients and Energy to Plants after Degradation:
During degradation, organic matters decompose into inorganic (e.g., N, P, S, microelements, etc.) and organic nutrients (e.g., amino acid, carbohydrate, etc.). Besides, the released CO_2 can diffuse in the air and provide substrates for photosynthesis.

Degradation Products can Promote the Mobility of Stored Inorganic Nutrients and Improve Their Availability:
Degradation products can directly improve the efficiency of soil nutrient release. Indirectly, the acidification of the degradation products or the decrease of redox potential can improve the availability and mobility of reduced-state elements, such as Fe, Mn, P and Mo.

Immobilization of Nutrients:
At the early stage of composting, when the C/N ration is larger than 30, microbial bloom will occur and absorb a great amount of nutrients from the organic matter. This is called the immobilization of nutrients, and often results in the competition of nutrients between crops and microorganisms.

Compost Contains Active Substances for Plant Growth:
During the decomposing process, various organic matters can release large amounts of different organic compounds, including crop growth stimulators and inhibitors or possibly substances similar to antibiotics that can directly affect crops or soil microorganisms.

Impact on Soil Microorganism:

- Compost provides nutrients and energy for soil microorganisms. Composts consist of various different components that can influence their degradation

rate. The degradation of lignin is slow, and very few microorganisms can utilize lignin.
- Humic substances with low content of N can improve the nitrogen-fixation ability of soil microorganisms, which can fix N_2 in the air and produce nitrogen compounds that can be utilized by organisms.
- Increasing soil benefits microorganisms in the soil which is an important task, as they can function as antagonistic agents towards pathogens.

8.9.2 Selection of Raw Materials and Methods for Composting

The Selection and Blending of Raw Materials:

The residues of various animals, plants and microorganisms can be used as raw materials for composting, as long as they do not contain toxic substances or pathogens that are difficult to eliminate. Due to the different compositions (refer to the Table 8-4 "Main Nutrients in Organic Matters" in the chapter "Application Guidelines for Organic Fertilizers" in this book), the raw materials require blending with other materials in order to balance the nutrients or fulfill particular purposes. The quality of the compost is affected by the materials for blending. The usual blending principles are (1) easy degradable materials should be mixed with materials that are difficult to degrade; (2) fine or sticky materials should be mixed with coarse or loose materials; (3) acid materials should be supplemented with alkaline materials; (4) plant materials should be blended with animal materials; and (5) low-nutrient materials should blend with high-nutrient materials.

Raw materials that can be used for compost include wastes of crop production, livestock industry, manufacturing and daily sewage. The degradation features, water holding capabilities and C/N ratio of these wastes are summarized in Table 8-7.

Selection of Composting Methods:

Based on different devices and facilities, composting methods can be categorized into horizontal type, rotary type, and spatial type each with their respective pros and cons. Horizontal type is the most common method applied by farmers and can be divided into track and non-track turning types. Rotary type uses revolving tubes while spatial type applies multiple layers for the replacement of compost. Selection of the composting method should be in accordance with particular requirements.

Composting Devices and Process:
- Process Flow:
  The process flow depends on the material source for composting. The general process is listed as follows with raw material shredding blending and adjusting moisture, pH value, C/N piling turning (for several times) sieving product.

Table 8-7. The Characteristics of Raw Materials for Compost

Categories	Microorganism Decomposition Features	Water Holding Capability	C/N Value
Wastes of Crop Production			
Grain Meals	light degradability	low	low
Vegetable Leaves	light degradability	low	high
Rice Straws	light degradability	medium	high
Hulls (Rice Hull, Peanut Hull)	persistent	low	high
Maize Pulp	persistent	medium	high
Banana Stem, Pineapple Peels, Bamboo Shoot Skin	persistent	medium	high
Bagasse, Residual Plants	persistent	medium	high
Mushroom Sawdust	strongly persistent	high	high
Wastes of Livestock			
Waste of Pig and Poultry	light degradability	medium	low
Waste of Cattle	light degradability	medium	medium
Wastes of Agricultural Manufacturing Products			
Fish Wastes	light degradability	low	low
Wood Wastes	persistent	high	high
Bamboo Wastes	persistent	medium	high
Wastes of Urban Daily Life			
Kitchen Wastes, Paper, Fibers	light degradability	medium	medium

Fine materials such as crop meals can be directly mixed and piled with other sources without shredding for this process.

- Composting Devices:
  Devices are important equipments to assist in the composting process. The more complete these devices are, the more efficient and automated the process is. In the meantime, with better devices comes higher investment and higher quality. Diverse equipments for selection are listed below:

(1) Storage Areas or Bins for Raw Materials:

Different devices serve different materials, as such that the main focus is to prevent massive loss of materials by rain, wastewater discharge or generating foul odor.

(2) Building Facility:

Building facility should be rain-proof, and with deodorizing and air suction, even negative pressure functions. There should be aeration devices under the building facility ground in order to provide sufficient oxygen for composting process.

(3) Turning Devices:

Turning devices can turn the compost inside out and perform the functions of mixing and shredding after a certain time of composting. The turning device includes track and non-track types.

(4) Drainage and Wastewater Disposing Plants:

Some raw materials contain much water and will flow out during the composting process, and thus it is necessary to collect the excessive water and sprayed back to the compost or drained out for further management.

(5) Sieving Devices:

Sieving may be performed via human labor or with machinery.

(6) Shredding Devices:

Different raw materials based on their hardness and size that require different shredding devices to shred the raw material or the product.

(7) Weighing and Packaging Devices:

There are automatic, half-automatic and manual weighing/packaging devices.

(8) Transmission and Transportation Devices:

There are various devices for the transmitting and transporting of raw materials and products.

8.9.3 Properties and Changes During Composting

Compost is an organic fertilizer produced from the degradation and maturization of organic substances (Figure 8-3). The changes in physical, chemical and microorganism properties during the composting process are specified as follows:

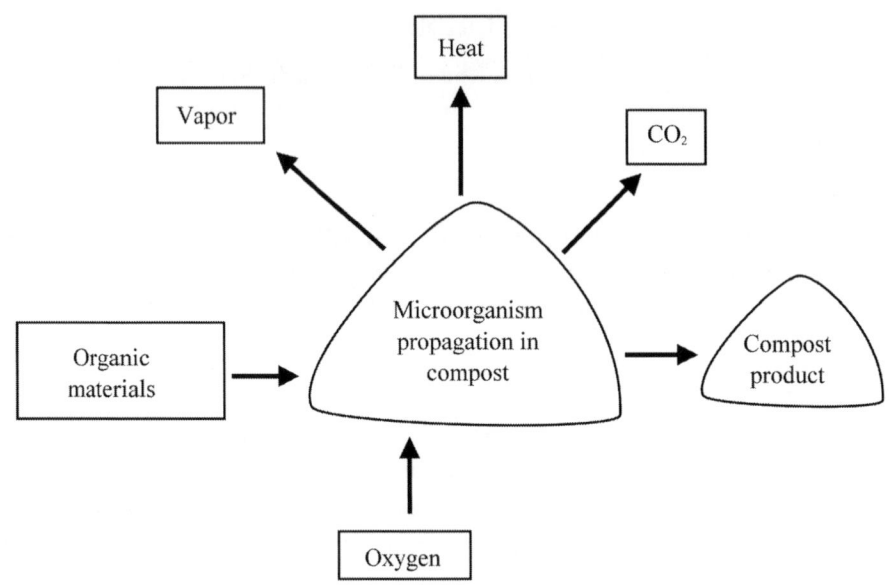

Figure 8-3. Illustration of the Composting Process.

Changes in Physical Properties:

Decomposing of Structure:
During the early period of composting, the structures of various materials based on their characteristics, will change from thick and long to slender and short, or even to powder particles. The longer period of the decomposition takes place, the smaller structures of organic fertilizer.

Loss of Water and Heat:
Sufficient water in compost can promote the softening of plant stalks and the activity of microorganisms, along with which temperature will be increased and heat produced. Heat emission will result in the water content decrease during the composting process. Therefore, addition of water is necessary whenever needed in order to complete the maturing process, or it will be slow down due to the shortage of water.

Temperature changes during composting process can be separated into three stages: high-temperature period (above 60°C), medium-temperature period (50 ~ 60°C) and low-temperature period (below 50°C). High-temperature period usually last about 10 ~ 14 days when pathogens, weed seeds or pest eggs are killed. The degradation of cellulose and lignin occur in the later two stages.

Decrease of the Volume and Dry Matter Content:
Degradation during the composting process will consume carbon and result in the release of CO_2, and thus resulting in the decrease of the volume and dry matter content.

Color:

Various materials for compost have different colors, and yet mature compost normally appear dark or dark brown in color.

Changes in Chemical Properties:

pH Value:

Due to different origins of raw materials, the original pH value of the compost varies. During the composting process, acids are initially produced and afterwards the pH value will increase gradually until it reaches the equilibrium point. Generally, lime is added into the compost to accelerate the maturing process, so that the pH value will be above 7. However, if the pH value is too high, ammonium in the compost will easily escape and lead to the loss of nitrogen.

EC Value:

Due to various origins of raw materials, the original EC values are also different. During the composting process, the EC value will usually gradually increase until it reaches a certain balance point. The more salts (such as fertilizers) or soil conditioners are added in the compost, the higher the EC value is.

Eh Value:

The formulations of materials in compost and its water content have a great influence on the Eh value. Under anaerobic conditions, the Eh value will be low if the water is in excess and will increase due to the loss of water. Additional massive water to the compost will lower the Eh value.

C/N Ratio:

A large quantity of CO_2 is released during the composting process which results in the decrease of the C/N ratio. Generally, when C/N ratio drops below 20, it indicates that the compost is matured. However, there are also exceptional cases. For example, the C/N ratio of fresh soybean meal with high nitrogen content is already below 20 before composting.

Soluble Organic Compounds:

Massive soluble organic compounds are generated due to their degradation during the composting process, such as organic acids, carbohydrates, fatty acids, amino acids, etc. At the later stage of composting, these compounds will gradually be degraded and mineralized into CO_2 or inorganic nitrogen.

Odor Substances:

During the composting process, different odors are produced from different materials with various water contents. Feces and crop meals can easily produce bad odor. Under high water or nitrogen content and anaerobic conditions, odors occurrences will increase.

Humic Substances:

During the composting process, humic substances are formed from the polymerization of organic matter and nitrogen substances or from modified lignin structures. Their compositions and structures are in fact different from the humic substances originally in the soil. Humic substances from composting can be degraded more easily than those are naturally formed over time.

Enzyme Activity:

Enzymes released by microorganisms or residues during composting process will increase after the initial stage, until they reach a stable state.

Changes in Microorganism Properties:

During the composting process, the varieties and quantities of the microorganism ecology change according to different materials and the types of compounds. Usually, carbohydrates, amino acids, fatty acids and proteins are first consumed by various bacteria and actinomycetes, and with actinomycetes as the major players at high temperature. At the later stage, celluloses, hemicelluloses and lignin are mainly utilized by fungi, actinomycetes and high-temperature anaerobic bacteria.

8.9.4 Conclusion

The changes during the composting process are complicated where a diverse organism contributes to the natural degradation of the organic material. Therefore, materials need to be as natural as possible to favor the survival of the microorganisms. We can adjust the formula, add beneficial microorganisms, control the moisture, turning time, temperature, and oxygen conditions to provide the best physical, chemical, and microorganism properties. By this mean, foul odors will be reduced and achieve the composting process with fast, efficient and high quality.

8.10 Stabilizing Technologies of Composting and Composting-Free Technology

8.10.1 The Stabilizing Process of Composting

Stabilization of organic wastes by widely known and applied composting technology was a long term task in agriculture studies. Composting is to turn solid wastes into stable and non-toxic product through the degradation process of organisms under controlled conditions. The compost product is stable and non-toxic to the environment and is easy to manage, store and apply to the farmlands. The stabilizing of organic wastes into compost is favorable to plants, as it can efficiently release nutrients, and be in accordance with sanitation and environment regulations.

During the composting process, the degradation process of microorganisms can break down large-molecular into small-molecular weight organic matters. Part of the organic matters can even be degraded into inorganic matters and become stable elements. This degradation process by microorganisms occurs at portions of the organic matter that is in direct contact with microbial enzymes, such as carbohydrates, proteins, fatty acids, etc. On the contrary, the degradation of lignin and cellulose are relatively slow as they are in the cell wall. Therefore, rich organic wastes in lignin or highly developed organic wastes take more time to degrade through composting. For example, sawdust will take at least half to one year to compost, unless other additives or microorganisms are added to accelerate the process. Generally, animal wastes have high contents of protein, so they can be degraded relatively fast. Reactions during composting include degradation, polymerization, energy release, microbial assimilation, metabolism and secretion. Organic wastes during the composting process produce small molecules, water soluble or insoluble organic compounds, and products from mineralization, such as CO_2, NH_4^+, N_2, N_2O, NO_3^- and other inorganic compounds. CO_2, N_2, N_2O or NH_3 will be emitted into the air while NH_4^+ and NO_3^- will remain in the compost. Among the emitted elements, CO_2 takes up the highest portion, and will gradually result in the C/N ratio reduction along the composting process. The produced heat will rise up the temperature and result in the water loss and CO_2 emission into the air, leading to the weight of organic wastes loss up to 50 ~ 60% during composting process. However, the composting process does not completely degrade from organic matters into inorganic matters. The applied-compost soil will continuously degrade.

Organic wastes of different compositions require different length of time to reach stabilization. The C/N ratio and lignin content are two main indicators for achieving stabilization in the required time. The reproduction, growth and degradation action of microorganisms are influenced by the organic materials and the composting method. Therefore, the matured compost will also exhibit different characteristics and compositions. It is rather difficult to evaluate the maturity degree of compost with a single standard or measurement. Regardless of the maturity degree of the compost, the purpose of recycling organic wastes by composting can be achieved as long as the application time, amount and methods are closely monitored.

8.10.2 Technologies to Accelerate the Stabilizing Process of Composting

The objective of managing organic wastes with composting approach is to avoid their damage to crops, soil and the environment when being applied in lands. Composting of organic wastes is necessary under the following conditions:

- Organic wastes, such as weeds, residues, and withered wood that contain pathogens could result in crop disease or pest issues.

- Organic wastes are common to both human and livestock that contain pathogens and diseases, such as feces.
- High in C/N ratio and low nutrients' release in organic waste materials, such as sawdust, bark, bagasse, etc.

To accelerate the stabilizing process of composting with various materials, the main mechanisms are to accelerate the propagation and metabolism of microorganisms, as well as eliminating pathogens. The propagation and metabolisms of microorganisms are affected by the physical (water, aeration, temperature, etc.), chemical (pH, EC, carbon content, nutrients, etc.) and biological properties (varieties and amount of microorganisms) of the raw material and composting conditions. When the compost temperature reaches 70 ~ 80°C, due to the energy release by microorganism metabolisms, it can be available in eliminating pathogens. The following are technologies to accelerate the stabilization process of composting:

Proper Adjustment of Water Content and Porosity:

Because organic wastes are different in water content, water holding capability, porosity (related to aeration) and the water content should be adjusted to about 55 ~ 75% at the highest, and to about 40 ~ 45% at the lowest with proper turning and mixing. The porosity of organic wastes should be of appropriate size in order to allow air to reach every corner and supply enough oxygen for microorganisms to perform the decomposition process. Composting under anaerobic condition should be avoided as it can easily generate foul odor, methane or nitrous oxide. Turning and piling during composting is usually conducted to achieve equal fermentation, optimal aeration and temperature reduction.

Optimal Control of pH Value:

Organic wastes often produce acids during the composting process, and thus it is suitable to add alkaline materials, such as limes, oyster shell powder to about 1 ~ 5%. Excessive addition of alkaline materials should be avoided as it will lead to ammonia evaporation and nutrient fixations

Adjust the C/N Ratio:

C/N ratio is an important indicator of the available nitrogen and maturity of an organic waste. High C/N ratio (> 30) indicates that the available inorganic nitrogen is insufficient, and the composting material is not matured that organic carbon have not yet decomposed into CO_2. Therefore, adjusting the C/N ratio of the organic material by addition of nitrogen fertilizers should be at priority, as it can accelerate the decomposition process.

Adjust Temperature:

The temperature of composting is determined by pile size, material condition, degradability, and environmental conditions. Heating devices are usually not necessary for temperature adjustment unless in the cold season or regions. Temperature of large compost piles can reach up to 80°C while the temperature gradually recedes from the inside out. Heat dissipation is apparent for small-sized compost piles as temperature accumulation is difficult. The temperature during composting fluctuates through medium temperature, high temperature and cooling stages and eventually reach maturity. When it is above 20°C, mesophiles start propagating and metabolism activates. They will be replaced by thermophiles when the composting temperature is above 40°C. And, up to 60°C is the most suitable temperature for composting. When the temperature is above 80°C, the activity and metabolisms of both mesophiles and thermophiles will cease.

The components and conditions of raw material compost influence the increase of the temperature. High content of ingredients that can be easily decomposed (e.g., carbohydrate, protein, cellulose, hemicellulose, etc.) will make the decomposition process easier and allow the temperature to raise up rapidly. Therefore, organic wastes that are difficult to degrade should supply with materials that are readily degradable (e.g., feces, meal, etc.) and alkaline materials in order to control the pH value and to create a favorable condition for the growth and metabolisms of microorganisms. The increase of the compost pile temperature is a phenomenon accounted for the decomposition process, which is helpful to eliminate pathogens contained in the compost.

Controlled by the Addition of Microorganisms:

Under optimal conditions, microorganisms present in the organic waste and the environment will begin to propagate at the early stage of composting. During the process, varieties of microorganisms undergo changes and eventually result in the maturity and stability of the compost. During this process, the addition of massive bacteria that are favorable to microorganisms will speed up the decomposition process and reduce the required time for maturation. These various microorganisms, generally including bacteria, actinomycetes and fungi, are called "composting microorganisms." In recent years, the sources of organic wastes have become diverse, and the additional microorganisms decomposition for particular harmful organic pollutants' degreadation can be beneficial for the process. Earthworms can also be used to assist the degradation process. Research and development on microbial fertilizers (biofertilizers) and multi-functional compost products will result in multiple benefits in compost quality.

8.10.3 The Stabilizing Process of Organic Wastes by Composting-Free or Fast Treatment Technology

Among organic wastes, crop meals are residual extracts that normally do not contain pathogens. There are over million tons of meal (e.g., soybean meal, rapeseed meal, peanut meal, etc.) applied directly to farmland as basal fertilizers, especially in orchards after fruit picking or in vegetable fields during land cultivation. Application of these organic wastes in soil will result in massive propagation of microorganisms and subsequent degradation activities. At the initial stage, the immobilization of nutrients by microorganisms or the production of organic toxic substances with high C/N ratio will result in apparent nitrogen deficiency. At this stage, nutrient immobilization will not cause problems as crops enter winter dormancy or are at pre-plantation stage. When crops require nutrients, organic wastes have already been degraded to an extent that they can provide nutrients for plants or the organic toxic matters have been degraded. Due to their high nitrogen content (> 5%) and readily degradable characteristics, meals can be directly applied without composting, which indicates that the stabilizing process of these sorts of organic wastes is conducted in the soil instead of composting.

After application in soil, organic wastes are degraded by soil microorganisms and undergo a series of biological and chemical reactions, producing many small-molecular weight organic substances and inorganic nutrients that may positively or negatively affect crop growth. For example, livestock wastes and plant residues may generate phenolic compounds after application in soil, and phenolic compounds have been reported to delay or inhibit seed germination and younger seedlings' growth. Acyl fatty acids can also exert similar inhibiting effects. Application of crops residues often produce serious toxic substances in soil and may be a cause of continuous cropping problems. Therefore, it is clear that organic wastes without composting will cause problems if applied directly to soil. Many technical issues need to be overcome in order to apply organic wastes without composting.

8.10.4 Innovative Concepts for Stabilization of Organic Wastes by Composting-Free or Fast Treatment Technology

The target of innovative technology development is to improve the composting process and reduce the required time. Even better, organic wastes could be rapidly transformed and blended. They could be directly applied without long-period composting and could achieve crop production and safety. Composting-free or fast treatment technology avoids composting, saves time, saves labors and avoids pollution issues. Therefore, this technology has been first developed in my laboratory.

Different kinds of organic wastes require different composting-free or fast treatment technologies. The process should be improved and accomplished with physical, chemical and microorganism methods according to the specific characteristics of organic wastes. Stabilization process in composting has its advantages, while stabilization without composting has its disadvantages and problems. Therefore, being aware of the particular shortcomings and making correspondent improvement can help to reach composting-free effects.

Technologies of composting-free or fast treatment are specified as follows:

Physical Method:

Water content, pests and pathogens, and odor controls are the main physical methods to improve composting-free or fast treatment process. Using temperature is the most available method to kill pest and pathogens sources. Odor is the most difficult to deal with. One can either apply microorganisms to eliminate the odor or use adsorption materials to eliminate the source of the odor.

Chemical Method:

Application of organic wastes that are not composted will have many negative impacts on the chemical properties of the soil, and thus it is necessary to overcome the disadvantages of the raw material beforehand with chemical methods.

Microorganism Method:

Direct addition of microbial enzymes with particular functions in the organic wastes can effectively ameliorate their negative impacts on the soil, including nutrients releasing rate, detoxification, root growth promotion, pest and disease control, etc. for the purpose of composting-free or fast treatment process

Among organic wastes, bark, sawdust and bagasse require composting as they contain large amounts of lignin and cannot be easily degraded. Degradation of these wastes can even produce toxic phenolic substances. Although the disadvantage of these wastes with high amounts of cellulose and low nutrient contents, they can be directly applied without composting by supplementing them with nutrients and the addition of favorable microorganisms.

For example, additional N, P and K fertilizers (refer to "crops fertilizer application manual") and selected high-quality beneficial bacterium *Bacillus subtilis* (bacteria solution 2.6%, w/w) to bagasse can be directly applied as a potting medium. The following describes a set of experiment that we have performed using *Brassica rapa chinensis* seeds. Seeds were germinated on different ratios of the above medium with soil (1%, 5%, 10%), with or without fertilization, and harvested after 40 days. The same potting medium was continuously cropped with *Brassica rapa chinensis* for three more harvests to 200 days. The amount of harvested plants for dry weight and nutrients absorption were measured after each harvest. The control group uses original soil without any bagasse addition. Results show that in the first harvest (40 days), the addition of bagasse (1%, 5%, 10%) result in retarded plant growth (2.3 ~ 51% of control). The amount of bagasse added was negatively related to the plant growth. When fertilizer was added into the bagasse medium (1%), plant growth was improved. The addition of fertilizer in higher concentration bagasse media (5%, 10%) did not ameliorate the plant growth. However, the addition of *B. subtilis* can improve the plant growth by about 350% for the second harvest (80 days). These results indicate that bagasse improved by additives can be directly applied without composting and can promote plant growth and nutrient absorption.

8.10.5 Conclusion

Utilization of organic wastes through stabilization processes can be either with composting, composting-free or fast treatment technology. Composting applies piling of the organic wastes to achieve maturity, while directly applied composting-free organic materials are mature in the soil. Composting has many advantages, such as allowing organic wastes to be matured for releasing their nutrients, eliminating pest sources, and being toxic-free. However, composting required space, takes lengthy of time, and requires generating leachate and foul odor. Therefore, improvements in composting stabilization technology need to be developed. In order to accelerate the stabilization process of composting, it is necessary to improve the physical, chemical and biological properties of compost materials. The common methods include controls in water, porosity, pH, C/N, nutrient contents, temperature and microorganisms.

There are several disadvantages for direct application of organic wastes to soil, such as immobilization of nutrients, production of plant toxic substances and residual effects of pests. Except for early application, organic wastes usually require direct improvement to become composting-free products. Direct improvement makes use of physical, chemical and biological approaches so that the reformed organic wastes can be directly applied and they can undergo stabilizing process in the soil without causing any problems.

8.11 Composting and Fast Treatment Technology of Kitchen Wastes

8.11.1 Disposing Problems of Kitchen Wastes

The method of disposing kitchen wastes is still quite conventional in many countries. It is common to dispose them in waste bins, garbage trucks, as pig feeds, or mixed together with other family wastes. The conventional methods will increase the volume of kitchen wastes and generate foul odor that cannot meet the requirements of environmental standards. Kitchen wastes can be further classified into raw and cooked kitchen wastes. The former is basic with mostly uncooked vegetable leaves, while the latter is more complicated with various leftovers of cooked food in either solid or liquid state, including meat and vegetables.

Direct application of kitchen wastes as feed for animals or pigs is considered unsanitary, because there may be pollutants or pathogens in kitchen wastes and also with food spoilage problems.

Kitchen wastes are the most difficult for composting as they consist of oil, water and food residues with water content as high as 80% ~ 90%. The treatment of kitchen wastes by incineration is costly. Treatment of kitchen wastes by burying will produce methane gas pollutants, which will affect environmental health, result in uneasy protests from nearby residents, and bring difficulty to acquire the land for waste disposal purpose. If dried by the heat, the wastes will produce sewage during filtering process and occupy the related device for 2 ~ 3 days and will consume much energy and cost. In order to manage massive kitchen wastes, it is necessary to develop fast treatment technology, which is an issue worthy to pay attention to by the public.

8.11.2 Composting Problems of Kitchen Wastes

Kitchen wastes cannot be directly applied to soil as a fertilizer is due their constituents, including:

- Oil will harm plants and cause poor aeration of soil and nutrient mobility issues.
- They can easily result in microorganism bloom, unbalance soil nutrient and toxicity issues.
- Large distinctions in C/N ratio.
- Problem of sodium salt.

Consequently, kitchen wastes should undergo composting and maturing before they can be applied as fertilizers.

Municipal kitchen wastes generally contain much oil and water. The main components are water (80% ~ 90%) and organic matters (10% ~ 20%), such as oil, carbohydrates (starch, cellulose, etc.), organic acids, proteins, lignin, bones, shells and inorganic matters such as sodium salt. Lignin, bones, shells and oil are hard to degrade, while the others are easily degraded and will produce acids and odor within a day.

Composting of Conventional Kitchen Wastes:

The composting of conventional kitchen wastes is conducted in composting factories. Common problems include:

- Kitchen wastes produce massive foul smell. Deodorization unit in composting factories is heavy-loaded and will be saturated in a short time, which tends to raise protest from residents.
- Water absorption requires large amounts of dry materials, which is over 50% of kitchen wastes and hard to acquire in recent years.
- It takes 2 ~ 4 months to complete the composting and maturing processes.
- Pathogens cannot be completely eliminated because of uneven turning.
- The water content of kitchen wastes is high at 80% ~ 90%. The filtrate water is large in volume and very stink and contains sodium salt.
- Composting of conventional kitchen wastes requires large space. The building and maintenance cost for the turning machine is very high. It is rather difficult to solve the above problems as composting of conventional kitchen wastes requires the combination of microorganisms and formulation technologies.

Composting of Kitchen Wastes in Tanks:

Composting of raw kitchen wastes in tanks for a small-scale is common in small communities and families. The common problems include:

- Poor aeration leads to the production of foul smell.
- Propagation of mosquitoes and fly.
- Poor draining and airtight conditions bring to high moisture.
- Compacting of waste leaves in the tank. To solve the above problems, it is necessary to combine with microorganisms and management technologies.

8.11.3 The Characteristics of Kitchen Wastes' Fast Treatment

Fast treatment of kitchen wastes is a subject of great technical difficulty. The author's research lab has developed fast treatment technology for kitchen wastes, which is the only advanced technique in the world that innovatively combines the technologies of microorganism enzymes, biology, organic chemistry and soil biotechnology. Based on our understanding of the compositions and problems of kitchen wastes, the novel technology overcomes each disadvantage and can produce organic fertilizer instantaneously with stabilization, detoxification and effectiveness.

The characteristics of the fast treatment technology for kitchen wastes are innovative and correspond to environmental protection concepts. Most importantly, the innovative technology can process kitchen wastes within 1 ~ 3 hours and can completely recycle resources with a small area and pollution-free (no leachate, no odor, and no toxic). This technology utilizes the entire nutrients in kitchen wastes and produces a product with high available, nutrient-balanced and complete organic fertilizer.

8.11.4 Effects of Kitchen Waste Organic Fertilizer Produced from Fast Treatment Technology

According to different requirements, fast treatment of kitchen wastes can produce liquid or solid organic fertilizers, as specified below:

Liquid Fertilizers from Kitchen Wastes:

Taking advantage of the fast treatment technology, 80 ~ 90% water content from kitchen wastes can be directly produced into liquid organic fertilizers. This process can completely utilize kitchen wastes, so no wastewater is produced. The fundamental formula composition of liquid fertilizers is a ratio of $N:P_2O_5:K_2O$ at 1%:1%:1%. Based on distinct origins of kitchen wastes, the organic matter and water content is about 10% ~ 15% and 80 ~ 90%, respectively. According to crop requirements, the N, P and K contents of the liquid fertilizer formula can be adjusted with organic and inorganic materials. Similar to ordinary organic fertilizers, kitchen waste liquid fertilizers can be applied as basal fertilizer or additional fertilizer for crops. Optimal adjustment of the application rate should also in accordance with the crop variety and soil condition. According to practical experiments, it is shown that 200 ~ 2,000 fold dilution of fast composting kitchen waste fertilizer can increase the growth of lettuce's younger stems up to 19% ~ 63%; 200 fold dilution can increase 25% of the younger roots' growth. Massive application of fast composting kitchen waste fertilizer in field practices has been proven to show prominent effects. Fast treatment kitchen waste fertilizer is as good as, if not better than, conventional compost in their function of promoting growth and mass plant for Chinese cabbage and green cabbage. Based on experimental results, it is recommended to use 200 ~ 600 fold dilution for direct application to crops and apply 2 ~ 10 tons/ha in land cultivation according to different soil fertility.

Solid Fertilizers from Kitchen Wastes:

Kitchen wastes contain large amounts of water, which has to be removed through filtering in order to produce solid organic fertilizers. Based on the distinct origins of kitchen wastes, the nutrient compositions in the solid fertilizers are very different. For instance, if the organic fertilizer contains of $N:P_2O_5:K_2O$ at a ratio of 2.5%:3.0%:2.8%, and 70% organic matter (dry matter) and 30% water, the application rate can be 1 ~ 5 tons per hectare. Excessive application on the other hand will result in crop or soil damages.

The performance effects of kitchen waste fertilizer are the same as organic fertilizers. With proper adjustment in formula based on different crops requirements, kitchen wastes fertilizer can provide nutrients after the organic matters are degraded, which can promote root growth and increase crop quality. The liquid fertilizer can be applied together with other fertilizers for particular purposes. Kitchen waste fertilizer has the effects of improving other fertilizers' function and can reduce chemical fertilizer applications. The particles and soluble fertilizers in the fertilizer can act as organic acids. Application of kitchen waste fertilizer can utilize deep-ditch method to improve soil properties, decrease soil degradation and achieve sustainable development. With its various effects, the application of kitchen waste fertilizers is worthy of popularization.

8.11.5 Conclusion

The quantity of kitchen wastes from our daily life is amazingly huge. If every ten people produce 1 kg of kitchen wastes per day, then a 23 million population of people will produce about 2,300 tons of kitchen wastes each day. Hence, it is really an important environmental protection policy to develop the multipurpose fast treatment kitchen waste fertilizer technology. In its raw state, kitchen wastes cannot be directly applied to farmland, but this newly developed fast treatment technology can solve the kitchen waste problems efficiently. Moreover, the organic fertilizer product can be directly applied to improve soil and crops. It is the only innovative technical breakthrough in the world.

The innovative fast treatment technology of kitchen wastes combines microorganism enzymes, biochemical organic reaction and the soil technology. It requires short reaction time (1 ~ 3 hours), occupies small space for devices, and produces zero secondary pollution wastes (no wastewater, smell, or toxic substances produced during the process). Fast treatment technology offers a most environmental friendly new choice for recycling kitchen wastes based on distinct requirements. The technology not only solves environmental issues, the fertilizer product can also be applied directly to farmland for improving soil fertility; meanwhile, the application of kitchen wastes fertilizer can reduce chemical fertilizer application and decrease soil degradation, so as to fulfill the goal of farmland sustainable development and achieve high production, environment and daily life quality.

8.12 Guidelines for Characteristics and Application of Bark Compost

8.12.1 The Origins of Tree Bark and its Value

Tree bark compost is a kind of commercial organic fertilizer. The main source material of compost is tree bark, while the largest amount comes from pulp mills. After years of fermentation the tree bark can be made into organic fertilizers. Tree bark is a kind of residues in wood industry, so there are various bark origins for commercial organic fertilizers. Only after decades, hundreds, and even thousands of years growing, trees in the forest can be logged for the use of human being. Xylem becomes wood and the phloem becomes bark. Bark is no less valuable than wood, so it is a pity to burn bark because hundreds of years growing will be ruined within a sudden. The organic carbon will go back to the atmosphere in the form of carbon dioxide. But if tree bark is made into compost, it can perform several effects in the same time. It can go back to, nourish soil, plant and delay the return of carbon element to the atmosphere. Besides, it can also improve soil and promote crop growth.

8.12.2 Characteristics of Bark Compost

Bark compost makes from natural tree bark. The organic matter content is high because there is little contamination and are few opportunities to be mixed with impurities. The containing lignin is more than other organic fertilizers. Lignin decomposes slowly in soil and is the main source of soil humic substances. The humic substances directly or indirectly transformed from lignin are good soil conditioner, with decomposition only faster than "peat" and good capabilities of water retaining and fertility maintaining. Mature compost is less likely to bark bred worm and odor smell. Generally, if nutrient content is high, the electric conductivity and fertility is also high. Immature bark should not be applied directly to crop roots. Bark contains large amounts of lignin and can release a lot of phenolic substances. Besides, it absorbs a lot of nitrogen and microelements in the decomposition process, which can easily lead to the lack of nitrogen and microelements in crops. During bark composting, part of carbon is released and forms carbon dioxide, and thus reducing total carbon content and increasing total nitrogen content. Correspondently, C/N ratio drops and crude ash content rises.

8.12.3 Good Qualities for Mature Bark Compost

- No odor smells, no mixed up with other materials.
- High maturity degree, low C/N ratio (below 20).
- No thick and bulky appearance, low water content (below 30%).

- No excessive heavy metals (according to the National Fertilizer Standards).
- No toxic substances harmful to plants.

8.12.4 Application Effects of Bark Compost

The biggest advantage of compost bark among organic fertilizers is that its decomposition is slower than other manures and bean meals. It belongs to slow-released organic fertilizer and soil conditioner. Lignin is an important raw material which can be transformed into soil humic substances. As a source of soil organic matter, the application effects of bark compost are listed below:

- As soil conditioner, it improves soil physical properties and promotes the formation of soil aggregate structure.
- It loosens up soil and increases water holding capability.
- It decomposes all kinds of essential nutrients that are provided for plants.
- It adsorbs and exchanges applied fertilizers, increases the fertilizer maintaining capability of soils, and increases the ability of slow-release of fertilizers.
- It provides the microbial carbon, energy and nutrient elements which are beneficial to the reproduction and activity of microorganism.
- It promotes the decomposition of man-made or natural toxic substances.
- A small amount of organic products or similar to the hormone growth from decomposition can directly help plants.

8.12.5 Application Range of the Bark Compost

There is a wide application range for mature bark compost. It can be applied on flat and slope land, can be applied to trees, vines, woody, agronomic crops, horticultural crops, grasses, and so on. Besides, it has specific potential in application to garden design, landscape planning, forest cultivation and lawn. It's high temperature and rainy in the region with clearly separated wet and dry seasons in a year, so that the decomposition of organic matter is quite fast. Therefore, it is necessary to apply the uneasily decomposed organic fertilizers and soil amendments (such as peat and mature bark compost), especially to the orchard on the slope land. Only in such way can the perennial orchard soil be taken good care of. Otherwise, if only labile organic fertilizers are applied, the abundant application will easily lead to nutrient loss and pollution, and the soil cannot be available and stay lasting preservation. Therefore, high-quality mature bark compost is a good soil improver for slope and flat lands.

8.12.6 Application Rate, Application Time and Attentions for Bark Compost

The application rate depends on the soil fertility. It can achieve the best effect to be applied as basal fertilizer. To reach the best effect, bark compost should be combined with appropriate added chemical fertilizer (N, P, K, and Mg fertilizers) in periods of basal fertilizer and top-dressing fertilizer. If bark compost with high nutrients should not directly contact with seeds, and should be mixed with soil beforehand. Otherwise, it may generate fertilizer injury or a temporary inhibiting phenomenon, just like why seeds fear chemical fertilizers. For the convenience of application reference, the respectively suggested application rates and application methods for different types of crops are listed in Table 8-8.

8.13 Housefly Propagation by Organic Fertilizer and Improvement Methods

8.13.1 Problems in Housefly Propagation by Organic Fertilizer

Warm season and polluted environment is very favorable for the growth and propagation of flies. With great adaptability, housefly (*Musca domestica*) accompanies human activity and lives on organic matter, such as discarded kitchen wastes, food and livestock waste. Housefly is presently one of the major pests because it not only disturbs human peace and damages city sight, but, more importantly, transmits diseases, does harm to human health, lowers daily life quality and carries great environmental health danger to human. Housefly can fly for 32 km, and for 12 km within 24 h, and their group can spread 5 ~ 6 km.

Serious propagation of flies often happens in garbage piling area, animal and chicken farms. Besides, organic fertilizer is also becoming a popular site for it. In recent years, the effectiveness of organic fertilizer has received approval, and farmers are gradually changing their concept of fertilizer application. Therefore, they apply much compost of livestock waste and organic fertilizer, such as compost from waste of chicken and pig, and meals. The application is also increasing year by year. Although this can improve soil fertility and reduce the bad influence of chemical fertilizer on farmland and the environment, the generated damage of flies on daily life and the environment is common in countryside and suburbs. Part of the reason is that serious outflow of countrymen results in the shortage of labor, and that application methods and specific fertilizers cannot get enough attention. The amount and density of flies propagation is closely related to organic matter and moisture of the environment. The best environment for the growth of fly larva is with a high content of water and organic matter; for example, rotten animal bodies. Moreover, the research on flies in garbage area shows that the propagation maggots are ones of the largest quantities in garbage before maturation. There are fewer amounts in garbage during maturing. No maggots in maturation exist in garbage because it can no longer provide maggot with nutrients as feed. In addition, climate can also

Table 8-8. The Application Rates and Methods of Bark Compost for Crops (only for general reference, actual application rate should be based on soil fertility and crop age)

Crop Category	Reference Application Amount of Tree Bark Compost	Application Methods
1. Citrus fruits: orange, Liu orange, citrus, pomelo, white pomelo, grape pomelo, lemon	Below 3 years: 2 ~ 5 kg/plant Above 3 years: 5 ~ 8 kg/plant	Applied as basal fertilizer or supplementary fertilizer: make ditches and apply the fertilizer to soil. It's better to mix soil and blend evenly after application.
2. Loquat, wax apples, star fruit, Mango, custard apple, banana, Litchi, guava, longan, etc.	4 ~ 10 kg/plant (based on the size of the tree)	As above basal fertilizer application methods
3. Deciduous fruit trees: apple, pear, peach, persimmon, plum, plum.	Below 3 years: 2 ~ 5 kg/plant Above 3 years: 5 ~ 8 kg/plant	As above basal fertilizer application methods
4. Flowers: (1) Floral Type (2) Medical Type (3) Plant Nursery	200 ~ 250 kg/0.1 ha 150 ~ 200 kg/0.1 ha 1/5 ~ 1/10 of mixed media	Basal fertilizer: land preparation after overall spread
5. Papaya	2 ~ 3 kg/plant	Applied as basal fertilizer or supplementary fertilizer: make ditches and apply the fertilizer to soil. It's better to add soil and blend evenly before application.
6. Grapes	4 ~ 10 kg/plant (based on the age of the tree)	As above basal fertilizer application methods
7. Leaf-flower vegetables: celery, mustard leaf, spinach, pakchoi, Chinese cabbage, broccoli, kohlrabi, etc.	High-attitude cold areas: 200 ~ 400 kg/0.1 ha Flat areas: 100 ~ 200 kg/0.1 ha	After overall spread or band placement, turn it into soil, prepare land, and make furrow and field planting.

Table 8-8. The Application Rates and Methods of Bark Compost for Crops (only for general reference, actual application rate should be based on soil fertility and crop age) (Continue)

Crop Category	Reference Application Amount of Tree Bark Compost	Application Methods
8. Fruit vegetables: Sweet pepper, hot pepper, tomato, eggplant, watermelon, strawberry, cucumber, watermelon, melons, bitter gourd, sponge gourd	100 ~ 200 kg/0.1 ha	Mix the fertilizer into soil in land preparation after application
9. Onion- garlic categories: onions, garlic, leeks	100 ~ 200 kg/0.1 ha	Prepare land and make headlands after overall spread of the fertilizer
10. Bamboo stem and root category: asparagus, bamboo, ginger	150 ~ 300 kg/0.1 ha	Applied as basal fertilizer: land preparation after spread. Mix it in soil.
11. Special crops: (1) tea tree (2) areca (3) lao flower	200 ~ 300 kg/0.1 ha 2 ~ 4 kg/plant 1 ~ 3 kg/plant	Applied under canopy between tree rows after tea picking. It is better to plow it in soil and cover it with soil.
12. Grasses: lawn (golf course and lawn)	200 kg/0.1 ha 100 kg/0.1 ha	Comprehensive blending with sandy soil or direct application of half quantity grass in construction. Use immediate spray to make it seep into soil.
13. Paddy fields categories: rice, taro, water bamboo	50 ~ 100 kg/0.1 ha	Land preparation after overall spread.
14. Grains and others: sorghum, corn, red sugarcane, peas, soybean and other legumes	200 kg/0.1 ha 50 ~ 100 kg/0.1 ha	Same above

significantly affect the fly's propagation, especially for temperature and rainfall. Fly growth session can last dozen days to several months, while it only takes 10 ~ 16 days in warm period or summer. Fly groups density tends to decline after continuous rainfall, yet it will raise again in dry period. Consequently, a large number of fly compose a common sight in countryside during basal fertilizer application period when winter has ended and temperature is raising. This problem was unexpected at earlier stage.

Immaturity and high water content of organic fertilizer is the chief conditions to attract fly laying eggs there. Organic fertilizer with low maturity degree (releasing smell) and water content as high as 65% which causes fly propagation most easily. Soybean meal is very likely to be chosen by fly for food and for spawning. The main reason is that soybean meal contains a pretty high content of organic matter and nitrogen. Besides, it is physically stickier than other fertilizers in the early fermentation period, and it also releases strong smell that lures flies. In application, dry animal wastes produce stinky smells with moisture and contain soybean meal feed that has not been completely digested, which is also quite likely to attract flies.

8.13.2 The Improvement Method for Housefly Propagation in Organic Fertilizer

To solve housefly propagation problems in organic fertilizer application, the following approaches are suggested:

Physical Approach:

Physical approach avoids organic fertilizer exposed on soil surface, so as to prevent the problem of housefly propagation in organic fertilization. The most available fertilizer management method is to cover applied organic fertilizer with soil of at least 4 cm deep and press it tight. Besides, the fertilizer can also be covered with plastic film or calcium cyanamide. In compost manufacture process or before packaging for sale, the compost should get mature and be disposed with additional material for prevention. Or by using high heat for killing fly eggs, this solves the problem of fly propagation.

Chemical Approach:

Chemical addictives can be used to kill housefly eggs or housefly larva in organic fertilizer. In organic cultivation, calcium cyanamide can only be applied to fruit trees during their vegetative period and its application rate should not be over 20% than the recommended amount. It is advisable to take advantage of 20% during vegetative period. Besides the nitrogen provided by 60 kg of calcium cyanamide in every 0.1 ha, the rest can be supplemented with other chemical fertilizers or organic fertilizers. If spreading evenly on the surface of applied organic fertilizer in ordinary cultivation, calcium cyanamide powder can not only be used as nitrogen fertilizer that saves pesticide and herbicide, but also promote the integration of organic fertilizer. In organic cultivation of rice, vegetable and tea, it is prohibited to apply any chemical fertilizer or chemicals, so organic fertilizer should be covered with soil, pressed tight together with turned soil, or covered with plastic film as soon as possible after application, so as to prevent fly propagation. Calcium cyanamide is a nitrogenous fertilizer with toxic that acts mainly as a skin and lung irritant. Toxicity of calcium cyanamide is relatively low but can be raised greatly and then can be combined with alcohol.

Biological Approach:

Biological materials can be used to prevent organic fertilizer form attracting fly spawning or to kill housefly eggs and larva. Plant materials (for example lemon grass) and microorganisms are such suitable biological materials. Although they cannot completely inhibit fly propagation at present, they are of practical significance when organic agriculture is actively promoted. Therefore, lemongrass and microorganisms demand further research and development as prevention materials for fly propagation. Besides, retail natural lemon grass oil and biological pesticides (such as *Bacillus thuringiensis*) also need further studies so as to expand their application.

8.14 Production Methods of Liquid Organic Fertilizer

The production of liquid organic fertilizer requires organic materials, devices and microbial inoculants, which are specified respectively as follows:

Organic Materials:

The raw organic materials for liquid organic fertilizer are mainly residues without toxins or products of various animals, plants and microorganisms, including commonly fish soluble, fish meal, meat and bone meal, dried blood, milk, meals, bran, yeast powder, humic acid, seaweed meal, eggs, animal wastes, and so on. The application rate and proportion of organic materials can be adjusted based on different nutrients demands and application purposes. Please refer to approximate composition of organic fertilizers (Table 8-9).

Production Devices:

The production devices of liquid organic fertilizer have different designs. The fundamental design takes fermentation tank as the main body, which has the function of aeration and turning. Bacterial sterilization can be achieved by steam method or high heat and pressure method.

Table 8-9. Approximate Composition of Organic Fertilizers

Organic Matter	Nitrogen (N)	Phosphorus Pentoxide (P_2O_5)	Potassium Oxide (K_2O)
		%	
Rice Bran	2.0	3.9	1.5
Rice Bran Oil Meal	2.5	5.8	2.0
Soybean Meal	7.0	1.3	2.1
Rapeseed Meal	4.6	2.5	1.4
Peanut Meal	6.3	1.2	1.3
Soy Meal	3.0	0.8	0.5
Cotton Seed Meal	3.4	1.6	1.0
Castor Seed Meal	5.4	2.2	1.5
Seaweed Meal	1.5	0.5	4.0
Steamed Bone Meal	4.0	23.5	0
Dried Blood	13.0	2.0	1.0
Feather Meal	13.4	0.3	0.1
Fish Meal	4.8	4.3	0.5
Bone Meal	0	35.0	0

Microbial Inoculants:

In order to accelerate production and get rid of bad smells, it is beneficial to add favorable microbial inoculants, which can bring out complete availableness of favorable microorganism and improve the quality of liquid organic fertilizer. The added bacteria need to be adjusted in accordance with various organic materials, so as to promote fermentation and regulate special functions of products.

Procedures and Time of the Production:

In the production of liquid organic fertilizer, the variety, amount and concentration of added materials should be determined by specific requirements and available cost. When deciding concentration, it is notable that osmotic pressure should not be over the tolerable degree of microorganism fermentation, or microorganism cannot perform the disintegrating function. The length of time for production depends on fermentation temperature, aeration quantity, microorganism amount, carbon, and energy sources. Usually, the production can be completed between 10 ~ 20 days.

8.14.1 Application Guidelines of Liquid Organic Fertilizer

The application guidelines of liquid organic fertilizer are similar to those of solid fertilizer, while the following points need to be paid close attention to:

Note the Maturity Degree of Organic Materials:

Liquid organic materials should not contain a large number of decomposable carbon materials, which can probably cause soil eutrophication and massive microorganism reproduction, and thus inhibit crops growth.

Note Dilution Factor:

If original liquid organic fertilizer with high concentration is applied directly to crops, it will cause damage. Therefore, it usually needs to be diluted before application. The dilution factor should refer to products specifications, and application to foliage and full plants needs to be paid special attention to. Concentration of liquid organic fertilizer applied before land cultivation can be higher than that of applied directly to crops, because plowing in land cultivation takes the effect of mixing and spreading crops.

Note Soil Characteristics:

The effectiveness of fertilizer application is influenced by soil physical and chemical properties. If harmful or limited factors exist in soil, they should be removed for improving fertilizer effects. For example, if soil is with high alkaline, the applied ammonia can easily spread out and disappear; if soil has great capabilities of fixing phosphorus fertilizer (such as laterite), the applied phosphorus fertilizer can be fixed very easily in soil with poor capabilities of retaining nutrients and water, applied fertilizer will suffer from much loss.

Pay Attention to Balanced Fertilizer Application:

Nutrients balance should be paid close attention to fertilizer application. It is not appropriate to apply simple liquid fertilizer excessively. Good fertilization should be considered comprehensively. Excessive nitrogen has bad influence on some crops. The content of major and micronutrients demands attention.

Note the Characteristics of Dilution Water:

Dilution water of liquid organic fertilizer also needs to be noted, because urban wastewater, for instance, has a high content of impurities, which are likely to produce insoluble substances or settlements in liquid fertilizers and affect fertilizer effectiveness.

Notice Application Time:

Liquid organic fertilizer is generally quick-released fertilizer. Liquid fertilizer with a rather high content of nitrogen is not suitable for full-bloom stage, or else the high nitrogen fertilization can probably cause flower and fruit drops. The growing period needs more fertilizer, but too much concentrated application to the roots will easily result in fertilizer injury. Therefore, this should also be noticed.

8.15 Application of Cultivation Media

8.15.1 Cultivation Media Expect Further Attention and Development

Since ancient times, human has always depended on the numerous kinds of crops cultivated in soil to extend life and develop civilization. Intensive farming and massive chemical fertilizer application can easily cause soil problems. In addition, the rapid development of industries and cities is more likely to produce soil pollution and cause sharp reduction of farmland. All of these problems are, unfortunately, common to modern agriculture. In recent years, the development of agricultural seedling cultivation and human focus on environmental greening has greatly increased the demand quantity of media for the crops cultivation. Therefore, the required amount of cultivation media has also relatively increased. Presently, application of cultivation media for crops cultivation tends to increase day by day, and it is the most for gardening crops.

Nowadays, there are various materials used as cultivation media, including organic and inorganic materials. Common organic matter includes peat moss, meals, compost, and so on; inorganic matter includes vermiculite, perlite, foam stone, river sand, and so on. Nowadays, the cultivation media sold in markets is peat moss, which is of high cost in application. Therefore, the development, application and promotion of local cultivation media made from local materials turn to be practically necessary (Table 8-10).

Ideal cultivation media should have good physical, chemical and biological properties. Many organic wastes, such as oil palm empty fruit bunch fiber, mushroom wastes compost, scrap of Cyatheaceae (a kind of tree fern), tree bark, bagasse, rice hull, corn cob, can be taken from local and can develop into excellent cultivation media. If they are skillfully utilized, properly regulated and blended, the cultivation cost will decrease and the industrial competition will be promoted, which is especially beneficial to seedling cultivation and pot culture.

8.15.2 Necessary Properties of Good Media

Various crops require different cultivation media for seedling cultivation and pot culture. For the reasons that different crops need specific media characteristics,

Table 8-10. Various Kinds of Media Formula and Applicable Crops (Adapted from Young, 1995. Handbook for Cultivation Media Made from Local Materials)

Media Name	Materials and Formula	Applicable Crops
Carbonized Rice Hull	Carbonized rice hull: other media (such as peat soil) (1:3)	Potting cultured plants, foliage plants
Compound Carbonized Rice Hull Media (I)	Carbonized rice hull: bark compost: peat soil: artificial soil for greening protection (1:1:1:1)	Chinese cabbage, rape, spinach, water spinach, lettuce, bitter gourd, loofah, crowndaisy chrysanthemum, tomato, etc.
Compound Carbonized Rice Hull Media (II)	Carbonized rice hull: bark compost: loam soil (1:1:1)	Tree seedlings culture
Rice Hull Compost Media	Compost: peat soil (1:1) Compost: peat soil: perlite (1:1:1)	Rose, lily, etc.
Compound Bagasse Media	Bagasse: pig wastes organic fertilizer (3:1) (compost)	Phalaenopsis (moth orchid)
Bagasse Media	Bagasse: soybean meal: cattle waste: crushed rice hull (6:1:1:2) (compost)	Anthurium (Flaming plant)
Charcoal Media	Full quantity of charcoal + small quantity of foam stone	Dendrobium Epiphytic orchids
Scrap of Cyatheaceae (#2) Media, Artificial Soil, Charcoal	Scrap of Cyatheaceae (#2): artificial soil for greening protection (v/v) (5:1)	Dendrobium Epiphytic orchids
Mushroom Cultivation Wastes Media	Mushroom cultivation wastes: clean soil: peat (2:4:4)	Grasses
Needle Mushroom Wastes Media	Needle mushroom wastes: Peat:rice husk (7:2:1)	Flowers
Peanut Husk and Shrimp Shell Media	Peanut husk: shrimp shell media (1:1.4)	Melons

Table 8-10. Various Kinds of Media Formula and Applicable Crops (Adapted from Young, 1995. Handbook for Cultivation Media Made from Local Materials)(Continue)

Media Name	Materials and Formula	Applicable Crops
Corn Cob Media (Carbonized)	Sand:corn cob:organic fertilizer (1:1:1) Perlite:corn cob:organic fertilizer (1:1:1) Organic fertilizer:corn cob:peat (1:1:1)	Flowers
Seedling No 1 Media (TSS-1 Media)	Fine peat: finland peat: HsiHu soil (1:1:2)	Tomato
Rice Husk Media	Husk (100%) + applying liquid nutrient	Vegetables, fruits and melons
Scrap of Cyatheaceae Medium	Scrap of Cyatheaceae (No. 3): sandy loam (1:1)	Ferns
Artificial Soil Media	Artificial soil: bark compost: Scrap of Cyatheaceae (2:1:1,v/v)	Phalaenopsis, four-season orchid (Cymbidium)
Corn Cob Media (Nursery Pot)	Corn cob: paper sheet:resin (1:0.18:0.018)	Water melon, melons, flowers

there are more or less differences in the nutrient requirements of different crops and seedling growth features. Therefore, the standards of excellent cultivation media are also different based on different demands of crops. It is specified to physical, chemical and biological aspects, respectivley:

Media should have Appropriate Physical Properties:

Cultivation media should possess appropriate physical properties, including porosity, texture, structure, and so on. Media physical properties affect their water holding capability, draining capability, roots adhesion capability, and germination rate. Water holding capability influences the frequency of water spray and irrigation; poor draining capability influences germination and root growth; adhesion capability influences on the adhesion of cultivation media with roots when being taken from nursery tray, and thus having an impact on the survival rate of farmland planting. Cultivation media are the basal materials to support plants, so their porosity and compaction will determine the seedling lodging. In brief, physical properties of good cultivation media should be with appropriate capabilities of water retaining, draining and aeration, should ensure high germination rate and fast germination, and should make seedling lodging uneasy. Besides, whether seedling can be easily and undamaged taken from nursery tray is related to the growth state of plants roots.

Media should have Appropriate Chemical Properties:

Appropriate chemical properties required by cultivation media include pH, electrical conductivity, cation exchange capability, nutrients and other chemical substances. Chemical properties of cultivation media influence the fertilizer holding capability, osmotic pressure and nutrients supply capability, and thus having impact on crops growth rate, height, hardiness, plant type, and so on. The nutrients supply amount and supply time in seedling cultivation play a decisive role in the transplant survival rate and adaptability after transplanting. If seedling is too tender and soft or grows too rapidly, it is symptom of a poor transplanting seedling. Cultivation media should not have abundant soluble fertilizer. Otherwise, high osmotic pressure will lead to fertilizer injury, or worse resulting in withering and death. In a word, the chemical properties of good cultivation media should be able to make cultured seedling or healthy plants growth, and should cultivate plants with proper growth, great adaptability and strong resistance to adverse conditions.

Media should have Favorable Biological Properties

Cultivation media should possess good biological properties, including performing the effects of beneficial bacteria such as rhizosphere protecting bacteria, mycorrhizal fungi, phosphate solubilizing bacteria, nitrogen fixing bacteria, detoxifying bacteria, and having the ability to reduce pest and disease. The biological properties of cultivation media influence crops germination rate, nutrient absorption, growth, transplant survival rate, resistance to pest and disease, and so on. Good microorganisms in cultivation media can reduce seedling pest and disease. If certain media have pathogens, worm eggs or others that can cause disease and pest, and then they are regarded as unhealthy cultivation media. Healthy microorganisms in media are another protection layer for plants. Therefore, excellent and healthy seedlings should not only have a excellent appearance, but more importantly, they need beneficial bacteria inoculation in the roots or inner plants, which is closely related to and can assist transplant survival rate and growth after transplanting.

8.15.3 Characteristics and Application Guidelines of Cultivation Media

There are various kinds of materials for cultivation media, including organic and inorganic matters, whose types and application guidelines are specified respectively as follows:

Organic Media:

Organic media are common materials. Organic media have good fertility and water holding capabilities as well as slow released function for nutrient supply. Different organic media have different physical and chemical properties, so that they can be used as materials for mixed media to meet various crops requirements. Peat and peat moss, which are applied abundantly at present, are both good organic media. Agricultural wastes such as mushroom cultivation sawdust wastes, rice hull, tree bark, bagasse, and corn cob need to be applied skillfully. Those local media need further development. Common organic media and application guidelines are specified respectively as follows:

Peat:

Peat has different types and composition based on its origins, formation period and places. Common types include peat soil, wood peat and peat moss. Because peat is formed by accumulation or alluviation of ancient plants through thousands of years, it does not have disadvantages of fresh plants, and is applicable as plant growth media. Peat is usually acid, so it needs nutrients and to adjust pH value before application. Acidic peat can also be used as amendment for alkaline media or soils.

Tree Bark:

The origins of bark are limited. Ordinary fresh bark need a long time (six months to a year, or even longer) to compost and to be mature for good media. Bark compost is a common commercial organic material. Bark which is applied as media should not be mixed with excessive chemical fertilizer for not making crops suffer from fertilizer injury.

Palm Fruit Fiber:

A great deal of fibers left from oil palms and coconuts can be used for cultivation media. In application to some crops, fresh palm fibers usually cause poor rooting. With appropriate disposition, maturing and adjusting, these fibers will become the optimal choice in place of peat media.

Mushroom Cultivation Sawdust Wastes:

The manmade packaging bags of modern mushroom production are usually called "Space Package," which often get discarded and pollute the environment. The reason is that the sawdust in "Space Package" is not properly treated. It has disadvantages if applied to farmland, such as lacking of microelements (such as Fe, Zn, Mn) and releasing phenolic phytotoxic substances, which influence crops germination or growth. Mushroom cultivation sawdust wastes without disposing often lead to leaves yellowing. In order to be used for cultivation media, mushroom cultivation sawdust wastes needs to be composted and they also require additional materials, such as rice hulls, carbonized rice hulls or detoxifying bacteria.

Rice Hull:

Rice hull can be used for cultivation media for its good aeration and resistance to rot. The application of carbonized rice hull is very common, but burning process used by carbonized is likely to cause air pollution. Raw rice hull needs to be combined with other materials in application, such as soil, sawdust and peat soil, and with liquid chemical fertilizer and biological fertilizer, in order to be suitable as crops seedling cultivation media.

Plants Residues such as Bagasse and Corn Cob:

Direct application of dried fresh plants residues tends to cause massive propagation of microorganisms and competition for nutrients, and thus this results in crops yellowing or death. Therefore, plants residues should not be applied until they have become mature from composting and mixing with other added materials.

Scrap of Cyatheaceae (A kind of Tree Fern, Snakewood):

The production of scrap of Cyatheaceae is becoming less and less. Snakewood sawdust is a kind of excellent media with good qualities of watering and aeration. If mixed with other organic materials or gravel, scrap of Cyatheaceae will produce good results. But, its single use can easily cause acidification and the problem of white fungi for a period.

Other Organic Media:

As long as its physical, chemical and biological properties are suitable, the organic matter can be applied as cultivation media. High electrical conductivity and fertilizer will probably cause damage or fertilizer injury for plants. Consequently, manmade organic fibers can also be used as media, which have great water absorbing capability and cannot be easily decomposed. It needs special attention in application whether the manmade organic fibers contain toxic substances. Those with toxic substances should not be used for vegetables cultivation and crops for food. Besides, those that will cause public damage or pollute the environment when being discarded are also not suitable for application.

Inorganic Media:

Inorganic cultivation media are also pretty common. Their most outstanding feature is that they cannot be decomposed. Inorganic media usually take natural minerals, gravel or soil as raw materials. Different media have different physical and chemical characteristics and are applicable to specific crops. Presently, the usual abundant inorganic media are vermiculite, perlite, rock wool, foam stone and other rocks. The commonly used inorganic media are river sand, sea sand, sandy soil, and other rocks. Due to the lack of nutrients or balanced nutrients supply, inorganic media generally should be combined with organic materials or fertilizers in application.

The characteristics of inorganic media need to be noted in application, such as water holding capability, nutrient holding capability, particle size (texture), hardness (grinding degree), structure, pH value and aeration and so on. When sandy soil gathered from riverbed is applied as cultivation media, it demands special attention whether there exist pathogens and factors that are likely to cause underground pests. It should not spread pests and diseases.

8.16 Introduction and Outlook of Organic Agriculture

8.16.1 The Development Outlook and Significance of Organic Agriculture

The development of science and technology brings along convenience and prospect for modern human. Agriculture development also enters the era of modern and industrial production from conventional ancient farming. The application of chemical fertilizers and pesticides in agriculture production has greatly improved crops output. However, excessive or improper application of chemical fertilizers and pesticides will not only pollute the environment or lower the quality of some products, but also cause the deterioration of agricultural ecological environment, which is worried by advanced and insightful people.

Human has relied on necessary substances from natural foods for cell metabolisms of bodies. Can they, after hundreds of thousands of years, adapt themselves rapidly to changes such as unbalance or shortage of substances supply? This is an issue worthy to be reflected on and studied. Another question is thus raised: can the cell metabolisms and living environment of human beings adapt completely under chemical and industrial food production and the environment?

The answer to the above questions, the proposal of "organic agriculture" is suggested. An organization called International Federation of Organic Agriculture Movements (IFOAM) has received immediate support from Germany farmers and relevant people, spread from Europe to America and Japan, and raised widespread attention from international agriculture fields. In 1986, IFOAM held the 6th international science conference in University of California-Santa Cruz. It can be seen that "organic agriculture" is a subject of scientific study instead of an entirely ancient and unscientific approach.

Organic Farming is defined in a paper of United States Department of Agriculture as a system in 1980 that excludes the use of synthetic fertilizers, pesticides, and growth regulators, and livestock food additives. In order to reach highest practicability and maintain soil productivity and farming, organic agriculture applies rotation farming, crops residues, animals fertilizer, beans, green manure, farmland organic wastes, mechanical farming, mineral rocks and prevention from harmful biological pests, so as to provide sufficient nutrients for plants and control insects, weeds and other pests. Therefore, organic agriculture adopts the concepts of "nature" and "ecological balance." In fact, organic agriculture also uses many methods which are the same as those of modern agriculture, while it excludes or strictly limits the application of manufactured (synthetic) substances and it pays close attention to comprehensive influences on the entire ecosystem.

There are many different types of management ways for organic farmland around the world. Common organic farmlands have mixed crops system, integrated crops and livestock system, vegetable garden and fruit tree system. Reports show that organic agriculture with mixed crop system, with or without livestock, has the most valuable immediate potential, because it will generate high economic profits in cooperation with crops and rotation system.

The following part lists out the collected literatures, and offers reference.

8.16.2 The Advantages and Disadvantages of Organic Agriculture

Organic Agriculture can Reduce Environment Pollution:

In recent year, agricultural chemical substances with strong pollution (including pesticides and chemical fertilizers) have been abundantly applied around the world, which results in ecological damage and unbalance. Organic agriculture is helpful to reduce environmental pollution, achieve ecological balance and lessen damage.

Organic Agriculture can Decrease Production Cost and Energy Consumption:

Organic agriculture adopts multiple approaches for the recycling of organic wastes, which can decrease production cost and input energy in entire system.

Organic Agriculture Increases Soil Organic Matter and Helps with Soil Productivity:

The increase of soil organic matter is beneficial to the physical, chemical and biological properties of soil.

Organic Agriculture can Make Nitrogen Self-Sufficient in Farmland:

Organic agriculture conducts intercropping and rotation system with legumes that can fix N_2, so as to increase nitrogen content in soil. Biological nitrogen fixing function is significant. The reports show that annual industrially fixed nitrogen for agriculture is about 50 million tons, while biologically fixed nitrogen in farmland soil is about 90 million tons around the world. In the middle 1960s, the cost of industrially fixed nitrogen fertilizers is about 1.5 million dollars, which is estimated to add about 20 times until 1990 and reach 300 million dollars.

Organic Agriculture can Reduce Soil Erosion:

The cultivation methods of organic agriculture allow water to seep into soil rapidly and preserve soil and water in proper methods.

Organic Agriculture can Reduce the Occurrence of Soil Pests and Diseases:

Organic agriculture values the balance of entire ecosystem, so that it can reduce the problems in the chemical resistance of diseases and pests. Biological control or natural enemy can be used for pest control; crop rotation system, intercropping system and different types of green manure can also be utilized for pest and disease control. For example, some soil diseases can be controlled by planting shallot or sesbania. Besides, special plants can be cultivated to get rid of certain pests. Such technologies are worthy of in-depth study.

Organic Agriculture may Generate Higher or Lower Production than Conventional Agriculture (with Application of Chemical Fertilizers and Pesticides):

Due to various soil types and distinct crop characteristics, there also exist differences in output of organic agriculture. After all, the cultivation backgrounds and application time length are not the same in different reports. However, organic agriculture has much potential because it protects soil productivity and decreases soil erosion. The problems of transformation from conventional farming to organic farming are seemingly related to nutrients supply. Long-term application of chemical fertilizers will lower the content of stable organic matter in soil. Soil that does not take legume as organic fertilizer or makes use of immature organic fertilizer may provide insufficient nitrogen, and thus result in poor initial growth of crops. In addition, some crops demand a great deal of fertilizers to generate high output, so the choice of crops is also very important.

The Application of Organic Agriculture Needs Rearrangement Based on Estimate of Present Available Nutrients Amount and Soil Fertility Measurement:

Organic matter is different in decomposition rate, so the nutrient release speed and release amount from decomposition are also different. Besides, short-term crops and perennial crop have diverse requirements for nutrients. Therefore, comprehensive experiments and studies are needed.

Study on Allelopathy is Likely to be Applied for Weeds Control:

In recent years, studies on allelopathy have emerged like mushrooms after rain. This kind of biological control follows natural principle. Organic agriculture excludes pesticides with high pollution possibility, so that application of natural allelopathic substances becomes an important method. To give crops the ability to remove weeds is also a study aim in the future. Before further achievement, organic agriculture control weeds with conventional grass coverage or sod cultivation under perennial crops. It's better to grow short types of grass for the sod cultivation.

8.16.3 Organic Agriculture Needs Further Experiments and Studies

It is necessary to lessen the damage on ecosystem, especially pollution on the land and water, for human survival and environmental improvement. Under the pressure of the abundant global pollution in recent years, the concept of organic agriculture is proposed. The new idea gives hope to the future agriculture like the first light in the morning. Either pure or comprehensive organic agriculture will decrease pollution and energy consumption. In terms of soil productivity protection, organic agriculture is beneficial to ecological balance of soil and the improvement of soil physical and chemical properties. Only healthy soil can generate good food and develop healthy body.

The area of Taiwan is located in subtropical and tropical zones. Its hot and rainy climate makes it naturally difficult for soil protection. High temperature accelerates the decomposition of soil organic matter, so the accumulated soil organic matter will decrease correspondently. A large proportion of farmland lacks organic matter (below 2%). Besides, slope land is likely to be short of organic matter due to the washing-away and erosion by rain. Sharp reduction of farm cattle makes animal manure and compost become very little, and it also decreases the sources of organic matter for soil conservation. Rice does not have high demands for organic matter unless that is in high economic agricultural products areas. Moreover, many farmers do not apply organic fertilizer. In this case, the fertility and productivity of soil with long-term intensive cultivation will be greatly damaged. In serious conditions, there will be "soil death," that is to say, "problem soil," including problems such as the accumulation of chemical salts, massive roots deaths, incapability of crop growth, and so on.

Although soil buffering capability is high, it requires a lot of investment to restore buffering once it is destroyed. Soil is the mother of crop production, so its continuous productivity protection needs to be emphasized and strengthened.

Organic agriculture demands further studies and experiments. This requires teamwork of specialists in study field of crop, soil, pest and disease control, and farmland management, which can provide multiple materials for the development of future agriculture.

8.17 Organic Agriculture Cultivation and its Specifications Foreword

In recent years, agricultural development is accompanied by the widespread application of chemical fertilizers and pesticides. Although crops production has been raised, excessive or inappropriate application of chemical fertilizers and pesticides will lead to the drop of farm production, quality, environmental pollution and ecosystem decline. This is concerned by advanced and insightful people.

To adapt with the cultural environment where ecosystem has been deteriorated by modern agriculture, it is necessary to propose a solution scheme of improving present cultivation management. According to natural ecosystem law, for those conform to natural law will last long. On the other hand, for those violate natural law and focus only on immediate interests will eventually fail. As the most advanced life on soil, human beings should be wise enough to understand this principle, whereas practice of natural law needs efforts from many aspects. The aim of agriculture is to satisfy human needs for foods and other things, but global grain insufficiency due to ever increasing population. For the welfare of longer future and in conformity to natural law, soil productivity needs to be maintained so as to achieve sustainable development. Therefore, "sustainable agriculture" requires further studies and promotion to ensure production, conservation and profits at the same time.

Organic agriculture is one of managements to achieve the goals of sustainable agriculture. Organic farming is defined as a system that excludes the use of synthetic fertilizers, pesticides, and growth regulators, and livestock food additives. In order to reach highest practicability and maintain soil productivity and farming, organic agriculture applies rotation farming, crops residues, animals fertilizer, beans, green manure, farmland organic wastes, mechanical farming, mineral rocks and prevention from harmful biological pests, so as to provide sufficient nutrients for plants and control insects, weeds and other pests. Organic farming is a kind of agriculture following the concept of natural law and eco-balance.

Organic agriculture also uses many methods with the same as those of modern agriculture, while it excludes or strictly limits the manufactured (synthetic) substances application and it pays close attention to comprehensive influences on the entire ecosystem. There are many different types of management ways for organic farmland around the world. Common organic farmlands have mixed crops system, integrated crops and livestock system, vegetable garden and fruit trees system. Reports show that organic agriculture with mixed crops system, with or without livestock, has the most valuable immediate potential, because it will generate high economic profits in cooperation with crops and rotation system.

It is difficult to recognize or identify products of organic agriculture only from the appearance. In order to meet the demands of the future, there should be at least certain standards for the production of organic agricultural products. Because the origins and characters of various substances in different places are diversified, the applied substances for organic agriculture cultivation in various areas may be different. However, the fundamental standards and principles should be the same, which are suggested here for reference.

8.17.1 The Objectives and Basic Production Principles of Organic Agriculture

The objectives of organic agriculture can be separated into short-term and long-term goals:

Short-Term Goals:

The goals are to develop farming patterns which exclude or strictly limit the application of chemical fertilizers and pesticides. These two may not be the same, according to the definitions in different countries and the healthy soil of obtaining the ecosystem.

Long-Term Goals:

The goals are to follow "natural ecological law" and develop farming and management patterns. In order to achieve beautiful countryside, healthy soil, safe products, happy farmers and independent agriculture, and to realize on agriculture, it is beneficial to balance the healthy ecosystem and to improve life quality for farmers.

The materials used for modern agriculture are diverse, so there should be certain standard for the organic farming application in order to construct a set of farmland management schemes for both production and conservation. To produce safe farm products with commercial quality, the basic production principles can be concluded below:

- Adopt the materials and farming approaches that are favorable to the environment and development of human health, and improve the quality of farm products.
- Use the materials that are suitable for soil organism activity and maintainable for lasting soil productivity.
- Cooperate with local agriculture system and recycle of renewable resources.

8.17.2 The Methods and Strategies of Organic Agriculture

The methods and strategies of organic agriculture cultivation are specified as follows:

Exclusion or Strict Limitation of Chemical Fertilizer:

Apply green manure, organic fertilizer and biological fertilizer and recycle organic wastes to improve soil productivity.

Exclusion or Strict Limitation of Chemical Pesticides:

Apply nonchemical pesticides and biological prevention to develop crop resistance to diseases and pests, and achieve ecological balance.

Reasonable Crop Rotation System:

Apply legumes and mutually supplementary rotation systems to achieve self-sufficient nitrogen fertilizer and nutrient balance.

Conservation of Water and Soil Resources:

Planting the land with forest and grass for the retaining of water and soil, so as to protect topsoil and water resources.

Preservation of Agricultural Ecological Environment:

Adopt pollution-free system that protects biological survival, so as to reach healthy soil and environment.

Found and Safe Authentication, Production and Marketing System:

Build the production and marketing system for organic agriculture in order to obtain profits.

8.17.3 Predictable Difficulties of Organic Agriculture

The following specifies the predictable difficulties of organic agriculture cultivation:

Insufficient Soil Productivity at Initial Stage:

Farmlands that have been applied conventionally with chemical fertilizers will suffer from production decrease at the initial stage after chemical fertilizers and pesticides are excluded. Usually, the farmlands need a period of soil conservation to be improved.

Unsatisfactory Appearance of Farm Products:

Formidable pests or diseases are very likely to occur when chemical pesticides are prohibited. Therefore, the appearance of farm products is mostly not satisfactory. It is requisite to seek natural laws or application of non-pesticide substances to overcome various pests and diseases. Besides, to enhance related education to consumers is also important.

High Cost of Farm Products:

The price of organic and biological fertilizers and labor cost for their application are both higher than those of chemical fertilizers, so that the production costs are raised. Except for rewarding organic fertilizer production, it is also necessary to develop fast treatment for organic fertilizer and reward biological fertilizer production. Although developing biological fertilizer needs a great deal of investment, its manufacturing cost is rather low. Therefore, if its application areas can be expanded, biological and organic fertilizers can lower in their production cost.

The Acceptance of Organic Farming for Farmers Needs to be Strengthened:

Farmers' knowledge of environmental protection is limited. They tend to repel unaccustomed ways of farming and management. Enhance relevant education to farmers and propose inducing factors to guide farmers to adopt the right conservation methods are needed.

It is Not Easy to Separate Farm Products of Organic Agriculture and Conventional Farming:

It is necessary to build safe authentication, production and marketing system.

8.17.4 Soil Quality Management in Organic Agriculture

Prohibited Substances:

The products of organic agriculture are difficult to be recognized only from the appearance. To set standards in production management is helpful to make sure that cultivation management follows certain specific standards. The prohibited substances in organic cultivation are mainly chemical fertilizers, chemical pesticides, synthetic adjuvant and substances with pollutants, lest they cause soil deterioration and influence the environment and human health. Different countries and certification bodies have more or less distinctions in terms of prohibited substances. For example, the "Organic Food Production Regulations" made in 1990 by National Organic Standards Board (NOSB) clarifies the prohibited substances, which cannot be applied for three years to farmland that produces organic farm products. The following sorts out and extracts those prohibited substances. Among them, the synthetic fertilizers and pesticides are all forbidden, and the rest are listed for reference in Table 8-11.

The Allowed Farming Conducts:

The production process of organic agriculture demands a lot of farming conducts or mechanical application. The allowed farming conducts are mainly the adoption of crop rotation system, plantation of favorable crops and control of harmful weeds,

Table 8-11. Substances Prohibited in Organic Agriculture Cultivation Suggested by Organic Foods Production Association of North America (OFPANA, 1990)

Fertilizers	Synthetic Substances	Pesticides and Chemicals	Oil and Others
Urea	Anti-gelling agent (synthetic)	Pesticides (such as pentachlorophenol, carbamate)	Carrot oil
Calcium Superphosphate	Wet spreader (synthetic)	Antiseptic	Creosote
Triple Superphosphate	Fumigant (bag) (synthetic)	Nematicide	Weed oils
Potassium Chloride	Growth regulator (synthetic)	Pesticides (such as the marathon, parathion, carbamate)	Petroleum solvents
Potassium Sulfate	Plant protection agents (synthetic)	Moth balls and crystals	(Aromatic agent)
Monoammonium Phosphate (MAP)	Bird bait and poison bait	Condensed Nicotine of organic phosphorus compounds	Sludge Emissions from ionizing radiation
Diammonium Phosphate (DAP)	Cryolite (synthetic)	Pyrethroids	
Nitric Acid	Spray regulator (synthetic)	Chlorinated hydrocarbons	
Calcium Nitrate	Evapotranspiration blocking substance (synthetic)	Third agent of pesticide (such as 3,4-methylenedioxy benzoic butanol-piperonyl butoxide)	
Magnesium Nitrate	Drop slot cleaner (synthetic)	Soil fumigant	
Ammonia Products		Methyl bromide	
Anhydrous Ammonia		Methyl sulfoxide	
Gypsum By-Product		Dimethyl sulfoxide	
Hide Power, Leather Barrel		Phosphoric acid	
		Methanal	
		Sodium hydroxideVitamin B1-plant growth regulator	

diseases, pests and other animals, so as to help with soil fertility and prevention from diseases and pests, and other animals. According to Organic Foods Production Association of North America (OFPANA), the farming and mechanical conducts are allowed as long as they are not related to synthetic additives. Such conducts are sorted and listed in Table 8-12 for reference.

Additives Applicable to Soil and Plants:

Crops need sufficient nutrients to grow healthily. Crops grow in soil, and soil provides the needed nutrients, water and oxygen. However, soil conditions are originally not necessarily good. Besides, soil under long-term farming also requires supplements from the outside. Organic agriculture cultivation needs natural substances, while synthetic and manufactured substances are not allowed. Additives that can be applied to soil and plants in organic agriculture cultivation mainly include natural residues from animals, plants, and microorganisms, beneficial microorganism inoculants, and minerals, and so on. Table 8-13 shows the additives applicable to soil and plants determined by Organic Foods Production Association of North America (OFPANA). Yet, some synthetic substances can also be applied in special situations and they are listed in Table 8-14 for reference.

8.17.5 Pest and Disease Preventives Applicable in Organic Agriculture

Organic agriculture does not apply pesticides to control pests and diseases. The permitted preventives should be natural favorable microorganism, plants, animals

Table 8-12. The Allowed Farming and Mechanical Conducts in Organic Agriculture Adapted from by Organic Foods Production Association of North America (OFPANA, 1990)

Farming		
Weed Control	Prevention from Pests and other Animals	
Crop Rotation	Cut grass	Remove pests by hand
Intercropping	Weed control by the goose	Balloon
Green Manure	Weed control by heat	Obstacle
N_2-Fixing Crops	Fire	Bird barrier and bird net
Cover Crops		Explosive devices
Plow		Gun
Grazing		Predators (Natural enemies)
Machinery and Faming Control		Mouse trap
		Sound device
		Vacuum device
		Health conducts

Table 8-13. Additives Applicable to Soil and Plants Adapted from Organic Foods Production Association of North America (OFPANA, 1990)

Animal Origins	Plants and Microorganisms Origins	Minerals	Salts
Animal Compost Dried Blood Bone Meal Feather Meal Fish Soluble Fish Meal Fish Soluble Matter Horn Or Hoof Meal Marine Animals Waste Wormcast	Plant compost Microorganism inoculant Residues of grape and other fruits Herbal products Humic substances Humic acid derivative Seaweed meal and its products Leaves, rotten leaves Covering (organic) Peat Plant products such as peanut meal, cottonseed meal, soybean meal, straw, rice hull, sawdust, vegetable residues and waste of canning factory and so on (under certain conditions) Coffee pod (without pesticides) Cotton gin fertilizer chips (without pesticides) Mushrooms compost (without pesticides) Beet pulp (without pesticides) Tobacco ash (only applied as fertilizer) Wood ash (without toxic substances)	Magnesium sulfate minerals Feldspar (potassium source) Granite powder (potassium source) Greensand (potassium source) Guano phosphate ore Rock phosphorus ore Gypsum Mine (calcium source) Kieserite Langbeinite Potassium sulfate mineral Uprooting grass: alfalfa Limestone: aragonite, calcium carbonate, dolomite (calcium source) Marl (calcium source) Oyster shell powder Perlite Vermiculite Pumice Stone chips Sand SulfurMinerals zeoilte containing microelements	Borate Soluble boron products (use only in need) Microelements from natural sources used for spray

and minerals and so on. Table 8-15 organizes and lists the suggested preventives determined by Organic Foods Production Association of North America (OFPANA). Yet, some synthetic pest and disease preventives can also be applied in special situations and they are listed in Table 8-16 for reference.

CH8 Soil Organic Matter and Organic Fertilizers   363

Table 8-14. Synthetic Chemical Substances Applicable to Soil and Plants in Special Situations Adapted from Organic Foods Production Association of North America (OFPANA, 1990)

Name	Specifications
Microelements	Limited to areas in short of microelements according to analysis of soil and plants.
Sodium Molybdate	Limited to areas in short and in necessary need for sodium molybdate.
Zinc Sulfate	Limited to areas that are undernourished according to analysis of soil and plants.
Ferric Sulfate	Application amount is limited because excessive sulfur will have adverse impact on soil.
Synthetic Microelement Substances	Limited to nourished areas according to records.

Table 8-15. Pest and Disease Preventives Applicable in organic Agriculture Adapted from Organic Foods Production Association of North America (OFPANA, 1990)

Microorganism	Plant	Animal and Others	Mineral
Bacillus thuringiensis (bt) Beneficial Microorganism Seaweed Animals Insects Nematode Protozoa Nosema Locustae Fungi Bacteria Virus	Melia azedarach and its extracts Pyrethrum Quassia Derris (not applicable within 5 days after picking) Sabadilla (pest control) Nox vomica (mouse control) Garlic Tobacco powder (limited to pest control) Herbal products Bug driving oil of plant properties	Insect extracts (insect juice) Bug driving oil of animal properties Deer or hare Expellant (limited to natural substances) Barrier with tackiness	Cryolite (limited to minerals and applied after cleaning) Sulfur (not applicable for after-ripening) Petroleum bug driving oil (limited to pest control)

Table 8-16. Synthetic Pest and Disease Preventives Applicable in Organic Agriculture in Special Situations Adapted from Organic Foods Production Association of North America (OFPANA, 1990)

Name	Specifications
Boric Acid	Used for inedible parts of plants
Bordeaux Mixture (Copper Sulfate Mixture Hydrated Lime)	It is good to take minerals as materials. This can add soil copper, so its application is constrained to soil and plants. Copper accumulation is especially toxic in acid soil.
Copper	Including those allowed substances that do not need to be determined by Federal Authorities, such as hydroxide, alkaline sulfate, hydrochloride, oxide and other fixed copper family. These substances can be considered to be used in Bordeaux mixture. See Bordeaux mixture.
Copper Sulfate	See Bordeaux mixture and copper.
Dormancy Oil	Limited to dormancy spray for woody plants.
Lime Hydroxide Lime	Only applied as fungicide spray on leaves. Refer to atmospheric control lime.
Calcium Polysulphide with Pheromone	Only applied as fungicide spray on leaves.
Foam Herbicide Soap	Limited to inedible agricultural products.
Sodium Carbonate	Applied for plant disease and pest control.
Suffocating Oils	Limited to woody and perennial plants. Applicable in winter and summer.
Sulfur (Natural Synthetic Type)	Used on leaves as pest control agents or antifungal agent.
Summer Oil	Prohibited to be applied to agricultural crops after harvest. Refer to suffocating oil.

8.17.6 Adjuvant Applicable in Organic Agriculture

The adjuvant needed in organic agriculture should come from natural and nontoxic substances. Table 8-17 organizes and lists the suggested production adjuvant determined by Organic Foods Production Association of North America (OFPANA).

Table 8-17. Adjuvant Applicable in Organic Agriculture Adapted from Organic Foods Production Association of North America (OFPANA, 1990)

Microorganism Origin	Plant Origin	Others
Alcohols (Natural Sources)	Plant extracts (application limited to plants)	Citric acid
Enzymes	Mandarin oil	Oxalic acid (limited to greenhouse disinfection)
Gibberellins (Limited to Acid Fermentation Sources)	Fruit wax	Natural chimeric substances
Inoculant	Tree seal (plant or dairy products)	Natural vitamin
	Plant oil spray adjuvant (including adhesive, Surface active agent and carrying body, 90% from plants)	Biological products
		Carbon dioxide
		Soda water (limited to Application after fruit picking)
		Nitrogen (application after harvest)

Yet, some synthetic adjuvant can also be applied in special situations and they are listed in Table 8-18 for reference.

8.17.7 Strategies for Sustainable Agriculture

Objectives:

Long-Term Goal:
The goals are to follow "natural ecological law" and develop farming and management patterns. In order to achieve beautiful countryside, healthy soil, safe products, happy farmers and independent agriculture, and to realize the on agriculture, it is beneficial to balance the healthy ecosystem and to improve life quality for farmers.

Middle-Term Goal:
The goal is to develop farming patterns that avoid the application of chemical fertilizers and pesticides, and to obtain healthy ecosystem.

Strategies:

Avoidance of Chemical Fertilizer:
Apply green manure, organic fertilizer and biological fertilizer and recycle organic wastes to improve soil productivity.

Avoidance of Chemical Pesticides:
Apply nonchemical pesticides and biological prevention to develop crop resistance to diseases and pests and achieve ecological balance.

Table 8-18. Synthetic Adjuvant Applicable in Organic Agriculture in Special Situations Adapted from Organic Foods Production Association of North America (OFPANA, 1990)

Name	
Alcohol	Methanol and ethanol made from compounds. Alcohol can be used in solution or medial products with brand. Isopropanol is prohibited.
Arsenious (Injecting Arsenic with Pressure)	It can be applied to existed trellis, but not to woody plants cultivated recently. It draws much attention that grapes and other plants absorb arsenic from forest that contains arsenic.
Chelating Agent	It can be applied to infertile soil in combination with microelements.
Atmospheric Control Lime	Carbon dioxide is removed from lime hydroxide through atmospheric storage method. It is limited to atmospheric control saving method after picking and harvest and application to fertilizer is prohibited.
Hydrogen Peroxide (Natural Molecular Compound)	Used for root soaking, foliage management; used as disinfectant, inhibitor to plant disease, and substitute for chlorine to purify water.
Petroleum Dispersion Adjuvant	Including adhesive and media. Dormont oils can be applied to perennial woody plants as dispersion adjuvant. Petroleum adjuvant should not contain synthetic pesticides.
Plastic Covering (Mulch), Furrow Covering and Disinfection with Sun Exposure	Applied only in the growing and harvest stage of plants. Plastic should be removed when it is no longer needed, instead of leaving it in soil and allowing it to disintegrate naturally.
Silicate	Limited to application to fruits after picking.
Tree Injury Protectant	Paints with plant or dairy properties are applicable. Other petroleum products should only be used with limitation when there is no substitute. It should not be mixed with substances containing antifungal agent or synthetic chemicals.

Reasonable Crop Rotation System:

Apply legumes and mutually supplementary rotation systems to achieve self-sufficient nitrogen fertilizer and nutrients balance.

Conservation of Water and Soil Resources:
Apply forest and grass for the retaining of water and soil, so as to protect topsoil and water resources.

Preservation of Agricultural Ecological Environment:
Adopt pollution-free system that protects biological survival, so as to reach healthy soil and environment.

Sound and Safe Authentication, Production and Marketing System:
Build production and marketing system for sustainable agriculture in order to obtain profits.

Predictable Difficulties:

Insufficient Soil Productivity:
Farmlands that have been applied conventionally with chemical fertilizers will suffer from production decrease at the initial stage after avoiding chemical fertilizers and pesticides. Usually, farmlands need to be improved during a period of soil conservation.

Unsatisfactory Appearance of Farm Products:
Formidable pest and disease is very likely to occur when chemical pesticides are prohibited. Therefore, the appearance of farm products is mostly not satisfactory. It is requisite to seek natural laws or application of non-pesticide substances to overcome various pests and diseases. Besides, to enhance instructions to consumers is also important.

High Cost of Farm Products:
The price of organic and biological fertilizers and labor cost for their application are both higher than those of chemical fertilizers, so that the production costs are raised. Except for rewarding organic fertilizer production, it is also necessary to develop fast treatment of organic fertilizer and reward biological fertilizer production. Although developing biological fertilizer needs a great deal of investment, its manufacturing cost is rather low. Therefore, if their application areas can be expanded, biological and organic fertilizers can lower the production cost.

The Acceptance for Farmers Needs to be Strengthened:
Farmers' knowledge of environmental protection is limited. They tend to repel unaccustomed ways of farming and management. It is needed to enhance relevant education to farmers and propose inducing factors to guide farmers to adopt the right conservation methods.

It is Not Easy to Separate Farm Products of Organic Agriculture and Conventional Farming:
It is necessary to build safe authentication, production and marketing system.

8.18 Microelements and Heavy Metal Elements of Organic Agricultural Soils

8.18.1 Crops Production Should not be Short of Microelements

Whether in organic agriculture or in conventional agriculture, crops have certain demands for nutrients, with various required amount from different crops. Macroelements and microelements included, crops need to take 16 essential elements from soils which are C, H, O, N, P, K, S, Ca, Mg, B, Fe, Mn, Cu, Zn, Mo, and Cl. Microelements refers to the elements that plants only need for minute quantities for proper growth. Although the required amount is less than macroelements and microelements, microelements are not unimportant at all. The confirmed essential microelements for higher plants include B, Fe, Mn, Cu, Zn, Mo, and Cl and so on. Other functional elements or favorable elements such as silicon, sodium, cobalt, selenium, and vanadium are likely to be added to the above list with future research and development. Some elements are needed only for a very small amount, but their significance should not be neglected. It is the same for crops of organic agriculture. Any shortage of the essential microelements for crops will directly affect their growth, output and quality. On the other hand, microelements should not be excessive, or else they will accumulate and become toxic substances in plants.

Due to years of massive and intensive farming, a great deal of microelements in soil is absorbed. On the other hand, some soils cannot provide sufficient microelements for crops growth. Besides, many farmers do not pay enough attention to the remaining quantities of microelements in soils and apply abundant chemical fertilizers for their convenience. The decline of organic fertilizer application greatly lessens the origins of microelements, and thus results in the lacking of microelements in soils in recent years. Nowadays, soils of organic agriculture need to be concerned for the amount of microelements. Otherwise, their shortage or excess will become limiting factors for crops growth.

8.18.2 Microelements Application in Organic Agriculture

There are various implement principles and standards for organic agriculture around the world. In terms of essential microelements application for crops, there are principles of suitable application and prohibited application. Some standards do not involve the issue of supplementary microelements. Materials for organic matter have somewhat microelements and can be served as the origins of partial microelements. However, if soils with severe shortage of microelements cannot obtain enough microelements, crops production and quality will be lowered. Under this condition, the only solution is to disuse the land in the way of organic farming.

Microelements application should mind the influence on soils and water quality, which is the same in organic farming. Organic Foods Production Association of

North America (OFPANA) stipulates that areas with shortage of microelements based on the analysis of soils and plants can be applied with microelements. For example, sodium molybdate should be applied only to areas where it is wanted and necessary. The federal government communiqué indicates in accreditation standards plan of organic produce countries, mineral nutrients applied as origins of major elements or microelements should be of low solubility and be non-synthetic. Highly soluble and synthetic microelements substances can also be applied to soils to mend nutrient shortage. In the meantime, the application should not result in deterioration of soil and water quality, including such as:

- Chelating compounds contains microelements.
- Products of soluble boron;
- Sulfates, carbonates or silicates of zinc, iron, manganese, copper, cesium, and cobalt can all be applied.

8.18.3 Heavy Metals and Factors of Organic Agricultural Products

Heavy metals are greatly considered in foods. Generally, organic agricultural products are considered to have low content of heavy metals. Heavy metals in chemistry refers to the metallic elements with high element density and their ions which mainly include the transition metals in the periodic table, such as chromium, nickel, copper, zinc, lead, cadmium, arsenic, mercury, and so on. The absorbing amount of heavy metals by crops is influenced by different crop types and soil environment factors. Usually, leafy vegetable products are more easily influenced than fruit products by the content of heavy metals in soils. The physical, chemical and biological properties of the soil can all have impact on crop absorbing of heavy metals. The chief influential factors are the content of soil heavy metals, pH value, beneficial microorganisms, and so on. The influences of crops, fertilizers and soil factors on heavy metals of agricultural products are specified respectively as follows:

Crop Factor:

Agricultural products that take different parts of plants, including leaf, root, stem, flower and fruit, have different capabilities of absorbing heavy metals and store heavy metals in respective parts. Among them, the accumulation in flowers and fruits is low, while heavy metal accumulation in leaf is generally high. Domestic organic agriculture has been through a decade and the results show that there is no obvious difference in heavy metals containment in vegetables between organic agriculture and conventional agriculture that uses chemical fertilizers. According to the study report on the field research of Sweden, Europe (soil pH 6 ~ 6.5), the cadmium content in wheat of organic agriculture is not necessarily lower than that of conventional agriculture with chemical fertilizers, and there is no obvious change for organic agriculture. The report also points out that organic agriculture may not reduce cadmium or other heavy metals in vegetables within a short term.

Fertilizer:

Factor:

Conventionally cultivated crops with chemical fertilizers often suffer from problems caused by high content of heavy metals in the applied chemical fertilizers and pesticides. For example, some ordinary superphosphate or triple superphosphates fertilizer contain much cadmium. Organic agriculture excludes such fertilizers, so that it can reduce potential pollution of heavy metals. However, the organic materials and compost applied in organic agriculture may also have excessive heavy metals. For example, pig feces compost may have a high content of copper, if copper element has been added to the feed. Long term application of such pig feces compost with high heavy metals will increase and accumulate heavy metals in soils, especially for copper and zinc. Therefore, application of compost with some problem wastes will add heavy metals in soils and also increase heavy metals in plants.

Soil Environment Factor:

Soil environment is influential to the heavy metals absorption by crops, because soil environment will affect the availability or solubility of heavy metals, which mainly decides whether heavy metals can be absorbed abundantly by crops. The main factor for the availability or solubility of soil heavy metals is pH value. If pH value is low, most heavy metals are highly soluble and can be easily absorbed by crops. On the contrary, if pH value is high, some heavy metals have low solubility and are not easy to be absorbed by crops. Long-term application of massive chemical fertilizers causes soil acidification, which is helpful to the solution of heavy metals and absorption by crops. On the other hand, organic agriculture uses organic fertilizer for a long time. In this way, soil pH can raise up, where the soluble quantities of heavy metals will decrease and the absorption of heavy metals will not increase. Besides, the added humic substances to soils can also adsorb heavy metals and thus should be able to decrease the heavy metal absorption by crops.

Soil Beneficial Microorganism Factor:

Soil beneficial microorganisms are capable of absorbing and chelating heavy metals. For example, it is proven that mycorrhizai fungi can reduce the heavy metals content in crops. The main reason is related to dilution function by promoting crops growth and the function of chelating heavy metals. The impact of other beneficial microorganisms on heavy metals in organic agriculture still needs further in-depth study.

To deal with the problem of heavy metals in organic agricultural products, the composition of the applied organic fertilizer should be paid closer attention to. Organic fertilizer or compost is not guaranteed to be problem-free. Therefore, it is necessary to control the total amount of soil heavy metals and the total amount of heavy metals in applied organic fertilizers, in order to ensure the quality of crops.

Chapter 9

Soil Microorganisms

9.1 Soil Microorganisms and its Requirements for Growth

9.1.1 The Importance of Soil Microorganisms

Soil contains three main components: inorganic materials (mineral, water, air), organic materials and organisms, which are all indispensable for crop production. Inorganic nutrients provide crops with basic nutrients, including macro nutrients and micro nutrients. Organic matter, on one hand, exerts great effects on physical and chemical properties, and on surface characteristics of soil; on the other hand, organic matter constantly supplies the crop with inorganic chemicals by means of microbial decomposition. Except decomposing organic materials, microorganism plays other roles in the soil, such as increasing the sources of soil nitrogen, improving the nutrient absorption of plants, enhancing the availability of nutrients, decomposing toxic substances, protecting the function of rhizosphere, increasing the structure of soil aggregates, etc. Different soil microorganisms play distinct roles in soil fertility and nutrient cycles.

Among soil microorganisms, both beneficial and harmful microorganisms exist, the two of which balance and benefit crop to the greatest extent. However, due to the inappropriate cultivation management, such as unbalance, crop cultivation problem or problematic soil might occur. In order to maintain soil fertility and provide plants with sufficient nutrients, a further understanding on features of soil microorganisms and transforms of various nutrients will contribute to a correct, flexible operation and management of the soil.

9.1.2 Growth Requirements for Soil Microorganisms

There are many types of soil microorganisms, and their demands for survival and propagation differ more or less from each other. In the following part, specifications will be presented in accordance with their demands.

9.1.3 Classification Based on Carbon Source

In accordance with the carbon source, microbial nutrient requirements can be divided into autotrophic and heterotrophic microorganisms. Autotrophy microorganism makes direct use of inorganic carbon dioxide, while heterotrophic microorganism can only use organic carbon source rather than carbon dioxide. Autotrophy microorganism contains algae, few bacteria, and plants; and heterotrophic microorganism contains bacteria, actinomyces, fungi, protozoa, etc. The decomposition of organic materials by heterotrophic microorganisms is one of the greatest importance in the nutrient cycle systems in nature. The decomposed function of organic materials in agriculture breaks down organic materials into inorganic nutrient ions, and this is so-called "mineralization".

Commonly, the microorganisms for the composting need organic carbon sources. If it is incorporated with other energy and appropriate environmental conditions, such as an increase in inorganic nutrients (for example, applying nitrogen fertilizer and phosphorus fertilizer, etc.), appropriate pH value (common applying lime) and moisture, the microorganism growth and propagation will speed up. Also, the decomposition of organic fertilizer in soil may need to add microorganisms that could conduct the fast decomposition, but should be accompanied with the applied organic fertilizer, in order to avoid the over-decomposition of soil organic matter.

9.1.4 Classification Based on Energy Utilization

Microorganism differs from each other not only in terms of carbon sources, but also energy utilization, and it can be divided into two kinds, i.e., one that uses solar energy, and the other uses the chemical energy. Algae and plants can use solar energy, and some photosynthetic microorganisms, such as green sulfur bacteria, purple sulfur bacteria, and purple non-sulfur bacteria, can also use solar energy. Most soil microorganisms can only utilize chemical energy instead of using solar energy, which is obtained from chemical matters' transformation, such as nitrifiers, anammox bacteria, sulfur oxidizers and reducers, methanogens, halophiles, and thermoacidophiles. In the mean time, some soil microorganisms can acquire energy from the transformation of nitrogen, sulfur, pyrite and CO_2 of bacteria to facilitate growth and composition within cells. Many bacteria, fungi and actinomyces obtain energy from the transformation of organic compounds. Therefore, the supply of organic fertilizer or nutrients will be extremely helpful to the growth and propagation of soil microorganisms, which can directly provide soil microorganisms with nutrients and indirectly facilitate crop absorption of decomposed products. The decomposition of organic fertilizer and crop nutrient supplies are indivisible in the soil system.

9.1.5 Classification Based on Oxygen-Dependent

According to the different requirement for oxygen, soil microorganisms can be classified into aerobic, anaerobic, and facultative microorganisms. Under good aeration condition of soil, aerobic microorganisms are more appropriate for growth and propagation. For example, under good aeration condition, ammonium ion in nitrifiers will nitrify into nitrite and nitrate. However, while under poor aeration or poor drainage conditions, along with a deficiency in oxygen, denitrifiers which are adapted to anaerobic environment can prompt the transformation of nitrate nitrogen into gas nitrogen (N_2O or N_2). Therefore, under anaerobic condition, the application of nitrate nitrogen is easy to be lost by denitrification. Soil for upland cultivation has good aeration condition, in which many organic decomposing microorganisms can decompose organic materials. Therefore, organic materials in the upland farming soil are easily decomposed, which lead to the decrease of organic matter content at a quicker pace in the soil. On the contrary, paddy field has a good maintenance of organic matter content, and the decomposition process of organic matters in paddy soil is quite slow.

9.2 Types and Change Problems of Soil Organisms

9.2.1 Types of Soil Organisms

Soil organisms include microorganisms and large-size organisms, several thousands of which cover all the five Kingdoms, with at least 11 Phylum of animal and all kinds of microorganism. In appearance, bacteria could be really tiny, from one micrometer to several centimeters, while the Australia earthworms could be as large as 1.5 meters. The microorganism includes virus, bacteria, actinomyces, blue-green algae, fungi, algae, protozoa, etc., and the large-size organism includes molluscs (such as earthworm), insect, etc. These soil microorganisms could live either in autotrophic way with inorganic carbon source as carbon dioxide or in heterotrophic way with organic materials as carbon source. Judging from its energy source, it could be divided into chemotroph and phototroph. Based on the influence to crops from soil organism, it could be identified as beneficial and harmful organisms, and can be divided into pest and non-pest organisms, or pathogens or non-pathogens. Pest organism exerts harmful influence on crops, such as nematode and other pests spreading diseases in soil, which have enormous negative influence on the quality of soil.

Soil microorganisms inhabit in residues, the surface or deep in the soil, or even moving in the profile of soil. Tiny microorganism could move a short distance, but it may also be brought to other soil location due to the movement of other organisms. The scattering of soil microorganisms is influenced by soil temperature, moisture and texture. An important factor on the activity of soil organism is the addition and content of organic matters. Generally speaking, the total biomass of soil organism accounts only for a small part of soil's total organic matter (about 1 ~ 8%).

9.2.2 The Inhabitation and Types of Soil Organisms

Soil organisms can be divided into four kinds of biota, and its effect on soil can be listed in the Table 9-1:

Microflora:
- Including bacteria, actinomyces and fungi.
- It is a major microbial biota and biomass.

Microfauna:
- Including protozoa, nematode, some mites and springtails.
- Such kind of organisms feed on microbial biota and organic matter, thus affecting other populations.

Mesofauna:
- Including mite, springtail, gnat and worm.
- It feeds on microorganisms, corrosion, omnivore and predation. Moreover, it could fragment residues, adjust microbial biota population and deal with excrement of large-size creatures.

Macrofauna:
- Including double foot insect, isopod, earthworm, scolopendra, millipede, larva and imago of insects, gastropods, etc.
- It has influence on the fragmentation and re-scattering of plant residues, and is helpful to the form of soil porosity and aggregates.

Virus:
- Virus is a small infectious agent that can replicate only inside the living cells of an organism. Virus includes two major parts: nucleic acid and protein. It infects, propagates, and grows only in living cells of bacteria, fungi, algae, protozoa, plants and animals. Virus of invasive bacteria is called phages.

Table 9-1. Influence of Soil Organisms and Types on Nutrient Cycles and Soil Structure

Type	Microorganisms Included	Effect on Nutrient Cycling	Effect on Soil Constitution
1. Microflora	Bacteria Actinomycetes Fungi	1. Mineralization and fixation of nutrition 2. Decomposing organics	1. Hypha entangles soil particles and form granule 2. Producing organic compound through decomposition and forming granule
2. Microfauna	Mite Protozoa Nematode Springtail	1. Adjusting bacteria, actinomycetes, and fungi 2. Transform nutrition	Affecting granule through microfauna
3. Mesofauna	Mite Springtail Gnat Worm	1. Adjusting fungi and microorganism group 2. Change nutrient transformation 3. Promoting fragmentation of plant remains	1. Promoting humification function 2. Producing pelletization of excrement 3. Creating biological holes
4. Macrofauna	Earthworm Centipede Myriapod Insect, larva, imago Gastropods (snail, limax)	1. Fragmentation of remains 2. Promoting the activity of microorganism	1. Blending soil organics and mineral particle 2. Promoting the re-distribution of microorganism 3. Enhancing humification 4. Creating biological holes 5. Producing pelletization of excrement

Soil microorganism can be divided into:

Bacteria:

Bacteria are a group of microorganisms whose cells lack a cell nucleus (prokaryotes). Commonly, there are $10^6 \sim 10^9$ bacteria per gram soil (Atlas and Bartha, 1998). In taxonomy, prokaryotic microorganism in the soil can be divided into the following types:

- Proteobacteria: Including α, β, γ, δ, and ε type, chemoheterotroph of gram negative.
- Cyanobacteria: Photosynthetic bacteria, several species of which possess nitrogen fixation.
- Planctomycetes: Parasitic organisms.
- Spirochetes: Spiral-type phylum.
- Bacteroids: It belongs to anaerobes.
- Fusobacteria: It belongs to anaerobes.
- Sphingobacteria: This kind of microorganism could decompose fiber and make movement through gliding.
- Firmicutes: Gram positive bacteria, including clostridiales, mycoplasmateles, bacillales, lactobacillaes.
- Actinomyces: In taxonomy, actinomyces belong to prokaryotic microorganisms. Because of its mycelia structure, it is usually specially illustrated by soil microbiologists. The difference between actinomyces and fungi is that actinomyces are prokaryotic microorganisms (i.e., without nuclei structure) with tiny profile ($0.5 \sim 1.0$ μm), and can form external spore. Among gram-positive bacteria, actinomyces is characterized by their hypha feature. Their hyphae are bifurcated and can produce spore, which occupy $10 \sim 33\%$ of the total amount of bacteria in soil (Alexander, 1997), and *Streptomyces* and *Nocardia* are two most common ones. Actinomyces are more tolerant to drought, and prefer alkalinity and neutrality, but sensitive to acidity. They are usually being used in agriculture.

Archaea:

Possessing special physical traits, archaea could especially make sulfur as its energy source. The *thermophiles*, halophiles and methanogens can be distinguished as follows:

- Crenarchaeta: thermoprotei, desulfuroccales, and sulfoloccales.
- Euryarchaeota: methanobacteriales, methanococci, halobacteria, thermoplasmata, thermococci, archaeglobi, and methanopyri.

In addition, bacteria can be classified by their different features and functions as follows:

- Carbon source: it can be further divided into autotrophic and heterotrophic bacteria.
- Energy source: it can be divided into photosynthetic and non-photosynthetic bacteria.
- Gram stain: according to the differences in cell wall, it can be divided into G^- and G^+ bacteria.

- Spore production ability: it can be divided into bacteria that are able and unable to produce spores.
- Demands for oxygen: aerobic, anaerobic, slightly aerobic and facultative anaerobic bacteria.
- Nitrogen fixation ability: symbiotic, associative, and free living N_2-fixing bacteria.

Eukaryotic Microorganisms:

These are microorganisms with nucleus in their cells. Eukaryotes in the soil include the following kinds:

Protozoa:

They are unicellular animals that include amoeba, flagellates and ciliates. Protozoa possess 80S ribosomes and organelles (containing mitochondria while part of protozoa contain chloroplast), and gets nutrients through surround food and engulf it. They include six types: archaezoa, microsporidia, rhizopoda, apicomplexan, ciliophora, and euglenozoa.

Algae:

They are the simplest eukaryotic containing chlorophyll and other pigment and usually scatter in wet soil and waters. They are eukaryotic with photosynthetic capacity, including five types: diatoms, brown algae, green algae, red algae, dinoflagellates, and so on.

Fungi:

It contains mold fungi and yeasts. The previous one, with hypha structure, is the summit of biological evolution. Its cell wall contains chitin and cellulose and could produce spore, including zygomycota, ascomycota, and basidiomycota.

In addition, lichens usually can be observed from the surface of soil, rock, and tree. Lichens are symbiont of fungi, and algae, green algae, or cynobaterium which can be divided into crustose lichens, foliose lichens, and fruticose lichens.

The ecosystem of natural soil is a system that has gone through long-term succession and become quite stable, and to some extent, it possesses biological balance. However, agricultural soil is influenced by cultivation management, especially the entrance of biological residues, which will promote the mass propagation of certain microbial biota. According to the benefits and detriments to crops, there are two functions, i.e., beneficial or harmful microorganisms.

9.2.3 Importance of Soil Microbial Types to the Soil

Since soil microorganisms include virus, bacteria, actinomyces, fungi, algae, protozoa, etc., we will discuss their importance to soils as follows:

Virus:

Virus can invade other higher organisms and bring about death or disease, but researches on its influence on the soil is still insufficient.

Bacteria:

Bacteria play an important role in soil microorganisms, including beneficial and harmful bacteria, which make great contribution to cycles of natural matters. Its importance to the soil is listed as follows:

- The major work of decomposition and conduct the mineralization process of decomposing organic matters to nutrients.
- Nitrogen fixation.
- The nitrification could transform NH_4^+ into NO_2^-, and finally NO_3^-.
- The denitrification could transform NO_3^- into NO_2 and N_2.
- Promoting the dissolution of phosphorus, sulfur, iron, manganese, etc.
- Closely connected to other soil microorganisms.

Actinomyces:

In taxonomy, actinomyces belong to prokaryotic bacteria. Because of its mycelia structure, it is usually illustrated specially by soil microbiologists. The difference between actinomyces and fungi is that actinomyces are prokaryotic microorganisms (i.e., without nucleus structure) with tiny profile (0.5 ~ 1.0 μm), and can form external spore. Few actinomyces may affect animals and plants. Their importance to soil is listed as follows:

- Decomposing organic matters and releasing plant nutrients.
- Exerting nitrogen fixation with leguminous plants.
- Secreting and releasing plant hormone, such as GA.
- Producing antibiotics.

Fungi:

Fungi belongs to eukaryotic microorganism (with nucleus), and with its hyphal structure, it also includes beneficial fungi and harmful fungi for agriculture. Its major roles are as such:

- Decomposing lignin and organic matter, and releasing plant nutrients through mineralization.
- Secreting plant hormone, such as GA.
- Their hypha binds up soil particles and exudates to promote soil aggregation.
- Mycorrhiza and root symbiosis.
- Producing antibiotics and inhibit the growth of other microorganisms.

Algae:

Algae are the simplest eukaryotes containing chlorophyll, and usually scatter in wet soil and waters. Its major roles are listed as follows:

- Carrying on photosynthesis, taking in solar energy and transforming it into chemical energy, as well as fixing CO_2 into organic carbon compounds, and increasing organic matters.
- Secreting organic acids, so as to conduct weathering of rocks.
- Cyanobacteria can fix nitrogen and transform it into NH_3, and finally transformed into amino acids and proteins.
- Aquatic algae can conduct photosynthesis and release O_2, absorb nutrients into plants and reduce nutrient loss.

Protozoa:

Protozoa is unicellular animal and includes amoeba and protozoa. Its important role includes:

- Engulfing and controlling soil bacterial biota.
- Speeding up the decomposition of organic residues.
- Promoting cycles of nutrients.
- Moving in the soil water and speeding up the provision of dissolved oxygen and nutrients.

In addition of the beneficial roles mentioned above, all microorganisms have virulent types, i.e., pathogens, which would bring about disease among plants or animals.

9.2.4 Problems of the Existence and Change of Soil Microorganisms

Soil microorganisms plays an important role in maintaining the high quality of soil, but today in many places, the index, quality diversity, ratio and other changes of soil microorganisms are insufficient. As we know the existence of microorganisms has a close connection with changes in soil environment and quality of soil. In long-term ecological equilibrium of soil, soil microorganism changes with time and seasons. Under certain principles, it exists for a long time in the circle of life. However, due to the impact of cultivation management, agricultural soil may be confronted with drastic environmental changes and faced with the following problems:

Death and Extinction of Soil Microorganisms:

Nowadays, it is common to see the overuse of various kinds of pesticides and fertilizers in agricultural management. Pesticides contain toxic substances, such as insecticides, fungicides and herbicides, which could be more or less harmful to the soil microorganisms. Researches on the changes of soil ecology, especially changes of soil microorganisms are scarce. Only those large-sized creatures could be observed, such as the decrease or extinction of earthworm may has enormous effect on upland soil, including water permeability, water-holding property and aggregation, etc. Therefore, the extinction of earthworm could be a visible index of soil organism changes, and thus it prompts us to think over the impact on ecology.

Changes in Soil Microorganisms:

With various kinds of type, soil microorganisms are the major part of soil organisms. Along with the influence of paddy rice-upland crop rotation on soil ecology, soil microorganism has a direct influence on biota changes of aerobic, anaerobic and facultative microorganisms. Upland farming is inductive to the growth of aerobic and facultative microorganisms; while during paddy, anaerobic and facultative microorganisms prevail. Nematode often appears in upland soils, while paddy cultivation will make it hard for nematode to survive in water, and thus it largely reduces nematode damage to upland crops.

There are many kinds of soil microorganisms, and the healthy soil possesses quite balanced ecological system. However, when soil environment changes (such as the addition of organic matters, alternation of aeralation, etc.), part of microorganisms may grow rapidly or stop growing. After a period of time, the balance revives again. The unbalance of microorganisms often leads to blockage or unbalance of nutrient supply and the deficiency in certain nutrients. During a long period, crops will be harmed, or toxic substance will be produced from some microorganisms that do harm to roots.

The deficiency of beneficial soil microorganisms is often associated with the deficiency in organic matters and inappropriate use of pesticides. Organic matter is the major food and habitat of microorganisms, which provides carbon sources, nutrients and energy. The deficiency in soil organic matters will lead to the deficiency of microorganisms, undesirable structure of soil particles, poor fertility and water holding ability, and thus the soil will naturally be barren. Inappropriate use or overuse of pesticides may lead to growth and decline of biota in the rhizosphere, and affect the roots.

Occurrence of Plant Diseases and Pests in the Soil:

Common problems in continuous cropping and plant diseases and pests include rotten roots, damping off, bacterial wilt, soft rot, nodule disease, *Phytophthora* disease, nematode, etc. Due to the inappropriate management in soil environment, plant diseases and pests will be easier to happen. Reasons for these may vary, but the major one is the deterioration of soil environment (such as soil acidification, and nutrient unbalance) and the increase of pathogens.

When there are various and excessive pathogens and harmful microorganisms in soils, along with undesirable condition of site (such as coarse texture and low pH value), the incidence of disease is high and the plant disease is serious. Reasons for excessive soil pathogens are often associated with continuous cropping, long-term upland farming, soil acidification, and inappropriate management.

Crops have a direct connection with soil through roots. When something has gone wrong with soil microorganisms, the roots will be firstly affected. However, root problems are hard to be observed until there are symptom on plant stems and leaves. Sometimes, when there are problems in soil microorganisms and the symptoms are quite obvious, the consequences will be as follows: the situation is irretrievable, the crop has got a second infection, the major cause is no longer easy to be detected, and the first detected chance has been lost. In this regard, the maintenance and preservation of soil microorganisms are more important than the remedy of soil problems.

The key for maintenance of soil microorganisms is to pay attention to maintain the physical, chemical and biological environment and carry on a long-term maintenance. More specific methods for maintenance of soil microorganisms will be introduced in the chapter of guidelines.

9.2.5 More Researches on Soil Ecology are Needed

Manifestation of agricultural soil is the integrated performance of soil physical, chemical and biological properties, which have a close interrelationship. If there is a limiting factor, it will exert negative influences on crop production. Therefore, the method of conservation of soil fertility requirements should be adapted to the different soil conditions. However, due to the fact that soil microorganisms play the important role in agricultural production, and that their problems are hard to be observed by our eyes, soil microorganisms are commonly observed from the extent of health and disease of crops. So far, researches on soil microbial ecology are insufficient in many places, we need to establish the following basic and applied researches:

- Establishing methods and models for soil microbial ecological systems, and investigating and analyzing the differences in different soil management.
- Establishing gene pool of soil microorganisms and conserving natural microorganism resources.
- Strengthening researches on pollutants and influences of pesticides on soil microorganisms. Imposing environmental standards on pollutants and use of pesticides, and establishing monitor system for the quality of soil environment.

- Making use of biological inoculants, so as to achieve the goal of soil conservation.

9.3 Soil Microbial Diversity and Its Conservation

9.3.1 Introduction of Soil Microbial Diversity

Soil organisms include microorganisms and large-sized microorganisms. Microorganisms include virus, bacteria, archaea, fungi, algae, protozoa, lichen, nematode, tiny arthropod, etc.; large-sized microorganisms include annelid (such as earthworm), insects, etc. They either live in autotrophic way that makes use of CO_2, or heterotrophic way that makes use of organic carbon source. They can also be divided into chemotrophic and phototrophic microorganisms.

As far as we know, there are many types of microorganisms: about 5,000 kinds of virus, 3,000 kinds of bacteria, 69,000 kinds of fungi, and 40,000 kinds of algae. In soil, there are about 5,000 kinds of microorganisms have been described and defined, but about 300,000 to 1,000,000 kinds of soil bacteria are still unknown, and only less than 1% microorganism bacteria are proved to be cultivated. About more than 99% kinds are unable to be cultivated and may disappear rapidly as the environment damaged, which include available types. Organism diversity includes gene, species, and ecosystem diversity, and scientific community all over the world that have limited understanding in these three aspects. Therefore, soil microorganisms are regarded as a black box, and its diversity is the most profound but most neglected part, which deserves our attention, investigation, exploration and application.

9.3.2 Resources of Soil Microorganisms

Knowledge in soil microorganisms is still quite limited, and we need an overall planning, investigation, collection and conservation. Due to continuous cropping and overuse of synthetic fertilizers and pesticides in many farmlands, soil microbial diversity is damaged. Under a poor natural balance, vicious cycle of problems could leads to worse plant diseases and pests. In the web of life in soil microbial ecology, if species constantly keep disappearing, are not planted and do not produce food, the overall operation and the balance of agriculture soil ecology will be affected. To increase soil microbial diversity will be positive to the health and balance of soil environment. Thus, in that way, ecology is getting conserved.

9.3.3 Functional Value of Soil Microbial Diversity

Soil is the major medium of crop growth and provides various ecosystem services. It is the basis for human existence and development, and the soil microbial diversity and ecological environment quality could directly affect crop growth and health, which exert inestimably enormous influences on life. Pedogenesis includes five major factors, in which the effect of soil microorganisms is necessary. There are six ecosystem functions of soil may offer to the Earth:

- "Growth place" for all organisms.
- "Treatment place" for cleaning residues, organic materials and wastes.
- "Regulation place" for cycles of elements.
- "Mitigating place" for hydrological cycle speed.
- "Renewing place" for soil fertility.
- "Conservation and transportation place" for plant nutrients.

The six functions mentioned above have the closest relationship with the function of soil microorganisms. Soil possesses many functions, and in terms of agricultural production, including production capacity, cleaning capacity, nutrient holding capacity, water holding capacity and the biological balancing ability, etc. If the diversity of soil microorganisms is decreased, it may affect soil quality and quality of agricultural productions, and produces groundwater pollution and problems of insufficient supply of nutrients, which impose threats to soil environment upon human being's lives and crops. There are quite complicated relationships between soil microorganisms, ecology and crop production capacity. In addition to crop adaptation and the influence of climatic factors, the production capacity of crop is also influenced by physical, chemical and biological factors of soils. Only when these three factors strike a good condition, the goal of enhancing soil fertility will be achieved.

In agricultural production, unhealthy soil may bring about physiological diseases and other plant diseases and pests in crops. Under such circumstances, more pesticides will be used, and this will again cause environmental and ecological problems. In this way, human beings will be the victim of overall vicious cycle of soil environment. There are close relationships between the quality of ecological environment of soil microbial diversity and food production we rely on, as well as a direct influence on our health. Therefore, it is an important job of agriculture production to pay attention to the protection of soil microbial diversity and the quality of ecological environment.

Nowadays, biodiversity is the focus of biological community, and biodiversity in farmland soil is an important standard for the maintenance of fertility and balance. The soilborne diseases in farmland soil are getting severe, which also is related to the continuous cropping. The soil microbial diversity of farmland has been threatened, and that leads to severe plant disease and pests. The question of soil microbial diversity needs our continuous attention.

9.3.4 Protection Methods of Soil Microbial Diversity

The protection and preservation of farmland soil microbial diversity call for long-term and reasonable attention. Only following the nature law can last; otherwise, it may fail. Protection methods of soil microbial diversity are listed as follows:

Do not Pollute Soil:

Common pollution sources of soils include irrigation water, acid rain, toxic matters containing in poured or removed soil from other places to other locations. Especially, inorganic and organic toxic substances in irrigation water, such as heavy metals, strong acids, strong bases, cleansers and other industrial sewages, even the untreated sewages from animal farms are all regarded as excessive nutrients and organic pollution. In addition to inorganic and organic toxic matters, there are still pathogen pollution sources. For example, septic tank water may pollute soil and do harm to soil microbial diversity. Therefore, irrigation water and quality of soil removed from other places to improve soil should be given enough monitoring and attention, so as to protect soil against pollutants.

Do not Overuse Agrochemicals:

Agrochemicals include pesticides, fertilizers and plastic materials, etc., and these should be used properly. Many pesticides could somehow affect soil microorganisms, such as killing or inhibiting microorganism, and even bringing about unbalance in soil microorganisms. All these could be regarded as reasons for soil microorganism damage. It is common to see plastic materials like plastic film being left and mixed into soil by turning plow. However, since plastic film is hard to be decomposed or even contains toxic substances, it would do harm to soil environment as well as the biodiversity of soil microorganisms.

Do not Conduct Continuous Cropping for a Long Period of Time:

When a farmland is used only for the cultivation of one kind of crops, plant diseases and pests associated with such crop will increase, bringing about continuous cropping problems. The increase of pathogens will affect the soil microbial diversity and balance, especially in the net-type and facility cultivations. Therefore, we should not carry on continuous cropping for a long period of time, but should do rotation cropping with various kinds of crops. In this way, rotation cropping will reduce problems caused by continuous cropping.

Different crops may differ from each other in terms of the seriousness of continuous cropping problems. For example, serious diseases of continuous cropping among ginger, asparagus, melons, etc., may have something to deal with pathogens and crop resistances or toxic substances. In addition to rotation cropping, intercropping may also be adopted, i.e., to plant different kinds of crops in the farmland at the same time, so as to avoid or reduce continuous cropping problem.

Applying Proper Organic Matters:

There is a variety of organic fertilizer, but those organic matterswith heavy metal, antibiotic or toxic pollutants should not be used. Organic materials that are not completely matured should not be applied to soil, unless they have been matured thoroughly or improved. In this way, the crops may not be damaged, or the eutrophication of soils caused by overuse of fresh organic materials could be avoided. As for organic materials in livestock manure, they should be treated after being added with other materials and gone through composting procedure or high temperature disinfection. This could avoid a large quantity of pathogens entering the soil, which may affect soil microbial diversity.

Improving Soil Conditions:

When soil condition is confronted with physical, chemical or pathogenic problems, we need to adopt various methods, so as to improve soil microbial diversity. For example, whether the soil is too sandy, too acidic, too alkaline, too salty, or is low in organic matters, soil microbial diversity could only be improved after various measures were taken.

Inoculating Beneficial Microorganisms:

When soil quality has been impaired, we can improve physicochemical property of soil and applying proper organic materials. If beneficial microorganism could be inoculated, the quantity of beneficial microorganism is increased and the soil microbial diversity is improved. But we should avoid using highly antagonistic microorganisms, unless they are used against massive pathogens.

9.3.5 Prospective

Soil microbial diversity is the basis of the protection and preservation of ecology, as well as the foundation of reducing agricultural disasters. If operation follows natural principles, agriculture could sustain. The benefit of biodiversity is sustainable, invaluable and intangible, especially in districts with special conditions, such as places with high temperature, rainy, and windy districts. In order to ensure a sustainable agriculture development, we should pay attention to soil microbial diversity. Only by doing these, healthy soil environment and human beings could sustain. This is a project that deserves the attentions from both government and people.

In order to achieve the goal of protection and prevention of ecosystem and reducing agricultural disasters, three measures concerning soil microbial diversity should be given at priority:

Strengthening Investigations in Soil Microbial Diversity:

We should learn about soil microbial diversity in various ecosystems, and make further judgment and decisions.

Strengthening Original Collection and Genetic Identification of Soil Microorganism Types:

While conducting investigations, we should also carry out original collection of soil microorganism types, and stock them in the soil microorganism gene conservation center. This center is still expected to be set up in the future, and genetic analysis will be carried out to identify the classification and relationship of species.

Enhancing Researches on the Utilization of Soil Microbial Diversity:

After investigation and collection of soil microbial species, we should analyze the characteristics and make feasibility evaluation, and explore the application according to different features of species, so as to improve the value of soil microbial resource.

9.4 Definition and Applying Purpose of Biofertilizer

9.4.1 Definition of Biofertilizer

Sustainable agriculture aims to establish an agricultural production system that is independent from massive application of chemical fertilizers and chemical pesticides, with low energy consumption and cost, as well as steady production, increasing income and reducing harm on environmental ecosystem. The method of making agricultural production independent from massive application of chemical fertilizers is to replace chemical fertilizer with organic, biological and mineral fertilizers. To increase the efficiency of chemical fertilizers with biofertilizer is a soil biological technology that deserves our attention.

"Biofertilizer" is also called microbial inoculant or microbial fertilizer. There are broad and narrow definitions for "fertilizer." Any material, regardless of its basis on the soil or plant leaves, if it can provide nutrients or improve physicochemical and biological properties of soil, and increase production or improve quality of crops, it can be called as fertilizer. In short, materials that are capable of increasing production and improving quality of crops can all be called fertilizer in a broad definition

(Sheng, 1971). Materials can only contain nutrients which is called fertilizer in a narrow definition. Biofertilizer starts with a broad perspective and refers to artificial cultivation of soil microorganisms, which makes the use of living microorganisms in soil to provide nutrients for crops, improve soil nutrient conditions or physical, chemical and biological properties of soils, thus increasing production and improving quality of crops. Since microorganism is the major source of biofertilizer, it is usually named as "microbial fertilizer."

9.4.2 Application Purposes of Microbial Fertilizer

The purposes of applying microbial fertilizer include direct and indirect purposes, which are described in the following parts:

Direct Purposes:

As described in the preface, the direct applying purpose of soil microorganisms preparation is based on a broad definition of fertilizer, i.e., biofertilizer, which makes living microorganisms as the source of nutrients (such as nitrogen-fixing microorganisms) or improves soil nutrient conditions. Therefore, the usage of beneficial microorganisms could reduce the application of chemical fertilizers, achieve the goal of equal production or increase production, and improve quality. Since the cost of microorganism production is lower than industrial fertilizers, not only the application of chemical fertilizers will be reduced and soil degradation will be decreased, but also the production cost will be reduced.

Indirect Purposes:

Such indirect purpose is to take advantage of other functions besides fertilizer effects; for example, the functions of protecting rhizosphere, promoting ability of growth, and absorption of water and nutrients of root system, increasing capability of root life span, neutralizing or decomposing toxic substances, improving survival rate of transplanted crops, and early flowering, etc.

The development of microbial fertilizer is the basis of sustainable agriculture, and its application is necessary for the success of sustainable agriculture. In the past few years, since the massive application of chemical fertilizers and pesticides has obviously affected soil and ecological environment, we should be alerted and reflect on it. Thus, the exploration and application of sustainable agriculture should be vigorously promoted. The development and usage of microbial fertilizer has proved that it could not only bring steady production of crops and increase farmers' income, but also mitigate damages on ecological environment caused by overuse of chemical fertilizers. Regulation on microbial fertilizer turns into a legal commodity, and the alternate usage of microbial and organic fertilizers could guarantee farmers' income and soil fertility, and lead agricultural management to a sustainable way. In this way, the agricultural objectives of superior, ecological, safe and healthy will be realized.

9.4.3 The Necessity of the Soil Microorganism Applications

Soil is one of the important production factors, and it is also the base of cropping. In subtropical and tropical zone, and slope fields, organic materials decompose rapidly in the soil. Together with the intensive cultivation, they bring about problem soil. Reasons for these include continuous cropping, overuse or improper usage of agrochemicals. Altogether, they lead to changes in soil physical, chemical and biological properties, unbalance of soil ecosystem, concurrent of plant diseases and pests as well as unqualified productions, etc. Soil is the most fundamental asset of human beings, and it is infused with vitality. To protect and preserve soil is the responsibility of every generation. For instance, damages on beneficial bacteria in soil could not be observed by our eyes. However, once there are in severe damages, it needs much time and money to recover. Therefore, soil is the asset accumulated after eon, and we should strengthen the protection and pay more attention to it.

Reasons for Usage of Soil Microorganism:

Applying Massive Agrochemicals:

In modern agricultural cultivation and management, in order to increase production and improve product quality, people tend to overuse pesticides and agrochemicals. In the long run, the inhibition of beneficial microorganisms or influence on ecological balances cannot be disregarded.

Polluting Soil:

Water or air pollution may acidify soil or increase pollutants. Under long-term pollution, soil microorganisms will be largely affected.

Changing in Cultivation System:

Rotation cropping of paddy and upland are common rotation systems in many places. In recent years, the transfer from rice field to upland farming is more common. Due to the different activities of microorganisms in paddy and upland farming, the proportion of aerobes and anaerobes could also vary. Beneficial bacteria in paddy field may not exert good influence on upland farming, and vice versa. Under the intended and unintended behaviors described above, the balance between beneficial and harmful biology will be altered. Therefore, under modern cultivation environment, we should pay more attention to the indigenous soil microorganisms.

Functions and Benefits of Soil Microorganism Application:

Improving Soil Fertility:

There are many kinds of beneficial soil microorganisms, and different microorganisms play different roles: supplying crops with nutrients, improving physical property of soil, increasing the availability of nutrients, etc., which are all very important. For example, nitrogen-fixing bacteria could increase nitrogen in soil; decomposing microorganisms could decompose organic residues, and provide nutrients; phosphorus-solubilizing microorganisms could dissolve insoluble phosphates, and turn it into phosphate ions so that could be used by plants; microorganism could secrete polysaccharide which may improve the structure of soil aggregates, increase physical property of soil, etc.

Assisting Plants in Absorbing Nutrients:

Plants mainly depend on root hair to absorb nutrients. With more root hair, the absorption area will be larger, and absorption ability will be stronger. The mycorrhizal fungi in soil could coexist with most plant roots. With hypha reaching out of root, its function is similar to root hair, i.e., to take in more nutrients, especially phosphorus. Many experiments on mycorrhiza have already increased the production of crops.

The form of nutrients may affect absorption efficiency. Not every form in soil could be taken in, but those organic acid secreted by certain soil microorganisms could form chelated structure, which is convenient for absorption or increasing nutrient availability, such as siderophores that could also do against pathogens.

Increasing Disease-Resistant and Drought-Resistant Ability of Plant:

Soil ecological balance could prevent pathogens from massive propagation, and thus we can make use of competitive microorganism to carry out biological control. The inoculation of mycorrhizal fungi or root protecting bacteria will take up the root surface to protect the rhizosphere from pathogens. Mycorrhizal fungi coexist in the rhizosphere, just like root hair could be conducted to absorb water and nutrients, which could resist drought. The increase of exudation in soil microorganisms will decrease the freezing point of soil water, which could protect crops from cold.

Saving Energy and Reducing Production Cost:

The production of chemical nitrogen fertilizer requires huge consumption of energy. For districts with limited and insufficient energy, the application of nitrogen-fixing bacteria may reduce usage of nitrogen fertilizers, and achieve the goal of energy-saving in agriculture. Phosphorus fertilizer production is based on acidification of apatite from phosphate rock. When it is spread into the soil, it should not be used up; while most fertilizers are fixed as insoluble phosphate salts with low efficiency. Nitrogen-fixing bacteria and mycorrhizal fungi could contribute to the source of nitrogen and phosphorus. With small usage amount of bacteria and low cost in bacteria production, we can lower production cost by reducing the application of chemical fertilizers.

Reducing Environmental Pollution:

Since enthusiasm is widely spread in environmental protection, agricultural pollution should not be neglected. Overuse of chemical fertilizers will pollute rivers, reservoir, and water resources. Eutrophication will promote massive production of microorganisms in water, thus affecting balance of water microorganisms, especially the pollution brought about by phosphorus and nitrogen fertilizers. Application of soil microorganisms (such as nitrogen-fixing bacteria, mycorrhizal fungi) can largely reduce the usage of phosphorus and nitrogen fertilizers, thus reducing environmental pollution to the lowest extent.

In a word, the application of soil microorganisms is a new technology, which can reduce environmental pollution, improve crop production and quality. Future researches will leap from several bacteria development to the compound or multifunctional chemical fertilizers and soil conditioners. A wonderful biological world in soil is waiting for our exploitation.

9.5 The Relationships between Plant Nutrient Absorption and Soil Microorganisms

There is a close relationship between soil microorganisms and fertility, because soil microorganisms could decompose organic materials in the soil and turn it into nutrients that could be used by plants. Soil microorganisms will dissolve minerals and turn them into soluble matters, and thus soil microorganisms plays a decisive role in the system of nutrient cycles. All these knowledge has been known to us, but it is only in recent few years that scientists become familiar with soil microorganism's effects on plant nutrient absorption. Soil microorganisms do not only affect the cycle of nutrients, but also plant nutrient absorption.

9.5.1 Plant Root System and Soil Microorganisms in the Rhizosphere

There are quite a number of soil microorganisms in the rhizosphere. It is reported that there could be as many as 10^{10} bacteria per gram, in average, 2×10^9 bacteria. The rhizosphere contains many exudates of plant root and residues of root cells, and thus soil microorganisms could obtain more carbon source, energy source, and nutrients for their propagation.

There are many kinds of soil microorganisms in the rhizosphere. Soil microorganisms in the rhizosphere mainly contain bacteria, fungi, actinomyces and protozoa, etc. Generally speaking, there are more bacteria on the surface of younger roots, while there are more fungi on the surface of old roots. The type and quantity of microorganism are often influenced by moisture, organic materials, nutrients, pH value, soil depth, as well as microflora.

In addition to microorganism decomposition in the rhizosphere, nitrogen-fixing bacteria, denitrifier, sulfate-reducing bacteria, and methane producing bacteria, etc., have their special functions and vary along with environments (Figure 9-1). In general, beneficial bacteria in the rhizosphere are the "guardians" of root, and some beneficial bacteria could resist to plant pathogens, thus protecting root system and forming an important link in the existence of nature. That is also the best example of mutual benefits of biological "co-evolution."

Rhizosphere microorganism exists around or in the surface of roots and its important role in promoting plant nutrient absorption deserves our attention.

9.5.2 The Influence of Microorganism on Plant Nutrient Absorption

In order to have further understanding in the effect of microorganism on plant absorption of nutrients, scientists make use of hydroponic and soil cultivations to observe the differences among sterilized, unsterilized or inoculated microorganisms.

For plants cultivated in water culture, it's simple to observe the influence of soil microorganisms. We can start from microorganisms to discuss factors affecting competition of inorganic nutrients, and the direct relations between microbiological production and plant nutrient absorption. However, for plants cultivated in soil, it is quite complicated due to the existence of microorganisms and soil organic matter decomposition and minerals. For example, phosphate-solubilizing bacteria can dissolve insoluble phosphates.

Hydroponic Cultivation:

During the experiment of microorganisms, the existence of microorganism in water culture could promote the absorption of phosphorus, sulfur, nitrate, potassium, zinc and iron, etc. It is reported that microorganisms could make absorption more than two times of nitrate as well as to enable its effectiveness in its transportation; but for the absorption of ammonium nitrogen, the figure is reduced to one third, and the quantity transported to the soil surface decreases. The influences of soil microorganism could also vary according to different nutrient forms for plants. In the future researches on hydroponics, we should conduct more with microorganisms.

Soil Cultivation:

In soil, ion concentration, its availability and interactions along with pH value, temperature, and aeration will all affect plant nutrient absorption. Therefore, by alternating these conditions, soil microorganisms can exert influences on plant nutrient absorption.

Nitrogen fixation
Nutrient availability
Nutrient absorption
Plant growth regulation
Phtotoxins
Siderophore
Antibiotics
Pathogens

Rhizosphere

Root exudation
Nutrient competition
Redox potential

Effects of microorganisms on plant growth and development

Effects of plant roots on microbial growth

Microorganism

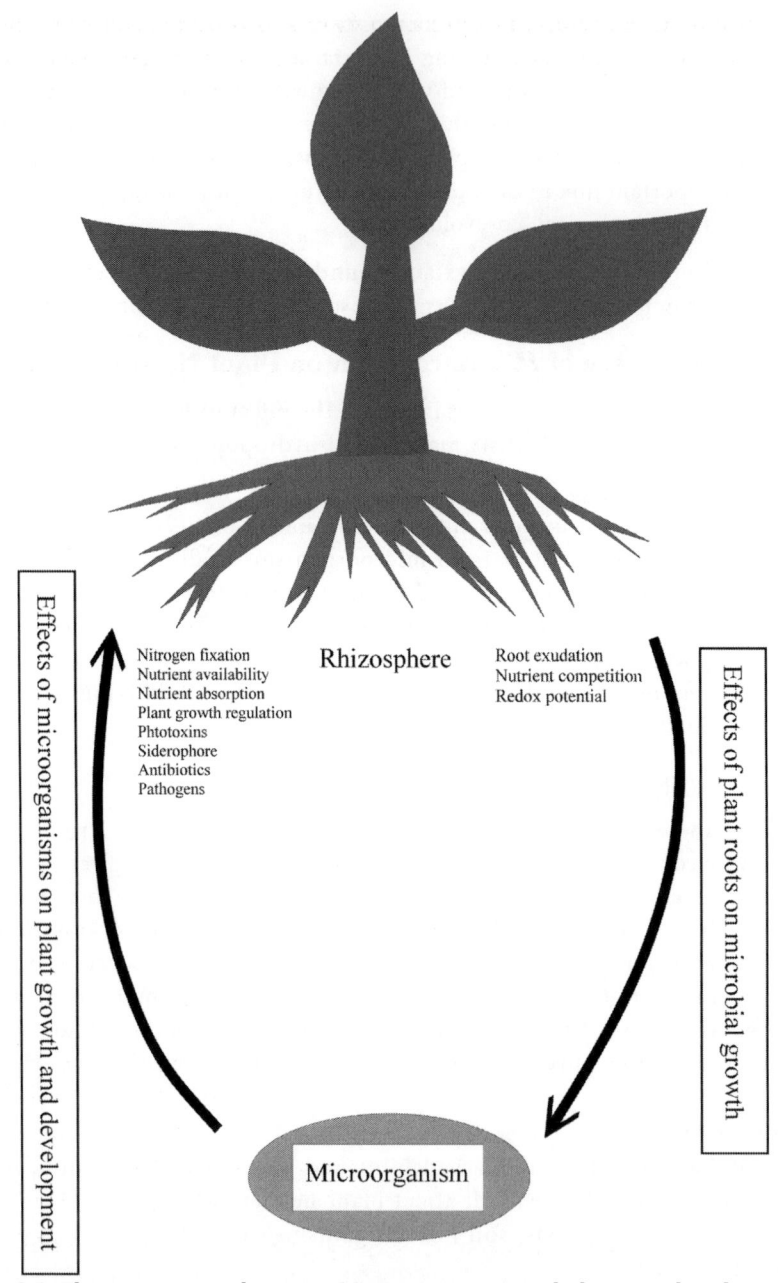

Figure 9-1. The Interactions between Microorganisms and Plants in the Rhizosphere.

Nitrogen-fixing bacteria in soil microorganisms offer the source of nitrogen. Under anaerobic condition, denitrifiers will transform nitrate into gaseous nitrogen. However, benefiting from the decomposition of microorganisms and organic materials in the soil could obtain more nitrogen and other elements. In this way, microorganism plays an important role in the loss or gaining of nitrogen in the soil.

Phosphorus and potassium fertilizers can be acquired through the decomposition of soil organic materials and dissolving of mineral by microorganisms. For a better satisfaction of plant requirements, phosphorus solubilizing microorganisms that have been discovered in recent year possess the best potential of development. When phosphorus fertilizers are applied into the soil, only a small quantity is taken in by plants, most phosphate fixed by calcium, iron or aluminum that are insoluble. Therefore, a proper usage of phosphate solubilizing microorganisms will be helpful to an effective exploitation of phosphorus fertilizers.

The biota of soil microorganisms may also expand due to the addition of organic materials in the soil. Therefore, soil microorganisms may take in a large quantity of nutrients in the soil during its mass production. If the soil is deficient in certain nutrients, crops will also be short of such elements. Commonly, a temporary deficiency in nitrogen or other microelements (such as molybdenum, manganese, etc.) is caused by the application of fresh residues to the soil, and this is also the reason why organic materials should be thoroughly matured. If fresh residues of plant are applied into the soil, it should be applied before planting, or we can add certain fertilizers with organic materials to prevent the composting microorganisms from competing with plants and taking away large amount nutrients.

9.5.3 Reasons of Microorganism's Influence on Plant Nutrient Absorption

The reasons of microorganism's influences on plant nutrient absorption include direct and indirect effects, and they are described in the subsequent parts.

Direct Effects:

Soil microorganisms may release substances that could directly affect root absorption ability of nutrients; for example, auxins transformed by soil microorganisms could stimulate plant nutrient absorption; gibberellins could promote plant absorption of potassium and enable its transportation in plant. Gramicidin D produced by microorganisms could also increase root absorption of potassium.

In addition to beneficial microorganisms, phosphate solubilizing bacteria could affect plant absorption and transportation of phosphate. Gibberellins could be as a regulator of membrane permeability. Gramicidin D could affect the function of cell membrane.

In addition to beneficial microorganisms, harmful microorganisms or pathogens in the soil may affect plant absorption ability of nutrients. Some even produce unique toxic substances, which react rapidly with membrane proteins in the root and thus inhibit the absorption ability of root. The worst may lead to the loss of electrolytes from root system.

Indirect Effects:

Auxins, gibberellins, cytokinins or other substances released by microorganisms in the rhizosphere could promote the growth of root system and enlarge the absorption surface of root, and thus increase root absorption. Also, the substances produced by microorganisms may promote photosynthetic rate and cell metabolisms, and improve the root absorption ability at the end. On the contrary, if harmful microorganisms also enter the cortex of root system, it will bring about absorption problems in the root system.

9.5.4 Pay Attention to Soil Microorganism Management

There are many types of soil microorganisms, including beneficial and harmful microorganisms. In nature, as things mutually reinforce and neutralize each other, it seems that we need not to carry out management under the natural principles of balance. But recently, due to human overuse of agrochemicals (including pesticides, fertilizers, etc.), and drastic changes taking place in intensively cultivated farmland, it is common to see communicable soil diseases, problems brought about by continuous cropping and other soil problems. Along with more attention on different methods of cropping and soil microbial management, these will become an important link to agricultural practice in the future.

9.6 Functions, Types and Characteristics of Nitrogen-Fixing Microorganisms

9.6.1 The Importance of Biological Nitrogen Fixation

Three elements of fertilizer are greatly demanded by crops, i.e., nitrogen, phosphorus and potassium, but only nitrogen comes from atmosphere. Though there are 80% nitrogen gas in the atmosphere, such kind of nitrogen (N_2) cannot be used directly by plants, unless nitrogen is transformed into ammonia. Such process of restoring nitrogen to ammonia is called "nitrogen fixation." In the world, nitrogen fixation includes "biological nitrogen fixation" and "industrial nitrogen fixation." "Biological nitrogen fixation" refers to the function of nitrogen-fixing microorganisms in natural world. After being transformed into ammonia by nitrogenase contained in nitrogen-fixing bacteria, actinomyces or cyanobacteria, ammonium is further transformed to organic nitrogen compound (such as amino acids). In agricultural application, the most frequently used microorganisms is rhizobia. Rhizobia could enter the root of beans, form root nodule, and fix the nitrogen as the source needed by plants. If beans could be inoculated with rhizobia, it could fix nitrogen and acquire quantitative nitrogen nutrients. However, the grass family and many other non-legume crops could only depend on nitrogen-fixing bacteria existing around rhizosphere to provide plants with limited nitrogen nutrients.

There are quantitative types of free living nitrogen-fixing bacteria in the soil. Though fixed nitrogen in unit soil area is less than symbiotic and associative fixed nitrogen, it makes contribution silently to the farmland, and become an important supplier in the natural soil ecosystem. So far in the agricultural extension for nitrogen-fixing bacteria, there are more applications of rhizobia and with positive effect.

9.6.2 Types of Nitrogen-Fixing Microorganism

There are many kinds of nitrogen-fixing bacteria, but all of them belong to prokaryotes, including bacteria (chemotroph bacteria, photosynthetic sufur bacteria, rhizobia), actinomyces, cyanobacteria, etc. (Table 9-2). In ecosystem, it includes free-living, associative and symbiotic nitrogen fixation. Free-living nitrogen-fixing bacteria scatter both in continents and waters, while associative nitrogen-fixing bacteria spread in the rhizosphere and on leaves. Since they have not formed special structure, their structure is quite different from that of symbiotic nitrogen-fixing bacteria (such as tuberculiform or cystidium). Rhizobia belong to the type coexisting with plants.

Nitrogen fixation possess of all microorganisms is the function of nitrogenase. Nitrogenase is specially protected within root nodules or free-living microorganisms, so as to avoid being destroyed by oxygen from air. What's more, nitrogenase could reduce many kinds of substrates; for example, it can reduce nitrogen (N_2) into ammonia (NH_3), acetylene (C_2H_2) into ethylene, cyanogen (CN^-) into ammonia and methane (CH_4), etc. Such kind of matter contains like $N \equiv N$, $N \equiv C$ or $C \equiv C$, as being listed in Table 9-3, but not CO ($C^- \equiv O^+$). Nitrogenase requires anaerobic conditions, and thus needs protection system in bacteroids or within root nodules, such as protection by polysaccharide, structure protein or leghemoglobin, or fast consumption of oxygen, etc. For example, increasing respiration could increase nitrogenase activity, or reduce photoreaction could reduce oxygen production, etc.

9.6.3 Taxonomy of Rhizobia

Rhizobia have been isolated more than one hundred years since 1889. At the early stage, the classification of rhizobia is based on host plants. Later, it is discovered that

Table 9-2. Types of Biological Nitrogen Fixation

Ecological Category	Bacteria Category	Examples
1. Free-Living Microorganisms		
Phototrophs	(1) Cyanobacteria	*Aanabaena, Gloeothece, Nostoc*
	(2) Purple non-sulfur bacteria	*Rhodopseudomonas, Rhodospirillu*
	(3) Purple and green sulfur bacteria	*Chlorobium, Chromatium*
Chemotrophs	(1) Aerobic bacteria	*Azotobacter, Beijerinckia, Rhizobium*
	(2) Facultative bacteria	*Azospirillum, Bacillus, Klebsiella*
	(3) Anerobic bacteria	*Clostridium, Desulfovibrio*
2. Associative Microorganism		
Chemotrophs	(1) Rhizosphere microorganism	*Azosprillum, Azotobacter, Bacillus*
	(2) Microorganism on leaves	*Beijerinckia, Klebsiella*
3. Symbiotic Microorganisms		
Phototrophs	(1) Gymnosperm and cycas	*Nostoc*
	(2) Litchen, liverwort	*Calothrix, Nostoc, Stignonema*
	(3) Hydropteridales, floating fern	*Anabaena*
Chemotrophs	(1) Leguminosae rhizobia	*Bradyrhizobium, Rhizobium*
	(2) Legume bacteria with stem nodule	*Azorhizobium*
	(3) Non-leguminosae rhizobia	*Rhizobium*

there are fast-growing and slow-growing strains, which finds the differences among various strains. In recent years, with advances in the analysis of molecular biology, new progress has been made in classification of microorganism. We mainly make use of 16S rDNA sequence and focus on the differences in separation and hybridization of DNA-DNA. Supported by other classification of biochemical analysis, new

Table 9-3. Functions of Nitrogenase

Substrates	Substrates structure	Production
Dinitrogen	$N \equiv N$	$NH_3 + H_2$
Acetylene	$HC \equiv CH$	$H_2C = CH_2$
Cyanide	$(C \equiv N)^-$	$NH_3 + CH_4$
Allene	$H_2C = C = CH_2$	$CH_2 = CH - CH_3$
Azide	$N \equiv N^+ - N^-$	$N_2 + NH_3 + N_2H_4$
Nitrous oxide	$N \equiv N^+ - O^-$	$N_2 + NH_3$
H^+ ion	H^+	H_2

types of classification come in one after another. The classification of rhizobia is booming, and so far in proteobacteria. Rhizobia belonging to type α includes nine genus (*Rhizobium, Allorhizobium, Sinorhizobium, Mesorhizobium, Bradyrhizobium, Blastobacter, Azorhizobium, Devosia* and *Methylobacterium*), and those belonging to type β include two types (*Burrkholderia* and *Ralstonia*).

9.6.4 Functions and Characteristics of Rhizobia

The application of rhizobia could cut down the usage of nitrogen fertilizers, enable drought-resistant ability of plant, reduce nitrogen pollution and soil acidification, increase production and improve quality of leguminosae, and increase fertility and health. Reduced application of chemical fertilizers could cause the consumption decrease of non-renewable resources. Schubert and Wolk (1982) pointed out that leguminosae nitrogen fixation only dissipates energy of plant itself, which approximately equals to the energy for absorbing and using the nitrate (NO_3^-) fertilizer, thus clearing up previous misunderstanding about energy consumption of leguminosae nitrogen fixation. Biological innoculant has low cost of production and convenient operation, while industrial production of chemical fertilizers (such as urea) consumes much energy and bring about transportation cost, etc. 1.3 ton petroleum are needed in the production of every 1 ton nitrogen fertilizer. However, rhizobia often possess the specificity or affinity of leguminosae host, i.e., soybean rhizobia could only be used in soybean, not the inoculation of peanut. Therefore, legumes that have a specific requirement for rhizobia need inoculation. Recently, we have adopted the method of screening phosphate-solubilizing rhizobia for the inoculants of non-leguminous and grass family crops, aiming at the function of dissolved insoluble phosphates in the rhizosphere.

In the application of rhizobia, the top priority is to select excellent bacteria strains. After isolated and purified, we carry out preliminary greenhouse and ultimate field screening work, to select the best strain in both adaptability and symbiosis, and make inoculant with mixed bacterial strains. It is a simple way to use the inoculant of rhizobia, and it could be sowed after it is mixed with seeds. The liquid inoculant used per hectare is about two to four liters, while it is common to use about 200 to 400 kg urea per hectare in green soybean field. Comparatively speaking, the rhizobial application can save time, money, and labor.

9.7 Necessity and Development of Rhizobia and Mycorrhizal Fungi

9.7.1 Soil Microorganisms and Crop Production

Climate, soil, crop species, cultivation management, and disease-pest control are five major factors that decide the production of crops, and are also popular with modern agriculture that pursues high production and quality. Agricultural production possesses integrity that requires the presence of all, and importance of five factors cannot be neglected. Soil and its management are one part of agricultural production. Healthy crops call for healthy soil, and soil are the basic asset of human beings. Therefore, the protection and preservation of soil is the responsibility and obligation of each generation.

Agricultural soil contains inorganics, organics and organisms, neither of which can be neglected for crop production, and they cooperate with each other. The cooperation is particularly important in perennial crops. Interactions among soil microorganisms may directly or indirectly affect agricultural production, like the unbalance or occurrence of soil disease and pests will decrease crop production and its quality.

In modern agricultural management, we cannot neglect those factors that affect soil microbial balance or promoting the growth of beneficial organisms, especially the overuse of pesticides, agrochemicals, environmental pollution of soil, and inborn soil microorganisms.

In agriculture, soil microorganisms could be divided into beneficial and harmful ones, and their proportion is influenced by environmental conditions. Intensive cultivation system commonly includes crop rotation of paddy field and upland, but recently many rice fields were changed to upland fields. Microbial system in paddy field soil is different from that of upland. Aerobic bacteria, rhizobia and mycorrhizal fungi are symbiotic microorganism of upland crops. Rhizobia could fix nitrogen, while mycorrhizal fungi could help plant with the absorption of phosphorus and other nutrients. Among soil nutrients, nitrogen and phosphorus are two of three macro elements of fertilizer. How to use microorganisms to fix nitrogen and assist the absorption of phosphorus fertilizer are important directions for the future agricultural development.

9.7.2 Necessary for Inoculating Rhizobia and Mycorrhizal Fungi

Plant Nutrient Requirements:

Rhizobia of Leguminosae:
There is about 80% nitrogen gas (N_2) in the atmosphere, but such kind of nitrogen cannot be used directly by plants, unless nitrogen is transformed into ammonia by symbiotic and non-symbiotic nitrogen fixing bacteria. Leguminous plants with rhizobia and form root nodules that take in nitrogen from atmosphere, and thus increase nitrogen source of plants. Such symbiotic nitrogen fixation has higher amount of nitrogen to supply for plants. Each year, nitrogen fixation per hectare varies according to different legumes. In soybean fields, nitrogen fixation could fix about 100 kg nitrogen/ha/year (about 217 kg urea), and some may reaches as high as 400 kg nitrogen/ha/year (about 868 kg urea). This nitrogen fixation is not only in favor of legumes absorption of nitrogen, but those residues are also effective for nitrogen needed by next crops, which can increase soil fertility.

Mycorrhizal Fungi:
Mycorrhizal fungi is a group of soil microorganisms, one type of fungi that colonizes in plant roots with symbiosis, including endomycorrhizas, ectomycorrhizas and endoectomycorrhizas. Crops are in great demand of phosphorus fertilizer, but phosphorus moves quite slow in the soil. The absorption of phosphorus mainly depends on the intercept absorption of roots. The infected hypha of mycorrhizal fungi with root symbiosis could largely increase contact with soil and absorption surface, so as to increase uptaking more phosphorus. Thus, among the absorbed nutrients by symbiotic mycorrhizal hypha, it is extremely obvious to increase phosphorus absorption. Actually, an increase in the absorption area may enable the absorption of other elements. Specially, mycorrhiza is helpful to the soil which contains little phosphorus.

Promote the Growth and Production of Crops:

In the soil, there are more or less indigenous rhizobia and mycorrhizal fungi, and the efficiency of their symbiosis with plants could be high or low. The infectious number of rhizobia or mycorrhizal fungi to the root is uncertain, and thus the selection for excellent rhizobia or mycorrhizal fungi as the inoculant is needed. Previous experiments show that the production of soybean in the inoculation with excellent symbiotic rhizobia at fields improves about 17%, and the highest reaches to 40%. Besides, inoculated mycorrhizal fungi could increase production by 15%, or as high as 65%. However, when soil contains too much nitrogen or phosphorus, the effect of inoculation will not be brought into play.

Save Energy:

Chemical nitrogen fertilizers used in agriculture will consume much energy in their industry production. In districts with limited or insufficient energy, the application of nitrogen fixation will achieve the goal of saving energy.

Reducing Cost of Production:

The production of chemical nitrogen fertilizers consumes much energy and adds cost. Phosphorus fertilizers are exploited from rock phosphorite, which is quite limited resource. After rock phosphorus was made into phosphorus fertilizer with acids, and applied into the soil at a later stage, it will not only be absorbed by crops, but also hard to be washed away. Over many years of phosphorus fertilizers' application, the soil may accumulate too much insoluble phosphates that are fixed by soil with low availability. The symbiosis of mycorrhizal fungi could contribute to taking up insoluble phosphates for crops. Nitrogen-fixing bacteria and mycorrhizal fungi are easy to be produced with low cost, and also need not to be used for too much amount in unit of area. Comparing with the application of chemical fertilizers, the application of mycorrhizal fungi could cost down the production to more than ten times, and thus increase farmers' income.

Reducing Environmental Pollution:

The era of environmental protection is coming, and environmental pollution that was brought about by overuse of chemical fertilizers cannot be ignored. Inappropriate application and overuse of fertilizers may pollute river, mountain, reservoir and water sources; and eutrophication function in water will result in the massive production of algae and also the balance among microorganisms. In particular, the management of nitrate and phosphorous pollution should be strengthened. Application of nitrogen-fixing bacteria and mycorrhizal fungi may reduce the use of nitrogen and phosphorus fertilizers, which reduces the pollution.

Increasing Disease-Resistant and Drought-Resistant Ability of Crops:

The mycorrhizal fungi symbiosis with plant root takes some space in root surface and endodermis, and thus reduces the infection of pathogens. Just due to the competition between mycorrhizal fungi and pathogens, plant root get protected.

Crops mainly depend on root hair to take up water, and when mycorrhizal fungi live in the root, hypha reaches out of the root surface, just like root hair, which is increase water absorption. When soil is deficient in water, hypha can assist crops in water absorption, and enable crop's drought-resistant ability. The importance of mycorrhiza is obvious in upland farming and dry seasons.

9.7.3 Effect of Biotic Soil Conditioners

The root of leguminosae has two major kinds of symbiotic systems: root nodule symbiotic systems fix nitrogen, and mycorrhizal systems promote absorption of phosphorus. If these two elements could supply for crop production in a product, it will become an ideal compound bio-fertilizer. Mixed inoculants with rhizobia and mycorrhizal fungi have developed and used in a very convenient way by mixing directly it with seeds. Test results in six experimental fields in Taiwan have shown an increase of 30% in production, and the highest could reach 90%. In some districts, the synergistic effect is obvious, but if part of the soil contains too much residual nitrogen and phosphorus, there is no need to inoculate. In the whole, the potential of developing biotic compound soil conditioners is great. In addition to certain types of bacteria that are related to the nutrient availability or absorption, we should also add phosphate-solubilizing bacteria, rhizosphere beneficial bacteria, disease-control bacteria, etc. This will better display for the biotic improvement of soil. The realization of increased production and improved quality of crops depend on the improvement of planting soil condition and disease control with additive effect.

In addition to practical tests, fast-growing soybean rhizobia grow about two times faster than slow-growing soybean rhizobia, which offer more help in rhizobia researches and soil microbial improvement.

During the research of mycorrhizal fungi, the method of massive producing spore of mycorrhizal fungi has been found. These spores could be made into inoculant and offer direct evidence for the usage to dissolve the insoluble phosphates in the rhizosphere by mycorrhizal fungi, which should be put forward for new ideas.

9.8 Introduction of Endomycorrhizal Fungi

9.8.1 Research History of Mycorrhiza Fungi

Mycorrhizal fungi were identified by German botanist A. B. Frank in 1885, and mycorrhiza is a symbiotic structure of root system and fungi. According to the shape and structure, it could be divided into ectotrophic and endotrophic mycorrhizal fungi. Ectotrophic mycorrhizal fungi can be commonly found in trees, like Pinaceae, while endotrophic mycorrhizal fungi include septum and non-septum. The previous category includes orchid, and the latter includes agronomy and gardening crops. Non-septum endotrophic mycorrhizal fungi and host root cortex cells form special structures like vesicles and arbuscules, etc., and thus they are called "vesicular arbuscular mycorriza" (VAM) or "arbuscular mycorrhiza" (AM). At the early stage, researches on endotrophic mycorrhizal fungi mainly focus on its classification and shape. It is not until the mid-1950s that mycorrhizal fungi are proved to have close relationship with nutrient absorption of the plants, which promotes relevant researches in academic community.

9.8.2 Distribution of Endomycorrhizal Fungi

Mycorrhizal fungi are found in moss, fern, gymnosperm and angiosperm. Among continental vascular plants, about 93% form mycorrhiza, and ectotrophic mycorrhizal fungi could only infect about 3% of phanerogam (flowering plants). Most of the plants (about 97%) are infected by endomycorrhizal fungi. Non-septum endomycorrhizal fungi scatter wider than ectomycorrhizal fungi and septum endomycorrhizal fungi, and so do host plants, such as major cultivated plants, grass, bush, forest and shade-living plants.

9.8.3 Classification of Endomycorrhizal Fungi

Endomycorrhizal fungi belong to Endogenaceae of Mucorales. At present, there are nine categories: *Acaulospora*, *Complexipes*, *Endogone*, *Entrophospora*, *Gigaspora*, *Glaziella*, *Glomus*, *Modicella*, and *Sclerocystis*. The classification of endomycorrhizal fungi are made firstly according to the different shapes of spores, and then to the hyphae. Commonly methods for root dyeing include fuchsin acid, trypan blue, aniline blue, or chlorazol black E. Spores separated from sieving mycorrhizal fungi often employ wet screening and sugar liquid gradient centrifugation. Presently, mycorrhizal fungi that have been found include many kinds, such as *Glomus*, *Gigaspora*, *Sclerocystic*, and *Glomus* is the most common.

9.8.4 Morphology of Endomycorrhizal Fungi

The endomycorrhizal fungi locate in the cortex cells of roots, but not in endodermis, stele and meristem. Mycorrhizal fungi form special structure, such as vesicular or arbuscular in root cortex. The part of vesicular forms sporangia. The abuscular is the place where exchange nutrient with host plant. Mycorrhizal fungi infect plant roots and the external hyphae extent to the outside of roots. The external hypha function as root hair that increasing the surface area to absorb nutrients and water.

9.8.5 Function of Endomycorrhizal Fungi

Many studies have proved that endomycorrhizal fungi could help promote the growth of plants. For example, it is most obvious that mycorrhizal fungi could help plants absorb phosphorus and other nutrient elements and water. Endomycorrhizal fungi could increase the resistances of disease, drought, water, and heavy metal of plants.

Endomycorrhizal fungi could increase the absorption of phosphorus fertilizer in the following ways: increasing contact area between root system and soil; increasing mitochondria of mycorrhizal root cell; enlarging nucleus, and lengthening life span, as well as increasing the activity of phosphatases and ATPase. In these ways, the ability of absorbing, dissolving and transferring phosphorus fertilizer is enhanced. In general, low content of phosphorus in the soil will be conductive to the infection of mycorrhizal fungi.

9.8.6 Relationship among Endomycorrhizal Fungi, Phosphorus Fertilizer and Environment

Phosphorus fertilizer has low solubility and mobility in the soil, and is easily fixed by soil. Soluble phosphoric acid tends to form insoluble phosphate compound with Fe, Al and Ca in the soil. Phosphorus is one of the three major elements required by the growth of plants, and when plants are seriously short of it, their growth may get stopped. Therefore, in almost each cultivated season, phosphorus fertilizers should be applied.

Generally speaking, with the increase of soluble phosphorus in soil, infection of root system by mycorrhizal fungi is decreased. The application of rock phosphate might not affect the infection of mycorrhizal fungi, and the inoculation of mycorrhizal fungi with rock phosphate powder could remarkably increase production. It has been shown that the inoculation of mycorrhizal fungi could help plants take in insoluble phosphates, and increase the utilization efficiency of rock phosphate in the soil. Different types of mycorrhizal fungi possess different sensitivities to phosphorus fertilizer.

Endomycorrhizal fungi could enable plants to take in water, resist drought, and reduce soil water conduction problem. Mycorrhizal fungi are aerobic microorganisms, which have close relationship with the aeration of soil. Neither too wet nor too dry condition will do better to the growth of mycorrhizal fungi. After the cultivation of paddy rice, the quantity of mycorrhizal fungi in the soil will drastically decrease. When soil moisture reaches saturation, the germination of mycorrhizal spores and the forming of hypha are restrained.

Temperature is one of the factors that affect the colonization of endomycorrhizal fungi. Under 25°C, the vesicules and arbuscules of mycorrhiza are formed. About 30°C, mycorrhizal fungi infecting soybean fails to form vesicules. If it is not infected under 41°C, this indicates that high temperature is harmful to the colonization of mycorrhizal fungi. Also, same harmful condition would occur at low temperature.

The light intensity and photoperiod are also one of the reasons affecting the colonization of mycorrhizal fungi. Light intensity exerts direct influences on plant photosynthesis and indirect influences on the colonization of mycorrhizal fungi. Strong light will promote the colonization process, especially in a long day.

The infection rate of endomycorrhizal fungi on host root system and the number of spore formation varies with the change of seasons, including the temperature in winter, solar intensity and photoperiod. The application of fungicides, pesticides and herbicides will all restrain the colonization and formation mycorrhiza. At present, one problem in researches on endomycorrhizal fungi is that scientists could not reproduce endomycorrhizal fungi by the synthetic medium so far, except for *Piriformospora indica*. Therefore, nowadays, many of the cultivation and reproduction still depend on symbiotic method of propagating it with plants. The physiology, zoology and heredity of endomycorrhizal fungi still need further and thorough researches.

9.9 Phosphate Solubilizing Microorganisms

9.9.1 Relation between Soil Phosphorus and Plants

The three nutrient elements absorbed the most by plants are nitrogen, phosphorus and potassium, which are called "three key-elements of fertilizer." Phosphorus plays a quite important role in plant nutrients and it deals with energetic and biochemical reactions of plants. Many biochemical processes need the cooperation of phosphorylation. What's more, since phosphorus is a component of nucleic acids, the biological genetic material has a great influence on cell division and the growth of meristem, and it is one of the indispensable elements needed in the production and reproduction of plants.

Among the major nutrient elements in the soil, there are distinct differences between phosphorus and nitrogen. Nitrogen tends to move or to be easily washed away from the soil, while phosphorus is just the opposite. An understanding of phosphorus forms and behavior will contribute to a proper application of phosphorus fertilizers and phosphate-solubilizing microorganisms, and thus it will increase the availability of phosphorus fertilizer.

Ordinarily, if the soil is not applied with phosphorus fertilizer, crops and, particularly, crops with high demands for fertilizer will be short of phosphorus. This tends to present deficient symptom. For some crops, symptoms of the deficiency in phosphorus can be judged on the purple color of stalk or leaves. But for some, it is hard to judge simply from the appearance, we need to judge from the analysis of the plant and the soil. When crop is deficient in phosphorus, the symptoms could be poor growth of plant, root, and leaves, or dark green with reddish-purple leaf tips and margins on older leaves of the grass family. For annual plant, once the deficiency in phosphorus is raised, a subsequent improvement would be a bit late. The shortage of phosphorus in fruit trees will seriously affect blossom, fruiting, fruit development and fruit sweetness. Phosphorus also plays an important role in potato tubers or other roots production.

9.9.2 Relationships of Phosphorus-Solubilizing Bacteria among Phosphorus of Soil and Fertilizer

Phosphorus is one of elements that mostly needed by plants, and its existence in the soil includes inorganic and organic forms. Inorganic forms often fix with calcium, iron and aluminum (Figure 9-2), which appear to be hard to dissolve and absorbe by plants. Organic forms exist in many organic compounds of phosphate bonding, such as phospholipids, phytins and nucleic acids, etc. These all come from soil organisms. Therefore, when soil contains many organic matters, there is rich in organic phosphorus. Commonly, organic phosphorus in the soil can provide plants with phosphorus nutrient after being decomposed.

The more inorganic phosphorus exists in a combined way, the less effective phosphorus exists in nature. Therefore, we need to apply phosphorus fertilizer in each cropping; otherwise, the production or quality will be influenced. However, the recovery rate of phosphorus fertilizer from soils is quite low. 100 dollars of phosphorus fertilizer with a cropping absorption are normally less than 20 dollars. The rests of the phosphate that are worthy of 80 dollars are almost fixed by soils. What's more, since phosphorus is hard to move in the soil, it is hard to be washed away by water or rain. Phosphorus fertilizer, which has not be absorbed in the last cropping, is fixed. Therefore, the plant cannot easily take in those fixed phosphorus, and that is the reason why each cropping should apply phosphorus fertilizer.

In districts with intensive crop production, most of the applied phosphorus fertilizer is fixed in the soil in each cropping season. Over a long time, it accumulates much insoluble phosphates, which is hard to be taken easily by crops. It is just like that we have deposited in each season, but these deposits cannot be drawn out. Nature is a wonderful world, where there is a kind of soil microorganisms called "phosphate solubilizing microorganisms," which can dissolve insolubilizing phosphates.

However, there is only a small part of phosphate solubilizing microorganisms existing in many soils. Scientific researches in recent years have isolated and identified phosphate solubilizing microorganisms, cultivated it in cultivation media, and applied to the soil. In this way, insolubilizing soil phosphates could be dissolved into soluble phosphate that could be absorbed by plants, and this is good news for the future agriculture.

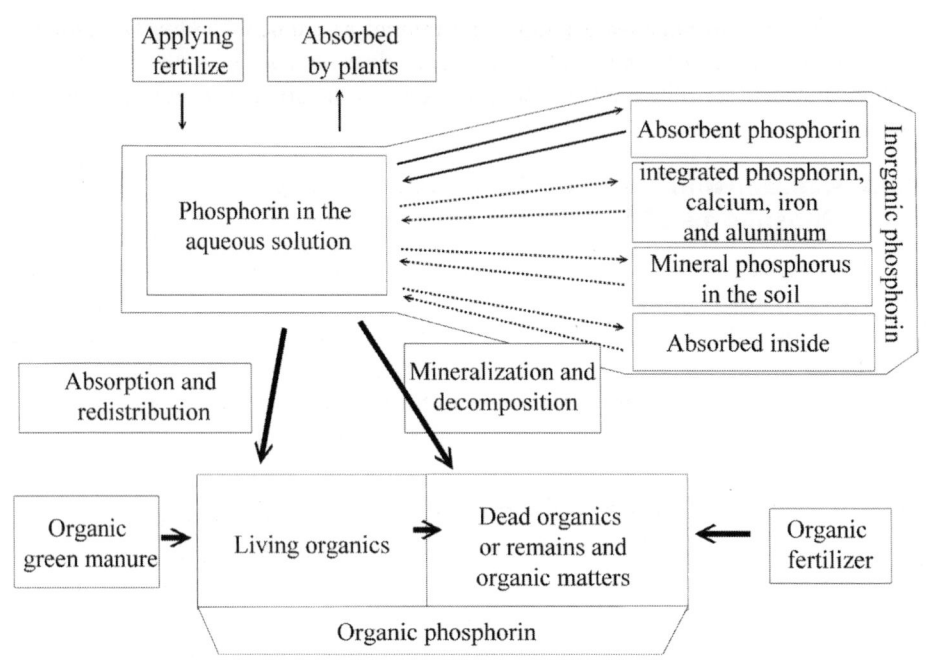

Figure 9-2. The Dynamic System of Phosphorus in the Soil.

The content of available phosphorus in phosphorus fertilizers is one of the determining factors of its quality. Phosphorus fertilizers made from acidic treatment products of rock phosphate powder, and its type and constitution may differ as the production district changes. The effect of rock phosphate powder for crop production is obvious only in acidic soils, and such kind of effect is limited and influenced by the content of available phosphorus. The coordination of rock phosphate powder with phosphate solubilizing microorganisms can improve the availability of phosphorus, which offers an example of the microbial fertilizer application.

9.9.3 Types and Characteristics of Phosphate Solubilizing Microorganisms

Among soil microorganisms with the ability of solubilizing phosphate can be called "phosphate solubilizing microorganisms," which can transform insoluble phosphorous into available phosphorus in the soil. It is one of the important research subjects in plant growth promoting rhizobacteria (PGPR). So far as we know, phosphate solubilizing microorganisms include bacteria, actinomyces, fungi, etc., which commonly includes *Bacillus, Burkholderia, Enterobacter, Pantoea, Pseudomonas, Rahnella, Thiobacillus, Aspergillus, Penicillium, Yarowia*, etc.

For one kind of phosphate solubilizing microorganisms, its ability of dissolving phosphorus differs with the different forms of phosphates (inorganic phosphorus fixed with calcium, iron and aluminum and organic compounds). For certain types of bacteria, they could only dissolve calcium phosphate, but might not dissolve the iron phosphate or aluminum phosphate. What's more, some types of bacteria could dissolve various phosphates, which can be applied in agricultural production.

Researches on phosphate solubilizing mechanisms mainly focus on the knowledge of calcium phosphate and iron phosphate's dissolve. One common mechanisms for phosphate solubilizing is that microorganisms may produce organic acids or inorganic acids indirectly, causing the phosphate release, or speed up the mineralization of organic phosphorus and phosphate release with the secretion of enzymes. Certainly, there are also many other phosphate solubilizing mechanisms that need to be discussed.

Commonly, phosphate solubilizing microorganism could dissolve calcium phosphate by secreting special matters, such as organic acids that includes formic acid, acetic acid, propionic acid, butyric acid, lactic acid, citric acid, succinic acid, malic acid, gluconic acid, oxalic acid, maleic acid and other five or six carbonic acids. These acids lower the pH value and change the solubility of phosphates, which is hard to be soluble. While certain secretions, along with chelated calcium and iron could promote an effective solubilization and uptake of phosphates. Besides non-symbiotic phosphate solubilizing microorganisms, the presence of mycorrhizal fungi, which develop a symbiotic relationship with plant roots and extend threadlike hyphae into the soil, can enhance the uptake of phosphorus from fixed calcium, iron, and, aluminum phosphates. Such solubilizing characteristics are related to the exudation of root and bacteria. Endomycorrhizal fungi could help plants with the absorption of phosphorus fertilizer in the following ways: expanding contact area between root system and soil; increasing the mitochondria of mycorrhizal root cell; enlarging nucleus and lengthening life; and increasing the activity of phosphatases and ATPase. In these ways, the ability of absorbing, decomposing and transferring phosphorus fertilizer is enhanced. In general, low content of phosphorus fertilizer in the soil will be increased to the infection of mycorrhizal fungi.

The effect of solubilizing microorganisms inoculation depends on the selection of microbial types. Excellent bacteria type could increase the production of crops. Good and bad microorganisms are intermingled in the soil. Despite of their phosphate solubilizing ability, their effects differ a lot. What's more, the cooperation of phosphate solubilizing bacteria with mycorrhizal fungi is proved to realize "additive or synergistic effect." Phosphate solubilizing microorganisms are also commonly applied with rock phosphate powder, so as to enhance the effect of increasing phosphorus fertilizers' absorption and production. With limited phosphorite resources and development of new technology, phosphate solubilizing microorganisms are greatly needed in future agriculture. The main goal is to improve the availability of phosphorus fertilizer, especially in the soil with strong fixation of phosphates. It is reasonable for perennial crops and crops with small root system to require phosphate solubilizing microorganisms; therefore, a sufficient providence of phosphorus fertilizer may satisfy the needs of crops.

9.9.4 Benefits of Phosphate Solubilizing Microorganism on Crop Growth

Phosphate solubilizing microorganisms could be used as fertilizer in a broad sense, i.e., one kind of "microbial fertilizer." It could make use of living organism to improve the availability of phosphorus in the soil, and to improve the nutrient condition. Therefore, the inoculation of phosphate solubilizing microorganisms could reduce application of chemical fertilizers, so as to realize the goal of increasing production and improving quality. Because the production cost of phosphate solubilizing microbial fertilizer is lower than industrial fertilizers, its application could largely reduce the usage of chemical fertilizers, reduce the possibility of soil degradation, and also bring down the production cost. Besides, the purpose of its indirect application is to make use of abilities except the fact that phosphate solubilizing microorganism affects on fertilizer; for example, protecting rhizosphere, promoting root growth and absorbing water and nutrients, expanding the life of root system, decomposing toxic substances, enhancing disease resistant and drought resistant ability, increasing the survival rate of transferring, early flowering, and so on.

All around the globe, due to the fact that phosphorus fertilizers have been applied over so many years, there are already accumulated large amount of fixed phosphates. However, fixed phosphates are unavailable and cannot wholly satisfy the needs of crops. During the past few years, it has been studied and developed the microbial inoculants of phosphate solubilizing microorganisms, aiming to increase the diversified application of phosphorus in the soil and reduce the usage of chemical phosphorus for decreasing soil degradation and environmental pollution, improving soil quality, and as well as increasing farmer income. In the agriculture development plan, it has been proved that the inoculation of phosphate solubilizing microorganisms could save 1/3 to 1/2 phosphorus fertilizers. Therefore, phosphate solubilizing microorganisms' development has its essence and environmental values. Phosphate solubilizing microorganisms in soils could increase the solubilization of rock phosphate powder, calcium phosphate, iron phosphate and aluminum phosphate. Researches on phosphate solubilizing microorganisms mainly focus on calcium phosphate dissolving microorganisms. Further studies and application of iron solubilizing microorganisms are expected in the future.

9.9.5 Effects of Applying Phosphate Solubilizing Microorganisms

The effect of phosphate solubilizing microorganisms' inoculation depends on selection of microbial type, and excellent microbial type could increase the production of crop. Good and bad bacteria are intermingled in the soil. Despite of their phosphate solubilizing ability, their effects differ a lot.

It is simple and convenient to use phosphate solubilizing microorganisms, and the method is just like that of root nodule nitrogen-fixing bacteria, i.e., by inoculating it to seeds or irrigating to the rhizospere and the soil.

In certain soil, the application effect of phosphate solubilizing microorganisms has been proved to have increased the production. It can also be applied with the additional usage of mycorrizal fungi to achieve additive effect. What's more, it can also be used in cooperation with rock phosphate powder, so as to increase the absorption of phosphorus fertilizers and increase production.

With the development of new technology, phosphate solubilizing microorganisms are greatly needed in the future agriculture. The main goal is to improve the effect of phosphorus fertilizers, especially in some soil with strong fixation of phosphates. It is reasonable for perennial crops and crops with small root system to require the presence of phosphate solubilizing microorganisms, and thus a sufficient providence of phosphorus fertilizer may satisfy the needs of crops.

9.9.6 Methods of Phosphate Solubilizing Microorganisms Application

Like other microbial fertilizers (such as rhizobia), there are two kinds of microbial fertilizer productions: liquid types and solid powder types, which can be specified as follows:

Application Methods of Liquid Types:

Methods of Inoculating Seedlings:
- Soak Seedling Method:
  Soak seedling in solution diluted with 5 ~ 10 times and taking it out immediately once it is coated solution.

- Spray Seedling Method:
  When there are many seedlings, we can just make the solution diluted for 5 ~ 10 times in the sprayer, and spray the solution on crop roots.

Methods of Inoculating Seeds:
- Soak or Spray before Planting:
  Seeds mixed with solution diluted with 5 ~ 10 times, and sow seeds after they have been mixed with liquid microbial fertilizer.

- Spray Seeds before Sowing and Covering with Soil:
  Spray seeds with solution diluted for 5 ~ 10 times on the unearthed seeds that are on the seedbed, moistening the seeds by the inoculant, and covering with the soil.

- Inoculation Methods after Planting or Direct Irrigation into the Soil:
  Put the 100 times diluted solution in the root or in the soil, or spray the 10 times diluted solution into the soil after the rain or irrigation.

Application Methods of Solid Inoculants:

Methods of Inoculating Seedlings:
Put solid microbial fertilizer in the caves or rows, plant seedling and let the root of seedling contact microbial fertilizer.

Methods of Inoculating Seeds:
- Inoculation of Big Seed
- Seed Coating Method:
  Put seed into agglutinant (such as CMC) and water, add more solid microbial fertilizer coating seeds, and thus attach solid microbial fertilizer outside the seeds.

- Contact in the Soil:
  Put solid inoculant into the soil, place seeds above it, and then cover with the soil.

Inoculation of Small Seeds:
- Mix:
  Mix equally small seeds with solid inoculant based on certain ratio, and then put the mixture on the seedbed for scattering seeds. The mixing ratio should be based on the amount of application and germination percentage. In theory, after being applied into soil, the seeds should contact the solid inoculant.

- Contact with Soil:
  As being described in the methods of big seeds, we just put solid inoculant into the soil, place seeds above it, and then do the covering with soil.

9.10 Roles of Microbial Application in Organic Agriculture

9.10.1 Introduction

Soil microorganisms in the farmland exert either direct or indirect influence on the growth of crops, these soil microorganisms include nitrogen-fixing bacteria, mycorrhizal fungi, decomposition microorganisms, and so on. There are many types of soil microorganisms with various functions; for example, increasing nitrogen source in the soil, improving the availability and solubility of nutrients, releasing the growth hormone of plants, speeding up the growth of root system and absorption of nutrients, decomposing organic materials and releasing nutrients, decomposing the toxic substances in the soil, resisting pathogens and pests, polymerizing and forming soil humus, etc. Different kinds of soil microorganisms play quite different roles, and they can exist in the soil. There must be certain reasons for its existence over millions of years, especially soil microorganisms of antagonism pathogens, and those growing on the rhizosphere for protecting the root system. It functions like the microorganisms existing inside the skin pores of human bodies, preventing the invasion of outside pathogens. In the cultivation management of modern agriculture, we cannot neglect factors affecting the balance of soil microorganisms or the growth of beneficial microorganism. Besides, due to the overuse of agrochemicals (including pesticides and fertilizers, etc.) and environmental pollution of soil, we should also pay attention to the natural soil microorganisms.

It is reported that organic agriculture refers to the mechanism of maintaining soil productivity by making use of rotation cropping, crop residues, animal manure, legumes, green manure, organic waste in farm, mineral rock, and biological pest control. From the application categories, we can get to know that soil microorganisms play a close relationship with organic agriculture. Organic agriculture is an agricultural production and operation mode focus on organic materials. In order to achieve the goal of increasing crop production, the application of organic materials has to be supported and coordinated by soil microorganisms. Otherwise, the cycle and supply of nutrients will be insufficient. In ancient agricultural production, production focused on organic materials and no special management on soil microorganisms, there must have been limitations in the performance of agricultural production potential. However, nowadays, in the farmland, because of the long-term application of chemical fertilizers and pesticides, many farmlands and hilly lands are deficient in organic matter. Thus, soil microorganisms group that lacks organic materials must be downcast. Therefore, microbial fertilizers are needed during the transformation from conventional agriculture to organic agriculture, so as to protect and cultivate soil as well as to maintain the production.

9.10.2 Roles and Application of Soil Microorganisms in Organic Agriculture

In order to develop organic agriculture, Liebhardt and Harwood (1985) believed that it is necessary to carry out following projects: lowering production cost and energy input, the self-sufficiency of nitrogen fertilizers, improvement of the cycle efficiency of nutrients, prevention of soil erosion, transformation from conventional agriculture to organic culture, allelopathic weed control, prevention of diseases and pests, etc. Therefore, only when microorganisms possess multi-functional types, organic agriculture could largely promote the development of agricultural production. In order to successfully achieve these new goals, we need to pay attention to the development of microorganisms. In the following part, by using the realization of agricultural development, descriptions on the role and application of soil microorganisms will be presented:

Self-Sufficiency of Nitrogen Fertilizers: Application of Symbiotic and Non-Symbiotic Nitrogen Fixation:

Leguminosae and green manure are important crops cultivated by organic agriculture, and its main purpose is to increase the source of nitrogen in biological nitrogen fixation, reduce production cost and energy consumption, and achieve the self-sufficiency of nitrogen in the farmland. However, in some farmland soils, if the conventional cultivation and fertilizing mode has not been cultivated with leguminosae, nitrogen-fixing microorganisms may be insufficient, good, and also bad bacteria are intermingled. In order to ensure the efficiency of nitrogen fixation among leguminosae, it seems obvious to obtain the needs for the with cooperation symbiotic application and non-symbiotic nitrogen-fixing bacteria.

There are many examples covering crops with leguminosae or leguminous help realize the self-sufficiency of nitrogen fertilizer in the farm. It has been known to us the inoculation of rhizobia could increase the nitrogen source of legumes. There has been the application of commercial inoculants in some place, the application and development of root rhizobia is already quite mature. We have promoted the inoculant of soybean rhizobia demonstration, but application study of non-symbiotic nitrogen fixating bacteria is quite rare, and the application goal has not been realized.

Improving the Efficiency of Nutrient Cycle: Application of Decomposing and Dissolving Microorganisms:

Application of Decomposers for the Nutrient Cycle of Organic Materials:

Fertilizers used in organic agriculture are mainly organic materials. Among the fertilizers, whether plant residues, green manures, composts, animal manures, animal residues, excrements, etc., these all depend on soil microorganisms for decomposing and releasing nutrients. Then, all these will be used by plants. Such efficiency of nutrient cycle is influenced by the characteristics and component of organic materials, environment, climate, soil nutrients, groups and types of soil microorganisms, etc. There are activities shown from directly controlling the decomposing rate. Organic material is quite complicated, and many types of soil microorganisms all take part in the decomposing process. Once animal or plant residues are decomposed by organisms and produce CO_2, nitrogen fertilizers can be used by plants (NH_4^+, NO_3^-), phosphate, sulfate and various kinds of microelements. The decomposition of organic materials to mineral nutrients is called "mineralization." In the process, some nutrients are used by microorganisms to conduct assimilation, which is called "immobilization." After nutrients are fixed for a while, microorganisms will die and new mineralization will be carried on by transforming organic cells into inorganic for the utilization of plants.

The utilization of organic decomposing microorganisms commonly includes cellulose decomposing bacteria, lignin decomposing microorganisms, anaerobic decomposing bacteria, and methane producing bacteria, and common decomposing bacteria, such as *Bacillus*, *Cellulomonas*, *Clostridium*, *Chaetomium*, *Trichoderma*, *Nocardia*, *Streptomyces*, *Penicillium*, *Aspergillus*, *Pseudomonas*, etc. The inoculation of decomposing microorganisms can speed up the decomposition of compost or organic materials, reduce the ratio C/N of organic materials, and the material weight, as well as intensify the humification process.

Organic wastes conduct decomposition based on the utilization of anaerobic fermatation, and produce methane. Such anaerobic decomposition microorganisms include primary and secondary microorganisms. Primary decomposition microorganisms (such as *Clostridium*) decompose cellulose under anaerobic condition and transform it into organic acids and alcohols, and then through secondary microorganisms (such as Methanococcus, *Methanosarcina*, *Methanobacillus* and *Methannobacterium*) which decompose organic acids into methane and carbon dioxide.

Application of Inorganic Dissolving Microorganisms:
Organic agriculture also uses rock minerals, which is slow to be soluble. There are different kinds of nutrients (such as phosphorus, potassium, magnesium, calcium, microelements, etc.) in minerals. After the application into soil, their utilization rate varies according to the characteristics of different soils. Some microorganisms may promote the solubilizing of mineral powder, and release potassium, calcium, phosphorus, magnesium, zinc, iron, etc.

Among those microorganisms that could dissolve inorganic, researches on phosphate solubilizing microorganisms, dissolving silicon microorganisms, mycorrhizal fungi, siderophore producing microorganisms are the most distinct. Phosphate solubilizing microorganisms is the most well-known and could enhance the absorption of phosphorus and promote the growth. Microorganisms that could dissolve phosphorus include bacteria, fungi, and actinomyces, etc. Mycorrhizal fungi are microorganisms that could symbiotic with roots. It could expand the contacting area with root system, so as to help crops take in various nutrients, phosphorus fertilizer from rock phosphates, thus promoting the growth and production of crops. Recent discoveries show that mycorrhizal fungi can not only expand the contacting area of root, but also help crops with the absorption of phosphorus that is hard to be dissolved (calcium-fixed phosphate, iron-fixed phosphate, etc). Mycorrhizal fungi also possess many other advantages, including disease-resistance and drought-resistance, which all deserve our attention.

Preventing Soil Erosion and Nutrients Loss: Application of Organic Polymer Producing Microorganisms and Transforming Microorganisms:

Organic agriculture could help reduce soil erosion, which can be regarded as one of its advantages in protecting and cultivation soil, because organic matters are conducive to the penetration of water or the stability of soil aggregates increase. Microbial promotion of stability in soil aggregates could prevent soil erosion; for example, certain soil microorganisms may secrete massive polysaccharides or organic polymers could all be used in organic agriculture.

Among nutrients in soil, nitrogen is most likely to be lost. In the different forms of nitrogen fertilizers, nitrate (NO_3^-) is most likely to be washed away by water, or get lost due to the denitrification by anaerobic microorganisms. Therefore, in organic agriculture, reducing the production of nitrate is an important method to prevent the loss of nitrogen in rainy seasons. Toxic substances in the soil (coming from microorganisms, root exudates, or residues) could inhibit the function of nitrification bacteria to some extent.

Weed Control: Application of Weed Controlling Microorganisms:

Since herbicides cannot be used in organic agriculture, it's naturally to conduct weed control in a natural way. Considering the labor cost of weeding, it is possible to make use of allelopathic plants or microorganisms for the weeding. There have been many studies on natural microbial herbicides, including host and non-host specific toxic substances, which have been developed by some studies.

Pathogen and Pest Controls: Application of Bio-Control Microorganisms:

Organic agriculture emphasizes not to use pesticides in crop cultivation. Therefore, how to carry out pathogen and pest controls without the usage of pesticide is the key to the success for organic agriculture. As an important way of developing organic agriculture, biological control could reduce plant diseases and pests, and one effective method is to kill or inhibit pathogens and pests by means of microorganisms.

Corp diseases often take place above or below the soil, and some diseases can be avoided by enhancing fertility and improving soil condition, while others cannot. Therefore, it is an ideal method of control by means of antagonistic microorganisms. Among crop diseases, some are mediated by insects, and thus soil microorganisms can be used in killing insects. There are some successful examples already existed.

Soil microbial application in biological control still needs further research and development by teamwork, because it is under many limiting factors, including climate, bacteria, types of diseases and pests, soil, ecosystem, etc.

The premise of organic agriculture does not use synthetic chemicals, such as fertilizers and pesticides. In order to increase the source of nitrogen, it is necessary to increase cycle rate of nutrients, prevent plant diseases and pests, and also the application of microorganisms. These are all crucial to the success of organic agriculture. In particular, under environmental pollution, silent microorganisms deserve the attentions to all sides. It is restricted that is should be depend on inborn organic agricultural production thoroughly. We have to develop organic agriculture featuring in modernity, science, naturalization and integrity, by upholding the spirits of teamwork. Thus, it will strive to make contributions to a better future for human beings.

9.11 Types of Microbial Fertilizers and the Quality of its Application

9.11.1 Purposes of Microbial Fertilizer Application

There is a closer relationship among the activity of soil microorganisms, soil fertility and the growth of plants. The biotic group and activity of soil microorganisms relate to the amount of carbon sources and energy sources in the soil, because soil microbial growth and propagation rely on the limited carbon material, including organic matters and carbon dioxide. Energies, such as chemical energy and solar energy, can also be used. Thus, soil microorganisms commonly have constituted to a quite stable ecosystem, and there are mainly two ways to increase the activity and biotic group of beneficial soil microorganisms: one is to increase carbon sources and energy sources; the other is to make use of additional microorganisms, or use both methods at the same time. Microbial fertilizer is the application of these principles, and in order to ensure an increase of certain beneficial microorganisms, the most effective way is to add such kinds of beneficial microorganisms.

There are many kinds of microbial fertilizers with different functions, such as a direct supply of nutrients (like nitrogen-fixing bacteria) or an increase in utilization availability, boosting the root growth and absorption. What's more, there are products resisting plant diseases and pests, or developing control pathogens. In the original soil microorganisms group, the efficiency of different bacteria type varies or even differs greatly due to the limitations of beneficial bacteria in the soil. Therefore, in order to ensure an explicit efficiency in the rhizosphere, it is necessary to inoculate beneficial microorganisms.

9.11.2 Types and Functions of Microbial Fertilizers

There are many kinds of beneficial microorganisms in the soil, and the four frequently used types include fungi, actinomyces, bacteria, and cyanobacteria. In ecosystem, these four types coordinate with each other and resemble in terms of function and effect. In the following part, it will illustrate types of microbial fertilizer by presenting the application effect of microorganisms.

Microorganisms that could Increase the Source of Soil Nutrients:

Bacteria of such kind mainly include "nitrogen-fixing microorganisms," which could increase nitrogen in the soil. There are symbiotic, associative and non-symbiotic nitrogen-fixing microorganisms, which are explained in the following part.

Symbiotic Nitrogen-Fixing Microorganisms:
It could be symbiotic with both higher and lower plants; for example, rhizobia and actinomyces may be symbiotic with leguminous and non-leguminous higher plants. Moreover, form root nodule could fix nitrogen from the air. Besides, there is also the phenomenon of cyanobacteria that is symbiotic with lower plants (such as Azolla, etc.). Except the symbiotic nitrogen-fixing microorganisms in the root, there are also symbiotic nitrogen-fixing microorganism in the stalk and leave, but their applications are quite rare.

Associative Nitrogen-Fixing Microorganisms:
Such kind of nitrogen-fixing microorganisms grow around the rhizosphere or epidermis without obvious relation of associative structure. The most famous one is the associative *Azospirillum* that exists in the root of grass family.

Non-Symbiotic Nitrogen-Fixing Microorganisms:
Many kinds of microorganisms belongs to such category, and examples could be given from cyanobacteria to bacteria. For example, in paddy field, nitrogen-fixing cyanobacteria and photosynthetic microorganisms all have non-symbiotic nitrogen microorganisms. Photosynthetic bacteria can make use of solar energy and CO_2, and increase organic carbon in the surface soil and soil in the paddy field.

Microorganisms that could Improve the Availability and Utilization of Nutrients in the Soil:

There are many microorganisms under such category, including fungi, actinomyces and bacteria, and the most obvious ones are "mycorrhizal fungi" and "phosphate solubilizing microorganisms." Mycorrhizal fungi belong to fungi that co-exist with the root of plant. Phosphate solubilizing microorganisms include fungi, actinomyces and bacteria, etc. These microorganisms could promote the dissolving of soil fixed phosphorus (calcium phosphate, aluminum phosphate and iron phosphate), and increase the phosphorus that could be absorbed by plants. Besides, there are still microbial types that could dissolve various minerals and compounds that is hard to be dissolved or utilized, and improve the availability of nutrient elements. What's more, some of the soil microorganisms could secrete "siderophore" (iron-loaded material), and possess the function of chelating iron, as well as promote the utilization of iron.

Microorganisms that could Promote the Growth and Absorption of Plants' Root:

In soil microorganisms, there are microorganisms that could produce or transform hormone of plant, which could speed up the growth of root system. There are also microorganisms that are capable of increasing in absorbing nutrients of plants and play an important role in the health of plants and the stretch of root.

Microorganisms that could Resist to Pathogens and Pests:

There are disease-resistant and pest-resistant microorganisms in the soil, and such kinds of microorganisms could protect the root system and kill pathogens. Rhizosphere-protecting microorganisms exist in the rhizosphere, and resist the invasion of pathogens. Microorganisms that kill pathogens can produce antibiotics, and thus kill or restrain the effect of pathogens. Inoculation of lighter and nonlethal virus to plant for resisting the invasion of pathogens is the developed application.

Microorganisms could Increase the Decomposition and Transformation of Organic Matters:

In the soil, organic matters could only be used when they are decomposed into smaller molecule or inorganic, and that is an indispensable part in the cycle system of the nature. When the decomposing process of organic matters is restrained, it could lead to the accumulation of fresh organic matters, bring about toxic effect, or the inability of utilizing these organic matters. Besides, soil microorganisms also play an important role in the humification in the soil. The products from humification could improve the physicochemical properties of soil, and become the advantage to agricultural production.

Detoxified Microorganisms:

In agricultural soil, there are artificial or natural toxic substances, including pesticides, pollutants, and phytotoxins. All these toxic organic substances rely on a group of microorganisms in the soil to be decomposed and detoxificated.

Microorganisms that could Prevent Soil Erosion and Nutrient Loss:

Some microorganisms could secrete quantitative polysaccharide and increase soil aggregation. There are also many fungi that could bind up soil particles and form steady aggregates. Thus, they live up to the expectation of preventing soil erosion.

What's more, some microorganisms could also take in nutrients or decrease the transformation of nutrients into forms, which is easy to be washed away and could reduce the loss of nutrients.

9.11.3 Quality Requirements for Microbial Fertilizers

Common standards and requirements for microbial fertilizers are as follows:

Maintain the Population of Microorganisms of Microbial Fertilizers:

In unit volume or weight of microbial fertilizers, it is better to have higher population of microorganisms. With a lower population of microorganisms, the application efficiency will decrease. In liquid, it is proper to have at least 10^6 bacteria per milliliter. While in solid, the number may vary with the weight of carriers. In theory, the more is the better. In general, the determined method for the population of microorganisms is to use culture plate count or microscope measurement.

Maintain High Bacteria Activity:

Besides the population of microorganisms, activity is also an important index. Active microorganisms could adapt to environment and display their functions or form the symbiosis. We could determine microbial activity by referring to the measurement of propagating ability, specific enzymes, or other active indexes.

Need High Adaptability:

Microorganisms have different adaptability to soil environment and crops, so we should pay attention to the effectiveness of microorganisms. Generally speaking, locally selected microorganisms could adapt to native soil and crops. Surly, we don't rule out other possibilities.

Need Little Contamination and Impurity:

We are afraid that contamination may enter during the production of microorganisms, which will lower the quality of microbial fertilizers. We need to pay attention to any possible pollution in the production process. The examination of which requires the microscopic examination, cultivation examination and other measurements.

9.11.4 Preserving Instruction and Noticing Items of Microbial Fertilizers

- It's best to store inoculants in shade place or in refrigerating chamber (above 5°C). Microbial fertilizers are active organisms with certain conservation period, so when the number of living microorganisms decreases, the effect will be weakened.

- Avoid using it with toxic pesticides. However, after seed sowing and covering with soil, pesticides could be applied.
- Use nitrogen-fixing inoculants. Do not mix with nitrogen fertilizers, although phosphate and potassium fertilizers are still need to use as basal fertilizer in the soil. For example, nitrogen fertilizers could be used as topdressing with a small quantity. The application of phosphate solubilizing microorganisms or mycorrhizal fungi should avoid the overuse of single superphosphate.
- When mixing with inoculants, if the seeds (such as soybean) cannot be soaked in the water, we should pour extra liquid product and avoid soaking seeds. Otherwise, the germinating percentage and viability of seeds will be decreased by liquid soaking.
- Sow immediately after seeds and mix with inoculant. Soil should not be too dry.

9.11.5 Application Methods of Microbial Fertilizers

In tillage farming, no-tillage farming, or nursery garden, the application methods of microbial fertilizers in sowing and managements for soil cultivation are in similar ways. That is to let roots of seeds or seedlings fully contact to microbial inoculants, so as to achieve the goal of inoculating microorganisms. The following application methods could be served as templates and thus can be adjusted to meet the specific needs of certain situations.

In terms of inoculants forms, inoculants consist of liquid and solid forms. Because seeds of crops vary in size, commonly, big seeds adopt the method of coating, and small seed (or seedling) are directly applied it to the soil.

Coating Method of Seeds:

Intended Items:
Large-size crop seeds (more than 3 mm in diameter).

Application Methods:
- Application Methods for Liquid Microbial Fertilizers:
  Put seeds into the container (if there are too many seeds, we can spread them on the floor), scatter agglutinants (such as CMC) into it and mix them evenly, then add liquid microbial fertilizers (we could use sprayer to add agglutinants in liquid microbial fertilizers), mix the seeds so as to make sure they are blended evenly and get wet (if there is too much water, pour out extra liquid microbial fertilizers; if there is little water, then add more clean water). At last, pour absorb powder (such as peat). After an even mixture, the seeds can be sowed (if the seeds are too wet, then blend them with peat but never immerse seeds in the liquid).

- Application Methods for Solid Microbial Fertilizers:
  Put seeds in the container or on the floor, add some agglutinants (such as CMC), pour in clean water (if use sprayer, can add agglutinants in liquid microbial fertilizers), and then inoculants could be added and mixed with seeds, as long as seeds are coated with a layer of powder.

Direct Application to Soil or Cultivated Media:

Intended Items:
Small crop seeds (less than 3 mm in diameter), nursery or seedlings.

Application Methods:
- Application Methods for Liquid Microbial Fertilizers:
  Application methods before and after seeds sowing, you could pour diluted liquid microbial fertilizers into the cave. Or after mixing liquid (or matrix), you could sow seeds or plant seedlings.

- Application Methods for Solid Microbial Fertilizers:
  Hole in the soil (or matrix), put inoculant powder in the cave, then you can sow seeds or plant seedlings on the microbial powder.

Application Rate:
The usage amount of liquid microbial fertilizers is determined according to extent that liquid could coat wet with seeds or soil, and powder is determined in terms of the extent that powder could coat well with seeds. Usage amount per hectare could be referred to the specification of various inoculants.

Application Frequency for Crops:

The multiple of dilution is determined by microbial content of the liquid microbial fertilizer, commonly it is about 100 to 500 times dilution. The application frequency varies with the types of crops:

- Short-term vegetable crops (harvest about 30 to 40 days): Apply one or two times at the seedling stage.
- Crops of fruiting vegetables (harvest about 60 to 100 days): Apply about one time at the seedling stage; apply two or three times during the middle age of fruit period.
- Perennial crop: applied 3 to 6 times per year, or even 10 times. In terms of fruit trees, they should be applied one month before the blossom, younger fruit period, and middle of the fruit period.

9.11.6 Notices of the Inoculation

- When mixing inoculants with seeds or water, you should pour extra water. Never immerse seeds that cannot be soaked in the water; otherwise, the germinating percentage and viability of seeds will decrease.
- Using nitrogen-fixing inoculants cannot mix with nitrogen fertilizer, although phosphate and potassium fertilizers are still used as basal fertilizer in the soil. If nitrogen fertilizer used as top dressing, a small quantity should be applied. The application of phosphate solubilizing microorganisms or mycorrhizal fungi should avoid overusing single superphosphate.
- Avoid using microbial fertilizer with pesticides. However, after seed sowing and covering with soil, pesticides could be applied.
- Mix seeds and inoculants should be sowed immediately. However, the soil could not be too dry or too wet, so as to avoid immediate irrigation in rainy days right after the sowing.
- It's best to keep inoculants in shaded places or refrigerating chambers. Microorganisms are active organisms in certain storage period. Therefore, when the number of living microorganisms decreases, the effect will be weakened.
- Highly acidic soil could be neutralized by lime materials, and then improved by microbial fertilizers.

9.12 Application of Microbial Fertilizer in Sustainable Agriculture

9.12.1 Application of Microbial Fertilizer is an Important Factor of the Successful Sustainable Agriculture

In recent years, the application of high amount of chemical fertilizers and pesticides in agricultural production have greatly affected soil condition and deteriorated ecosystem, which deserve our attention. Therefore, we should greatly promote researches and applications of sustainable agriculture.

Agriculture production should establish a production system that is independent from chemical fertilizers and pesticides, and with low energy input. Under the principle of reducing production cost, we should stabilize the production, increase farmer income, and reduce damages to the environmental ecosystem. How could sustainable agriculture reach its goal? In order to be independent from chemical fertilizers, the method is to make use of microorganisms and organic fertilizers. Due to various types of plant nutrients, no fertilizer is omnipotent. If we replace chemical fertilizers with microorganisms and organic fertilizers, we should let microorganisms cooperate with organic fertilizers. In this way, we could achieve twice the result with half the effort.

With limited agricultural labor and reduced labor cost, a proper cooperation of organic fertilizer and microbial fertilizer will form the basis for the success of sustainable agriculture, due to the fact that organic fertilizers are in large volume and the application cost is high, while microbial fertilizer has many functions and cost less. Reasons for using microorganisms as fertilizers are listed below:

- They could increase the source of nutrients (nitrogen source and decomposition of organic matters), and improve the result of nutrient availability (absorbable type).
- They could protect the rhizosphere (disease-resistant function), increase absorption of water and nutrients, lengthen the life span of root system, neutralize and decompose toxic substances, and increase survival rate of transplanting.
- They could reduce environmental pollution and soil acidification.
- They could improve fertility, including the improvement of physical, chemical and biological properties of soils. These functions are indispensable requirements of sustainable agriculture.

9.12.2 Why could Microbial Fertilizer Replace Part of Chemical Fertilizers?

There are many kinds of microbial fertilizers. Judging from their functions, they could be categorized as: nitrogen fixation, microbial fertilizer that could promote soluble ability of unavailable nutrients, absorption of nutrients, the growth of root system, decomposition of organic matters, soil physical and chemical properties, detoxification, plant resistance, and protection of root system, etc. Microorganisms may have more than one function, and some microorganisms are multifunctional activity; for example, certain nitrogen-fixing microorganisms may have fix nitrogen, dissolve insoluble phosphates and promote the growth of roots. In the following part, we will specify the functions of microorganisms to replace chemical fertilizers:

Nitrogen-Fixation Function:

Nitrogen-fixation microorganisms include symbiotic, associative and non-symbiotic microorganisms, which could fix nitrogen gas to transform into ammonia for available nitrogen compound. Such function could directly increase nitrogen source in soils, such as rhizobia, nitrogen-fixing facultative bacteria in the root of grass family and free-living nitrogen-fixing microorganisms. The function of nitrogen fixation could replace or reduce the application of chemical nitrogen fertilizers.

Beans and leguminous green manure are important crops cultivated in organic agriculture, and their major purposes are to increase the source of nitrogen, bring down production cost and energy consummation, and achieve the self-sufficiency of nitrogen. However, farmland soil in conventional fertilization modes or soil that has not be used in the cultivation of beans, nitrogen-fixing microorganisms may be insufficient, good, and also bad bacteria are intermingled. In order to ensure bean efficiency in fixing nitrogen, the cooperation of symbiotic and non-symbiotic nitrogen-fixing microorganisms seems to be extremely crucial.

There are many examples of achieving self-sufficiency of nitrogen fertilizer by making use of legume crop or cover beans. We have already laid certain foundation and gained achievements in the application and development of rhizobia.

Solubilizing Function:

In soils, there are many fixed nutrient elements that could not be utilized by plants; for example, phosphorus, calcium, and iron, etc. could be used by plants only after they have been dissolved by microorganisms in the rhizosphere. Therefore, phosphate solubilizing microorganisms could enable reuse nutrients to replace or reduce the application of chemical fertilizers, such as solubilizing calcium phosphate microorganisms, solubilizing iron phosphate microorganisms, and solubilizing organic phosphate microorganisms.

Sustainable agriculture makes use of mineral rock that is not easily to be dissolved, because mineral rocks contain various kinds of nutrient elements (such as phosphorus, potassium, magnesium, calcium, microelement, etc.). After being applied in the soil, their utilization efficiency by plants varies with the different characteristics of soil. However, some microorganisms still could promote the solubilizing of mineral powders, and release potassium, calcium, phosphorus, magnesium, zinc, iron, etc..

Among microorganisms that could dissolve inorganic matters, research on phosphate solubilizing microorganisms, silicon solubilizing microorganisms, mycorrhizal fungi, and siderophore production microorganisms are the most distinctive. The most well-known one is phosphate solubilizing microorganisms, and microorganisms include bacteria, fungi, and actinomyces, etc., and solubilizing microorganisms could promote plant absorption of phosphorus and the growth of crops.

Mycorrhizal fungi are microorganisms that could symbiotic with roots, expand the contacting area of root system, increase and promote crop absorption of various nutrients. Also, it could enable crop to absorb the phosphorus fertilizer in the phosphate ore, and promote the growth and production of crops. Mycorrhizal fungi could not only expand the contacting area of root, in recent years, people find that it could also promote the absorption of phosphorus from not easy to be dissolved forms (calcium phosphate, iron phosphate, etc.). Other merits of mycorrhizal fungi, like enable disease-resistant and drought-resistant abilities, all deserve our attention.

Function of Promoting Nutrients Absorption and Growth of Root System:

In soil microorganisms, there are microorganisms that could promote the absorption and growth of root system. By enhancing the absorption ability of root system and expanding surface area, we could reduce the application of chemical fertilizers, and increase the supply efficiency of nutrients in the soil, such as the application of beneficial microorganisms and mycorrhizal fungi.

Decomposition of Organic Matter:

In sustainable agriculture, organic matter is the major fertilizer, and organic matters like plant residues, green manures, composts, animal manures, animal residues, and excrements, etc., all rely on soil microorganisms to be decomposed and to release nutrients, and then plants will reuse these materials. Efficiency of such nutrient cycle is affected by the characteristics and components of organic matters, environment, climate, soil nutrients, biotic groups and types of microorganisms. Among these, soil microorganisms, environment, and nutrients could be altered by human, and the activity of decomposition rate can be directly controlled, because the characteristics of organic matters is quite complex and many types of microorganism participate in the decomposition process. Whether animal or plant residues, once being decomposed by microorganisms, they could produce carbon dioxide and nitrogen, phosphorus, potassium, sulfurs and various microelements that can be used by plants. And, such decomposition function of organic matters is called "mineralization." Also, in part of the process, some nutrients are utilized by microorganisms, and assimilation function is carried out, and such kind of phenomenon is called "immobilization." After nutrients being fixed for a period of time, when microorganisms die, it will perform the mineralization again, and then transform organic matters into inorganic matters which could be used by plants.

Common utilization of decomposing microorganisms in organic matters include cellulose, protein, and lignin decomposition microorganisms, anaerobic decomposition microorganisms, methane producing microorganisms and common decomposition microorganisms, etc. The inoculation of decomposition microorganisms could speed up the decomposition of compost or organic matters, and decrease the C/N rate and weight of organic matters, and some fungi can intensify the humification.

9.12.3 Guidelines for Getting the Best Effect of Microbial Fertilizers

In order to get the best effect of fertilizers, we should pay attention to the coordination of soil and condition of crops. This is true for microbial fertilizer, we need to notice the cooperation condition described in the following parts.

The Soil should not be too Acidic or Too Alkaline:

Whether the soil is too acidic (lower than pH 5) or too alkaline (greater than pH 7.8) will influence the absorption and availability of various nutrients, as well as the performance of microbial fertilizers. Highly acidic soil should be neutralized by lime materials (like limes, oyster shell powder, silicate slag and dolomite, etc); and strongly alkaline soil could be neutralized by acidic materials (such as sulfur powder or acidic peat).

Materials and Location for Coordinating the Propagation of Microorganisms:

Microbial fertilizers are viable microbial products. After they are applied in the soil, they need to survive through propagation, and the best surviving place is the rhizosphere of crop. Therefore, application of fertilizer reach to the root could achieve the best and direct results. If diluted liquid of microbial fertilizers are added with little (0.1 ~ 0.5%) humic acid, molasses, amino acids, soluble seaweed-extract powder or nutrients, it will be helpful to the propagation and survival of microorganisms.

Satisfy for the Needs of Crops:

Different crops in their growth periods will have diverse requirements. For microbial fertilizers, it should be applied the inoculation the early the better, especially during the time of younger seedling. In terms of perennial fruit trees, during the growth period, we should emphasize on increasing nitrogen and enhancing the effect of phosphorus in microorganisms. Moreover, during the middle size period of fruit, we should emphasize on strengthening the function of phosphorus for crops in microorganisms.

Requirements for the Quality of Microbial Fertilizers:

Since microbial fertilizer is active microorganisms, requirements for the quality of inoculants are as follows: the living microbial population should be maintained, the activity of microorganisms should be high, they should be capable of adapting to soil environment, and with few contaminations in the product.

9.12.4 Reasons for the Application of Microbial Fertilizer could Hardly Lead to the Residual Problem in Large Quantity

The application of microbial fertilizer will not lead to the residual problems or accumulation in large quantity, such as brought about by chemical fertilizers and organic fertilizers, and the reasons are illustrated in the following parts:

Variations in Cubical Space and Environmental Conditions of the Soil:

Ecological niche of the soil is quite complicated in cubical space and environmental conditions, and their physical, chemical and biological properties vary greatly. Therefore, it is not easy for extraneous microorganisms to adapt to such changeable environmental conditions.

A Counterbalance of Enormous Biotic Group of Microorganism in Soil:

Soil is a complex microbial ecosystem. These microorganisms exist in their own niche and counterbalance various biotic groups of microorganisms in accordance with nature rules of mutual promotion and restraint as well as predation. Therefore, it is difficult for extraneous microorganisms to keep a foothold and conquer niches of different microorganisms.

Carbon and Energy Sources are in a Constricted System in the Soil:

Once carbon and energy sources appear in the soil, they will be immediately used by the original microorganisms until the end. Thus, it is not easy for extraneous microorganism to obtain enough carbon and energy sources, and reproduces at a large scale.

In a word, soil possesses a complex cubic space condition, counterbalancing ability and limited carbon and energy sources. Any extraneous microorganism could hardly surmount all the predicaments described above. Therefore, applied extraneous microorganisms are not likely to remain residual in large quantity in the soil, nor bring about problems as overuse and residues by the application of microbial fertilizers.

9.13 Interactions among Soil Microorganisms

9.13.1 The Importance of Soil Microorganisms

Soil microorganism is the most basic organism in nature and environment. In recent years, with the rise of biotechnology, people start to use soil microorganisms in agriculture, industry, medicine and the treatment of environmental protection. Therefore, the role of soil microorganisms becomes more and more important. Soil microorganism includes virus, bacteria, archaea, fungi, algae, and protozoa, etc. There are many kinds of soil microorganisms, but only less than 1% could be cultivated. Thousands of microorganisms could symbiotic with animals or plants, but they could not be cultivated, and they exist in wide environmental conditions, including special environmental conditions, such as special temperature, acidity, alkalinity or salinity, etc. Soil microorganism plays a crucial role in maintaining the high quality of soil. However, nowadays, in many places, knowledge of the standard of soil microorganisms, diversity of essential biology, proportion of soil microorganisms, etc. is still insufficient. So far as we know that close relationship exists between soil microorganisms and the changes and quality of soil environment. Under the challenge of agricultural management, agricultural soil undergoes drastic environmental changes and some common problems.

Soil microbial diversity is the basis of the ecological protection and cultivation, also the foundation of reducing agricultural disasters. Only with emphasis on soil microbial diversity could we have healthy soil environment and healthy human beings. It is a project that calls for the attention of both government and all the people. In the following part, by giving the example of the interaction among soil microorganisms, we hope to provide insights into the world soil microorganisms.

9.13.2 Interactions among Soil Microorganisms

Since ancient time, soil microorganisms inhabit in soil, born and dead in the changeable nature, and interact with each other, which includes the function of both promotion and inhibition. Therefore, in different soil niche, the physical, chemical and biological properties all related to such changes.

Relationships among soil microorganisms include the relationships of +, -, 0: + represents positive effect, - represents negative effect, and 0 represents no effect each other. In the following part, their interactions are listed in Table 9-4.

No Relationship -- Neutralism:

Neutralism refers to that microorganisms co-exist in the same habitat with neither harm nor benefit to each other. In soil, if the microorganism is rich in nutrients, or two organisms have asynchronous requirements for growth, the phenomenon of close relationship come into being. If the environment changes or nutrients are insufficient, contiguous microorganisms may start to interact with each other.

Positive Relationship:

Positive relationship of soil microorganisms indicates that one or two of the microorganisms gain benefits in the habitat. There are three kinds of situations: commensalisms, mutually beneficial coexistence, and mutualism. When applying microbial fertilizer in agriculture, in order to realize the positive relationship, we need to explore the conditions and application in depth.

Commensalisms:

It means that in the environment, the growth or operational effect of one organism results from other organism's work. Therefore, one side gains benefit, while on the other

Table 9-4. Interactions among Soil Microorganisms

Type	Microorganism A	Microorganism B
1. No relationship:		
Neutralism	0	0
2. Positive Relationship:		
(1) Commensalisms	+	0
(2) Mutualisms		
a. Protocooperation	+	+
b. Symbiosis	+	+
3. Negative Relationship: (antagonism)		
(1) Competition	–	–
(2) Ammensalism	0	–
(3) Parasitism	+	–
(4) Predation	+	–

"+": positive effect; "–": negative effect; 0: no effect.

side, though without benefits, it is still no harm. This situation could happen frequently, but not necessarily. Let alone a long-term relationship, and there exists no mutual specificity. Microorganisms of commensalisms have close relationships with each other, just like the neighbor.

Common commensalisms in the soil include:

- Transformation of nutrients and media.
- Supply of growth factor.
- The decomposition of toxic substances.

For instance, outside of soil particle, aerobic bacteria consume oxygen, and exert positive influence on the anaerobic oxygen inside, which is commensalisms.

Among organic decomposition microorganisms, the splitting decomposition conducted by cellulose decomposing microorganisms will assist the latter in utilizing other components of plants for decomposition.

In soil, some microorganisms are auxotrophic microorganisms that need vitamin to maintain growth, while some are prototrophic microorganisms that may be helpful in supplying vitamin for the growth of microorganisms in need. Besides, sulfate reducing microorganisms in their growth produce poisonous hydrogen sulfide, while photosynthetic microorganisms use hydrogen sulfide as its source of electron, transform to sulfuric acid, and reduce the poison brought about hydrogen sulfide on other microorganism. Commensalism in soil pollution problems could decompose toxic substances, and protect and cultivate soil.

Mutualism:

It indicates the mutually beneficial relationship among microorganisms, including protocooperation and symbiosis.

- Protocooperation: Mutually Beneficial Coexistence

  It means that microorganisms gain benefits from each other and exist by itself in its previous growing environment. However, the relationship is not symbiotic, and the cooperation is not fixed. Therefore, they could be replaced by microorganisms with similar effects. The coexistence of two microorganisms could produce different effects or reactions that could not come into being when microorganisms exist alone. First, they could compose products that could not be formed if they live apart; second, they could form special structure; third, they could decompose toxic substances or remove pathogenic reactions.

  For example, two microorganisms need A and B, two growth factors which the soil lack, but one microorganism could produce A, and the other could produce B. In this way, both could obtain the growth factors in need. Here is another example: algae could produce oxygen through photosynthesis that could be used by microorganism. Through respiration, bacteria could release carbon dioxide that could be utilized by algae in photosynthesis, and thus forms a cycle. The partnership among microorganisms is quite prevalent. However, so far, our knowledge in this aspect is insufficient. Therefore, many soil microorganisms could not be cultivated.

- Symbiosis: Mutually Beneficial Symbiosis

  It means that the relationship between two microorganisms is specific, even permanent, including absolute and facultative symbiosis. Absolute symbiosis is totally dependent and inseparable, while facultative symbiosis symbolizes the independent existence and growth in proper environment.

  For example, green algae and blue green algae (cyanobacteria) could form symbiont with fungi. While fungi take in water and minerals for the utilization of algae, through photosynthesis algae provide organic carbon source, thus they realize the goal of mutually beneficial coexistence. There are also examples of algae symbiotic with protozoon; for example, in the body of empennage paramecium, there are hundreds of symbiotic green algae conducting photosynthesis, and without illumination the algae will be digested by insects. The symbiosis of microorganism, advanced animal and plants is quite prevalent, such as the nitrogen fixation formed by rhizobia and beans, as well as by Azolla and Cycadaceae with cyanobacteria. Others like the symbiotic microorganisms of animal intestines and stomach. Some microorganisms could decompose matters that could not be decomposed by animals, such as cellulose.

Negative Relationships: Antagonism Effects (- -; 0 -; + - Relationships):

The negative relationships among microorganisms refer to the fact that in soil environment. One or both of the two microorganisms may bring negative effects, such as competition, antibiosis, parasitism, predation and antagonism. They are separately discussed in the following parts:

Competition:

With limited environmental resources, two microorganisms may compete with each other, so as to maintain existence, snatch nutrients or take spaces. In the competition, negative effects come into being and the weak one may be sifted out. The competition could be divided into inside species competition and outside species competition. The previous one is more prevalent than the latter. Environmental factors may affect the competition process, such as physical and chemical properties of soil, like pH value. Additional carbon and minerals could alter the results of competition.

Antibiosis:

Antibiosis indicates that products of one microorganism will kill another microorganism. An alternation in pH value or the production of toxic substance such as antibiotic, hydrogen sulfide, and organic acids, etc., of one kind of microorganism may affect the growth, restraint and death of another microorganism. For instance, *Thiobacillus* may transform sulfur into sulfuric acid, and restrain microorganisms to be sensitive to acidic.

Bioinhibitors or toxic substances include two type substances that require high concentration (such as organic acids, chelating agents), and low concentration (such as antibiosis). Such toxic substances are often used to control pathogenic microorganisms. Generally speaking, the previous one is more likely to be accepted, while there are still some problems for the application of antibiosis in the environment.

The importance of antibiosis in ecologic systems is still assumed. However, in lab, green house and fielding researches, its potentials have been inferred. For example, *Bacillus subtilis* can produce antibiotics to resist fungi in the field, or make use of streptomycin or other production bacteria in antifungi to enhance the ability of symbiotic rhizobia of beans and increase the production of plants.

Besides, siderophore could reduce plant diseases, but should be with high concentration. Siderophore is a kind of low molecular weight (500 ~ 1,000 daltons) produced by microorganisms in soil with low content of iron. Siderophore could chelate iron, such as *Pseudomonas fluorescens* and *Pseudomonas putida* could effectively prevent diseases.

Parasitism:

Parasitism means that parasite choose larger organisms as its host, and take the cells, organs, or body fluid as the source of nutrients. In the process of parasitism, the host causes harm, but the parasite will avoid harms from outside competitors. There are two kinds of parasitism: absolute parasitism and facultative parasitism. Absolute parasitism refers to the fact that parasites display the effect of parasitism in the whole life circle; for example, parasitic vibrio bacteria adsorb Gram negative bacteria, while facultative parasitism does not absolutely rely on any host for its life cycle completion; for example, bacteriophage parasite in the cell of bacteria.

There are specific parasitism and non-specific parasitism. The previous one is highly specific, while the latter choose to host widely, and different species or genus could be its host. Also, parasite can become the host of another organism, and such phenomenon is called "hyperparasitism."

Among soil microorganisms, the major parasitics are bacteriophage, parasitic vibrio bacteria, and fungi. The hypha enters the body of host, stretching and growing, destroys the host.

In nature or agricultural soil ecosystem, parasitism effect could control the soil microbial community and effectively limit the invasion of outside microorganisms, striking a balance among microorganisms in the ecosystem.

Predation:

Predation refers the phenomenon of larger microorganism feeds on the smaller microorganisms. By feeding microorganism with smaller size, predator obtains the carbon source and growing factors needed during the growth. In the soil, bacteria are preyed by euglena and amoebae; therefore, after bacterial inoculants applied to the soil, populations of these two organisms will increase.

The phenomenon of partial predatory exists in the predation, and the existence of predator is often influenced by the number of prey. For example, Parameciidae in each split reproduction will consume about 18,000 bacteria. The more it preys, the fast the reproduction of body, and a balance will also be struck.

9.13.3 Application and Correlation of Microorganism

The application of microorganism in agriculture could partly replace chemical fertilizers and pesticides. It exerts effective influences on the fertility and the environmental protection, and becomes an indispensable method of sustainable management of agriculture. Thus, it is a project deserving our great attention.

Microbial fertilizer and microbial pesticide are the focal work of future development of agricultural biotechnology, which make microorganisms as its major preparation. Specifications are listed as follows:

- Microbial fertilizer refers to active microorganisms or resting spores, such as bacteria (including actinomyces), fungi, algae and other specific cultivation of metabolic substances. When they are applied in the crop production, they could provide plant nutrients, including an increase in the nutrients of plants, or stimulation on the growth of plants, or plant absorption of nutrients promotion. Besides utilization of beneficial microbial fertilizer, there are still other functions, such as the effect of protecting rhizosphere, promoting the growth of roots and absorbing water and nutrients, expanding the life span of root system, neutralizing or decomposing toxic substances, increasing the survival rate of planting and flowing in advance, etc.
- Microbial pesticides consist of a microorganism (such as bacteria, fungi, protozoan or virus) as the active ingredient that could be used in preventing diseases of plants, pests, weeds and harmful animals.

Therefore, when applying in environmental ecosystem, microorganisms will form close relationships with each other. In order to display the effect of certain microorganisms to the utmost, we need to impose the effect of bacteria. As for how to enhance their existence and behavior in the environment, in case the effect of beneficial bacteria will be weakened, it is best to form a mutually beneficial relationship to avoid the competition and prey cycle between the antagonism and other beneficial microorganisms.

9.14 The Possible Impact of Biotechnology and Transgenic Crops on Agriculture and Environment

9.14.1 21st Century is the Century of Biology and Environment

The development of biotechnology and transgenic organism science will exert earthshaking influence in the 21st century. By making use of the modification of DNA fragment in genetic engineering, biotechnology achieves the goal of realizing new phenotype or new function in organisms. Originally, "biotechnology" is a noun used early in scientific development, indicating using the treatment of biology to realize specific effects, which may not appear in genetic engineering. But, later biotechnology belongs to genetic engineering. Due to the fast development of chain reaction of taq DNA polymerase and gene sequencing instruments, knowledge on gene has been quite fast accumulated. In recent years, the development of protein sequencing, the structure and equipment largely speed up the growth of biotechnology, which will become one of the important characteristics in the 21st century.

Any development result could have both positive and negative influence on human beings and the environment, and biotechnology is no exceptional. That is also a subject that deserves our attention in 21st century. We need to emphasize on how to weaken its problems, and this is a scientific challenge. Due to the development of new biotechnology and transgenosis technology, including medicine, natural science and agriculture, various application effects on humans and environment are still being observed, and we cannot completely judge from its goodness. In this article, based on discussing possible attacks on the environment brought about by the application of biotechnology and genetically modified organisms in agriculture, the author provide forward prospective thinking and thoughtful questions, so as to present reference for research, development and application.

9.14.2 Uncertain Effects of Agricultural Biotechnologies and Transgenic Crops

Agricultural application of biotechnology and novel organisms of transgenosis include in animal, plants and microorganisms. All over the world, foods associated with such two kinds of technology are treated differently. Many consumers, most noticeable in Europe, worry about genomic modified organisms (GMO) foods contains hidden health risks. For example, whether there are new toxic matters in the new food of genetic engineering? How about the nutrient content? Would there contain allergic factors or unidentified substances? Or would there be some possible unexpected and side effects? All these may take place in the transgenosis.

In agricultural production, in terms of crop cultivation, types of crop and the system, the crop may be changed or replaced by the food industry system, in which the cell cultivation is accelerated by biotechnology and transgenosis. That may lead to a decrease in the crop choice in agricultural production as well as consumers' choice. Some transgenic agricultural products (for it may have special effect or ingredient) could not be judged from the appearance, so people may eat them by mistake or over-eat them. Therefore, we should pay attention to prescribe strict standards on the production and consumption of special agricultural production.

9.14.3 Possible Impact of Novel Organisms on Agricultural Ecosystem

The main theoretical basis for the attack of biotechnological and transgenic organisms on overall environment includes competition, mutation, gene hybridization and transformation, induced resistance against chemicals, and toxins, etc. They will influence the dynamic equivalence of nature and agricultural environment, including ecosystem balance, competition among species, food chain system, cycle of materials, etc. Attacks on agricultural soil condition mainly focus on the physical, chemical and biological properties of soil ecosystem. We will specify from the possible problems on agricultural ecosystem:

Reduce Biodiversity:

Due to the cultivation of unitary transgenic plants, it may cause the problem of simplex ecosystem.

Transgenosis may Intensify Species Competition:

The genes of transgenic organisms may hybridize naturally or be transferred to wild plants or weeds, increasing the competition of wild plants and weeds, and imposing attack on environmental ecosystem.

Induce New Virus:

Transgenic anti-virus plants may induce more serious virus.

Changethe System of Food Chain:

New anti-pest crops may lead to a change in the biology of food chain due to decrease of certain insects, and thus affect the balance of overall ecosystem.

Increase Organisms of Resistance:

For example, utilization of transgenic biological insecticide could enhance pest ability of resistance. Or the use of transgenic biological germicide may enhance microbial ability of resistance, and thus affect the balance of agricultural system.

Enhance Chemical Effect:

For example, utilization of transgenic crops that resist to herbicide may produce chemical effects that strengthen other plants.

Changing the Important Metabolisms of Crops:

For example, utilization of transgenic biotechnology may strengthen leguminous ability in the nutrient absorption, and thus weak the N_2-fixation of symbiotic root nodule.

Change the Metabolisms of Matters in Cycle:

If special novel transgenic organisms could strengthen ability of absorbing oxygen and resisting bacteria, it will lead to a change in soil or aquatic ecology. As a whole, it will affect the metabolic system of matters in cycle and the ecosystem of aquatic environment.

9.14.4 The Development of Biotechnology should be Careful for the Possible Crisis

In terms of the researches on biotechnology and transgenic biology, due to the deficiency in a complete global controlling system, one problem may lead to a global crisis. Therefore, in case any novel organism goes wrong, we should prevent for problems, and ensure to decrease the occurring problems. Otherwise, we need to cost extra labor, energy and money to solve these new problems. Of course, with new development, there is new hope. If we watch out, we may reduce more crises. These are also the control and management prescription made by many advanced countries in terms of biotechnology and transgenic organisms, which call on scientists to comply with the safety rules.

9.15 Advantages and Disadvantages of Soil Sterilization and Pesticide and Guidelines for Preventive Management

9.15.1 Major Functions of Soil

Is it necessary to apply the soil sterilization and pesticide? Will the function of soil be affected? These are questions that we should pay attention to. Therefore, first of all, we should get to know the function of soil. The major function of soil includes productivity, cleaning, ecological balancing, buffering, retention of water and fertilizers, and cycle of matters, etc. They are to be specified as the following:

Productivity:

In agricultural application, soil could mainly provide crops with producing media, nutrients, and common needs of crops such as nutrients, water and (or) oxygen. Nutrients in the soil include some are easy to be used, but some are hard to be used. Soil with high productivity could provide sufficient nutrients, water and oxygen. Beneficial microorganisms in the soil, such as rhizosphere microorganisms, are closely related to the production of crops. In addition to protect rhizospheres, they are also associated with the health of root system and nutrient absorption.

Cleaning:

Soil contains diversified biotic groups of microorganism and adsorptive ability, which could clean up various organic waste or toxic substance placed by human. They could decompose and adsorb toxic substances and decrease its toxicity. They could not clean up heavy mental which could not be broken down, only with accumulative effects. The application of organic fertilizer in agriculture also depends on the decomposition capacity of soil microorganisms to provide nutrients for crops.

Ecological Balance:

There are many microorganism biotic groups in the soil, and they have mutual balancing abilities. No microorganism may control other microorganisms. For agricultural crops, a few parts of microorganisms could lead to diseases, and thus is harmful to microorganisms. However, most of the microorganisms are beneficial ones that could exert direct or indirect influences on others.

Buffering Capacity:

Soil possesses the buffering capacity of stabilizing acidity and alkalinity. When adding acidic or alkaline matters, soil will be entitled with such function of mitigation and changing resistance.

Retention of Water and Fertilizers:

Inorganic and organic matters contained in soil are with porosity and adsorbability, which could maintain water and fertilizers in the soil. An increase of organic matter content in agricultural soil will enhance the ability of maintaining water and fertility.

Cycle of Matters:

Substances in nature switch among solid, liquid, and gas states. Soil could decompose these matters and play the cyclic function of various elements (such as carbon, nitrogen and sulfur, etc.).

Among the major six functions stated above, five of them have the closest relation with agricultural production, i.e., productivity, cleaning, ecological balance, buffering capacity, retention of water and fertilizers.

9.15.2 Advantages and Disadvantages for the Application of Anti-Pests and Anti-Pathogens in Soils

In general, the application of anti-pests and anti-pathogens in soils include three kinds: physical method, chemical method and biological method. They are going to be illustrated separately in the following parts:

Physical Method:

Physical methods such as temperature, insolation, and water control are part of this approach for anti-pests and anti-pathogens in the soil. Common physical method has little side effect on the function of soil. However, it takes a long time for the effects to be completely displayed, so the physical method has to be carried on for a long time.

Temperature Controlling Method:
Temperature may affect the existence of soil microorganisms. Commonly, high temperature is frequently adopted, and we can use the method for increasing the temperature, such as covering the soil with plastic film and steaming, etc. The method of covering plastic soil may have limited effect, but it is simple to operate. While the method of steaming commonly conduct to against pests and pathogens, and sterilization through steaming, and the method is quite laborious and takes a long time.

Insolation:
We can turn the soil and bask the soil in the sun, so as to reduce the occurrence of diseases and pests. This method is simple, with limited effect, and it may be unavailable to certain pests.

Water Control:
Various pathogens and pests differ in the adaptability to the water content in the soil. When soaked in water, some biotic groups may decrease, such as nematodes. If incorporated with the loosening of soil and insolation, the moisture content may decrease and pests are reduced.

Others:
For example, soil removed from other place is to improve the local soil. Or the dilution method of adding minerals, or the change in the physical environment could both decrease the biotic group of plant diseases and insect pests.

Chemical Method:

Soil sterilization which is against pests and pathogens based on chemical principles includes pesticides, acid-base control agents, toxic gas created agents, chemical improver of soil, etc.

Pesticides:
Chemical agents are common doses of soil sterilization and anti-pests and anti-pathogens, and these chemical doses are all poisonous or hypertoxic. They could even kill all the harmful and beneficial organisms in the soil with high side effects on the functions of soil. Under severe situation, the productivity or dependence of crops may decrease and vacant patch of seedlings will take place. Therefore, when using liquid of soil

sterilization which is against pests and pathogens, we should refer to the prescriptions on the instruction book of pesticides.

Acid-Based Controls Agents:

Soil pathogens and pests may adapt to certain pH value. If they prefer acidic soil, we should apply lime materials to improve pH value of soil, and the propagation and problems of pests will be reduced. For instance, nodule disease is likely to take place in acidic soil. If soil particles are to completely take control of the pH value, they will be influenced by soil properties. For example, hilly land soil is not easily to be completely controlled; therefore, the effects of acid-based control agent are limited, and if they are not over-dose, there will be few side effects.

Toxic Gas Created Agents:

The method of using the produced toxic gases to kill diseased microorganisms is quite effective, because poisonous gas has good aeration in the soil, such as methyl bromide, ammonia, and other chemical matters. But we have to pay attention to chemicals with strong toxicity and its high side effects. Common ammonia has low side effects, and it is produced by ammonia through alkaline matters.

Chemical Improvement of Soil:

Such as the application of organic fertilizer, its composition may change the soil chemical property, decreasing the biotic groups of pathogens and pests, and reducing the occurrence of problems.

Biological Method:

Mutual promotion and restraint exist among soil organisms. Soil sterilization which is against pests and pathogens could make use of such kind of principle. For example, antibiotics could be produced by microorganisms that resist other microorganisms, or the natural enemy of pests, enhancing microbial function of sterilization and against pests and pathogens. In biological method, when the bacteria come from local soil, its side effects are quite low. When the microorganisms of sterilization which is against pests and pathogens come from other places, we should notice its possible defects and of attacking environment.

9.15.3 Notice the Possibilities the Application of Soil Sterilization for Anti-Pests and Anti-Pathogens

Healthy soil possesses ecological balancing power. Once the soil gets sick, it means that soil get unbalanced. First of all, we should diagnose the soil and explore reasons in the soil analysis, and ask for improvements. Do not immediately apply soil sterilization which is against pest and pathogen agent in the soil, because various medicaments may have side effects and destroys the function of soil.

In terms of the application of soil sterilization and anti-pests and anti-pathogens, we should especially pay attention to the harm that done to the function of soil. Harm that is done to the beneficial microorganisms will lead to unhealthy root system, low nutrient absorption and the availability of fertilizer. Decomposition ability of organic matters will be weakened. After sterilization, some of the microorganisms will significantly reproduce due to the absence of balance, and thus cause more problems with soil microorganisms. Under vicious circle, soil could maintain crops production only when soil gets frequent sterilization. This means that soil has lost the balancing ability.

9.15.4 Guidelines of Soil Sterilization Management and to Fight against Pests and Pathogens

In order to reduce soil sickness and the application of sterilization and against pest and pathogen agents, in terms of soil management, we should do the following work to maintain healthy soil:

Reduce Continuous Cropping of Soil, and Adopting Rotation Cropping:

Soil sickness is often associated with problems in continuous cropping, which could bring about soil physical, chemical and biological problems, and the plant diseases and pests, or soilborne diseases, such as root rot, root tumor, damping off, etc. We could adopt rotation system, and avoid continuous cropping in similar crop. Or the adoption of interval upland-lowland farming and rotation cropping in paddy field may reduce the problems of pests and pathogens.

Apply Organic Fertilizer and Organic-Improved Material:

Organic fertilizers and materials (such as shrimp shell, crab shell, peat soil, etc.) could improve the physical, chemical and biological properties of soils, which could improve the balancing ability of microorganisms. We should choose compound organic fertilizer with various raw materials, instead of single material or overuse, and try to avoid overusing organic matters in excrements of single material.

Reduce Soil Acidification, Avoid Using too Much Chemical Fertilizers, and Neutralize Soil with Lime Materials:

Overuse of chemical fertilizers may lead to soil acidification, especially use acid-forming fertilizers. Therefore, we should try not to use too many chemical fertilizers. Acidic soils could be neutralized by lime materials. Lime materials belong to alkalinity, such as magnesia lime, dolomite, oyster shell powder, silicate slag, etc.

Avoid Using too Much Nitrogen Fertilizer and Notice the Supplement of Nutrients:

Commonly, the overuse of nitrogen fertilizer may lead to acidification and weaken leaves of plants, reduce the disease-resistant ability of plants, and thus bring to diseases. We should pay attention to the balanced supply of various kinds of nutrients.

Pay Attention to the Water Source and Prevent Pollution of Irrigation Ditches:

Many soilborne diseases have their origin in the irrigation water, such as Phytophthora disease. Thus, ditches should be independent so as to avoid passing by complex farmland and crop pollution.

Increase Application of Microbial Fertilizer to Rhizosphere:

Application of microbial fertilizers (such as mycorrhizal fungi, phosphate solubilizing microorganisms, etc.) could enhance the protection of rhizosphere, reduce the occurrence of diseases and pests, increase utilization of nutrients in the fertilizer, improve the quality of corps, and thus achieve many things at one stroke.

9.15.5 Conclusion

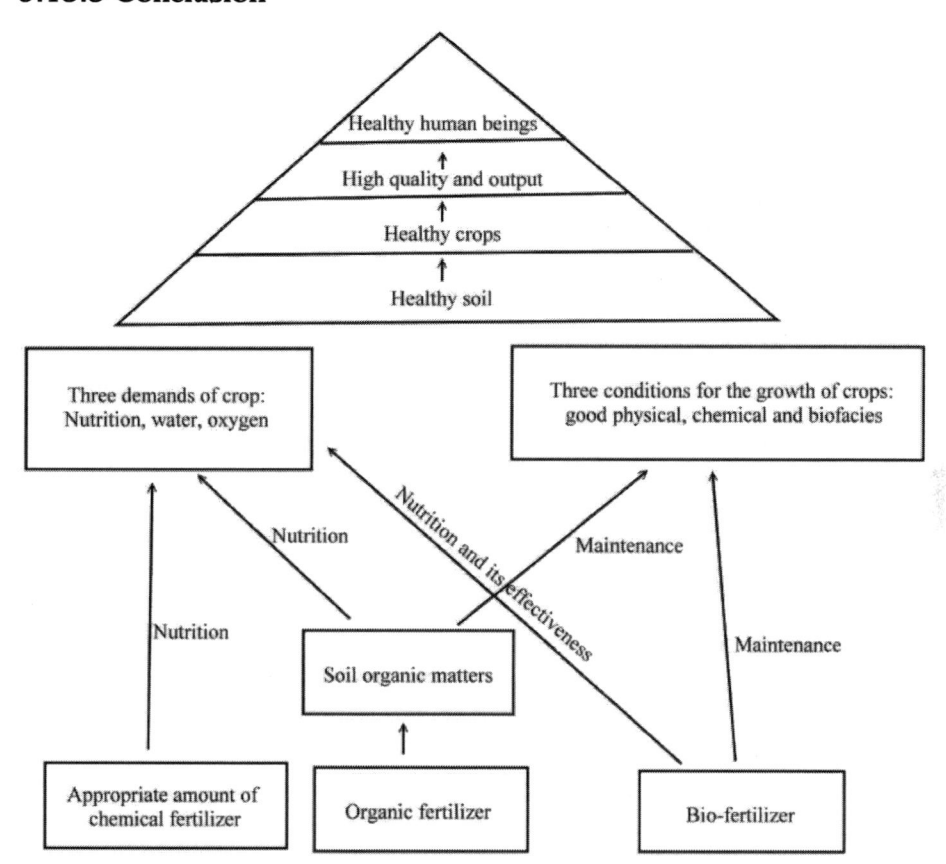

Figure 9-3. Human Health Related to Soil Quality and Fertilization.

442 Soil and Fertilizer

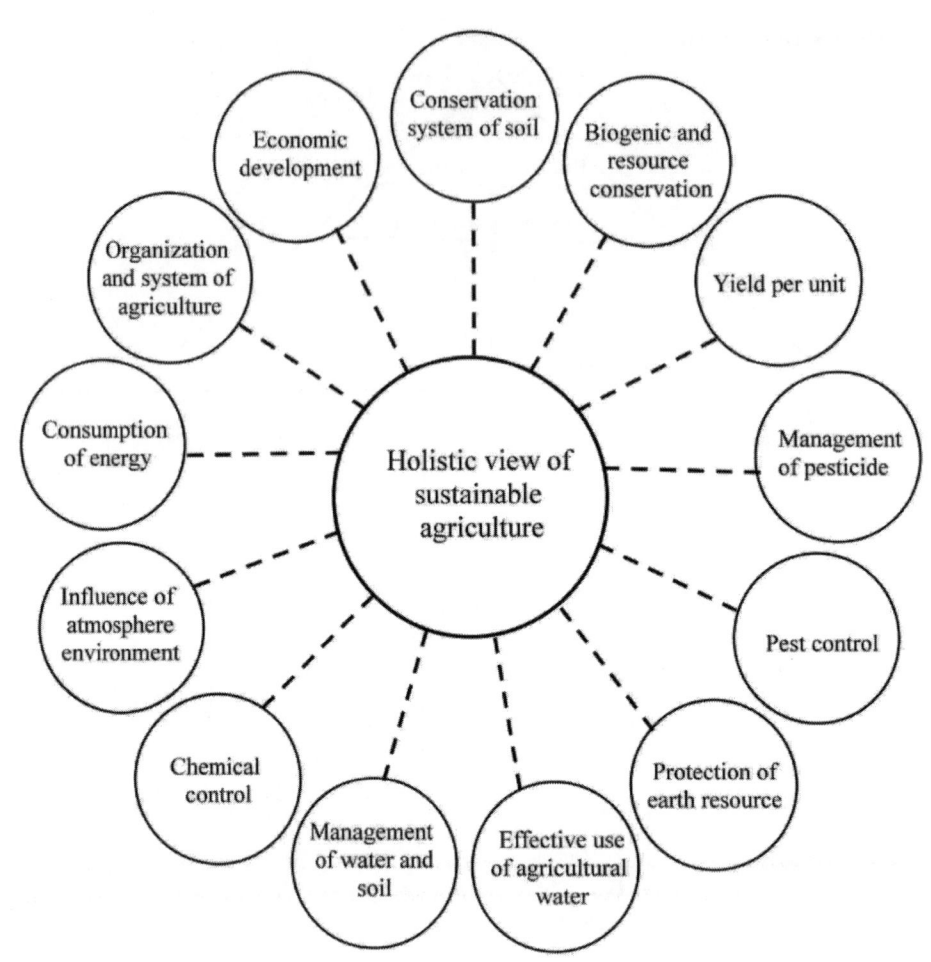

Figure 9-4. Overall View of Sustainable Agriculture.

Chapter 10

Q & A: Mailbox for Soil and Fertilizer

10.1 Q & A for the Diagnosis of Problem Soils

Q1: What is the appropriate frequency for testing the properties of the soil? How to collect the soil sample from the farm for the diagnosis of soils?

A1.1: The frequency for testing the properties of the soil depends on its usage and the ways of cultivating and managing. In principle, the more often the cultivating system of the crops develops, or the soil is cultivated and fertilized many times in a year, the more necessary it is to test the changes of the soil. In this case, the test should be done at least every two years. However, the soil in which the perennial crop grows should be tested every three to five years, so as to avoid soil degradation, or to provide records for the right fertilization.

A1.2: The method of soil sampling is based on the principle of acquiring the "representative" soil. So, attention should be paid to the typical soil sample. That is to say, the soil should not be taken from the fertilization spot in the field; instead, it should be taken from a wide range. Moreover, the depth of the soil samples should correspond with the depth of plant roots.

10.2 Q & A for Plant Nutrients and Guidelines of Fertilization

Q2: Can calcium superphosphate be mixed with urea?

A2: Urea can absorb water and become wet, so it is not appropriate to be put aside for a long time when mixed with calcium superphosphate. However, the mixture can be fertilized right away. Otherwise, it will get lumped severely, the powdery calcium superphosphate in particular.

Q3: Do hydrophytes (aquatic crops), such as water spinach, taro, and lotus, have something different from the common crops on fertilization? Or, what kind of things should we pay attention to?

A3: Hydrophytes grow in the aquatic or paddy field. The water flow and the soil aeration of the aquatic or paddy field are worse than that of the upland. When hydrophytes are fertilized, the loss and effectiveness of fertilizers should be considered seriously. The nitrogen fertilizer in the aquatic field can be easily lost in the running water and in denitrification, but the availability of the phosphorus is improved in this situation. Things that should be paid attention to in the fertilization of the aquatic or paddy field are listed as follows:

A3.1: Nitrate Fertilizers should not be Used in the Aquatic or Paddy Field:
Nitrate nitrogen should not be used to fertilize in the hydrophytes, because in anaerobic conditions, nitrate nitrogen will be used by the denitrifying bacteria to produce nitrogen gas or nitrous oxide. Because of this, it will scatter and disappear from soil into air.

A3.2: Applying the Fertilizer should be after the Drainage of Water:
Soluble fertilizers are easily dissolved in water and then be lost. If the fertilization is done after drainage, fertilizers can easily be moved into soil, or it is proper to place fertilizers into the depth of the soil.

A3.3: It is Appropriate to Use Slow-Released or Granular form Fertilizers:
It can increase the recovery of fertilizers, and lose less.

Q4: *A crop season of pepper or bitter melon can be harvested for half a year. In order to avoid wage consumption in fertilization, while additional fertilizer with covering the plots with plastic film, full amount of fertilizers will usually be applied once before crop planting into the field with covering plastic film. When taking this approach, what kind of things should we pay attention to? Is it appropriate to foliar dressing by using materials, such as soluble fish meal? Or, what should be improved in order to keep the harvest period over half a year for good practices?*

A4: When taking this approach, we should pay attention to the enhancement of the maintenance of soil fertility, the foliar dressing and the drip irrigation.

A4.1: The Maintenance of Soil Fertility:
The soil needs the ability to supply sufficient nutrients and the capacity in nutrient holding. Use matured organic fertilizers and slow-release fertilizers or fertilizers with long-term effects.

A4.2: The Foliar Fertilization:
Liquid organic fertilizer with different proportions of nutrient can be sprayed on the leaves by foliar fertilization. But the proportion of nutrients should be adjusted according to the growth period of crops. The proportion of nitrogen fertilizer should not be too high, because the high proportion of it will cause problems like excessive growth, flower falling, fruit dropping, or less blossoms. Whether nitrogen is excessive or not can be seen from the size and thickness of leaves.

A4.3: The Drip Irrigation:
A drip irrigation system can be set under the plastic film. Liquid fertilizers with different formula can be used to supplement nutrients. The drip irrigation can be more long-lasting and effective than the foliar dressing.

Q5: *What are the advantages and disadvantages of rock wool cultivation and hydroponic cultivation? Is there anything we should pay attention to regarding the supply of the nutrients?*

A5: All methods of cultivation have their advantages and disadvantages. The advantages and disadvantages of the rock wool cultivation and the hydroponic method are shown in the following Table 10-1:

Table 10-1. The Advantages and Disadvantages of the Rock Wool Cultivation and the Hydroponic Method

	Rock Wool Cultivation	Hydroponic Cultivation
Advantages	1. The crops grow fast. 2. The controlled environment has less pathogenic problem, if in good control. 3. The aeration and the stability of roots are good.	1. The crops grow fast. 2. The controlled environment has less pathogenic problem, if in good control.
Disadvantages	1. The rock wool waste should be dealt with seriously or be recycled; otherwise, rock wool dusts will cause severe pollution. Dust and fibers are a health risk. 2. Technicality should be paid attention to. If not carefully adopted, the cultivation can turn out to be a failure easily. 3. It is easy to fail once the diseases in the water occur. 4. The production of food crops requires a research on long-term consumption of it. The accumulation of toxic elements or lack of nutrients should be avoided.	1. The dealing with the waste water should be careful; otherwise, it will pollute water. 2. Technicality should be paid attention to. If carefully adopted, the cultivation can turn out to be a failure easily. 3. It is easy to fail once the diseases in the water occur. 4. The production of food crops requires a research on long-term consumption of it. The accumulation of toxic elements or lack of nutrients should be avoided.
Attention	1. The concentration and proportion of all nutrients and adjustment of environment and weather should be paid attention to. 2. The value of pH and EC should be measured frequently or controlled automatically.	1. The concentration and proportion of all nutrients and adjustment of environment and weather should be paid attention to. 2. The value of pH and EC should be measured frequently or controlled automatically.

Q6: Do different crops have their own preference of nutrients?

A6: Different crops have their own selection, preference, sensitivity, and tolerance towards the nutrients. Preference results from the difference of the nutrient absorption of the different crop root systems. Tea trees are NH_4^+ and Al-preferring. Alfalfa has a higher proportion of calcium and magnesium than general grass. The coastal plants have a preference for salts. Rice plants prefer NH_4^+ to NO^{3-}, and grow well with NH_4^+ as sole N source. However, most terrestrial plants could significantly reduce growth and develop NH_4^+ toxicity. All these are best examples. The differences of biochemical process of cell membrane of roots and qualities of ion carriers among crops will affect the absorption of nutrients.

Q7: Will the type of fertilizer affect color or flavor of fruits?

A7: Factors that affect color and flavor of crop fruits include weather, species, soil, cultivation management, and pathogen prevention. The reason why types of the fertilizer in the cultivation management affect color and flavor of fruits is that they affect biochemical metabolism, the distribution of nutrients and the storage and preservation of crops. There are three types of fertilizer: chemical fertilizers, organic fertilizers, and microbial fertilizers. All of them will affect the amount and ratio of the absorption of nutrient elements and organic compounds. For example, excessive nitrogen fertilizer will make crops and leaves grow rather fast. Therefore, when the amount of products of photosynthesis distributed to fruits become less because of the high nutrient demands of new shoots and leaves, the color and flavor of fruits will be affected. In order to have good color and flavor of fruits, we should pay attention not only to all nutrient elements (macro and micro elements), but also to the cooperation with organic fertilizers and microbial fertilizers. In this way, it will turn out to be fruitful easily.

Q8: I have learned the cultivation of grafting pears for three years. My knowledge of the fertilization is from the seniors. How can I judge whether I have fertilized too much or not, especially the amount of the organic fertilizer applied after the harvest every year?

A8.1: Two ways for judging whether the fertilization is over or not: identify by visual inspection and analysis method:

Visual Inspection:
   The appearance of the crops can be observed by seeing whether the branches and leaves grow too fast or not, and whether plants have the deficient symptoms or not. If there is something wrong with the old leaves, it may be relative to nitrogen, phosphorus, potassium, and magnesium. If there is something wrong with new shoots or younger leaves, it may be relative to calcium, boron, iron, zinc, and manganese. If there is an excessive use of fertilizers, the leaves will curl or droop in the afternoon.

Analysis Method:
The nutrients and features can be analyzed and determined through plants and soil analysis, so as to research and judge the amount of fertilizers. And, research results of soil and fertilizer named as Crop Fertilization Manual (published by the Council of Agriculture of Taiwan) can be referred to judge whether the ratio is adequate or not.

A8.2: The Supplement of Organic Fertilization after the Harvest:
The qualities of nutrients and organic fertilizers should be concerned. The organic fertilizer with excessive nitrogen should not be used especially before the leaves fall from deciduous fruit trees. Otherwise, it will result in excessive growth, which is not good for the blossom next time. If plant branches grow too fast without trimming, soil nutrients will be consumed and the storage capacity of them in stems and branches will become less after harvesting and become wastes at last.

Q9: *Nowadays, the fruit growers often use humic acid, soluble fish meal, and seaweed extract to increase the sweetness of fruits in order to improve the quality. Should soluble fish meal and seaweed extract be sprayed on leaves or irrigated on the roots?*

A9: Humic acid, soluble fish meal, and seaweed extract are organic matters. Humic acid and seaweed extract are peat and seaweed plant extracts which can increase the growth of root system. Soluble fish meal is made from the soluble matter of fishes. The main contents of those materials are nitrogen fertilizer and amino acids. The organic matters that are not decomposed completely are better to be irrigated into soils. In this way, the macromolecules can be decomposed into small molecules to be absorbed by roots. Therefore, humic acid, soluble fish meal, and seaweed extract can all be irrigated into soils. However, they contain many inorganic nutrients, and trace elements and organic matters can absorb small molecules and hormone directly. They can be diluted and sprayed on leaves, but it is not proper to spray them on the leaves when the concentration is high.

Q10: *After heavy rain, the margin of old leaves of grape plants will get yellow, burned, and even withered. Is this called "fertilizer injury" or "virus"? As growers have applied potassium sulfate, is it proper to judge it as "potassium fertilizer injury"? If it is, what can be done with it? Some growers haven't applied the "potassium" fertilizers, but their crops have the same symptom. So, will the "potassium" fertilizer be accumulated and stored in the soil? And, will it be released easily after heavy rain?*

A10: After the rain or a long-time rain, the margin of the old leaves of the grape crops gets yellow, burned, and even withered. The reasons might be as the following:

A10.1: The Water Damage of the Root System:
The water saturation will lead to the deficiency of oxygen in soil and damage root system caused by suffocation. When it is sunny after rain and water loss of leaves is heavier than that of roots, namely, the leaves lose more water and become soft, the symptoms of leaves turn out.

A10.2: The Roots or Leaves have Diseases:
The heavy rain results in rotten roots and the lack of water absorption. Then above symptoms will appear.

A10.3: The Fertilizer Injury of Root System:
Applying soluble fertilizer will hurt the roots because the fertilizer will be easily dissolved into the soil with the rain.

A10.4: The Acid Rain:
Strong acid can lead to the damage of leaves.

A10.5: The Damage Caused by Pesticides:
The pesticides are absorbed by plants so that plants are damaged.

All above reasons are possible. The analytical research and judgment should be made depending on the situation. If potassium fertilizer is accumulated and stored in the soil, there would be a large amount of salts, and the long time rain and excessive organic fertilizer will cause the root rot easily, to summarize the first, second, and third problems.
Remedial methods and matters that need attention are listed as below (remedial methods can be selected according to the above reasons):

- The liquid wax or the protective agent can be sprayed -- they can prevent the problem brought by loss of water when sprayed on leaves.

- The bactericide can be sprayed to prevent the occurrence of diseases.

- If roots are damaged by the fertilizer or water, stop fertilizing, breaking ground or moving roots. Nutrients can be supplemented through the foliar application.

Q11: Is calcium cyanamide (black fertilizer) an alkaline fertilizer? What's the relationship between alkalinity and the pH value of calcium cyanamide?

A11: Calcium cyanamide is an alkaline fertilizer. Calcium cyanamide can absorb water to produce calcium hydroxide. The value of pH is above 7, which means that the H^+ concentration is lower than OH^- concentration in solution. Alkalinity can be shown by the value of pH.

Q12: What are the functions of calcium cyanamide (black fertilizer)?

A12: Calcium cyanamide has the following functions:

A12.1: Function as the Nitrogen Fertilizer:
Cyanamide contains CN^- group as the nitrogen fertilizer. It can be transformed into ammonium nitrogen fertilizer when it is decomposed.

A12.2: Control Diseases and Pests:
It is toxic and can kill some microorganisms and insects.

A12.3: Removal Weeds:
It can kill the seed of some weeds.

A12.4: The Defoliator:
It can lead to falling of leaves of some crops, such as cotton, and can be used as defoliant.

Q13: What does EDTA belong to? Why does it work only when it is mixed with magnesium, iron, calcium, zinc, and sodium? Does the purity of EDTA have a direct relationship with its price?

A13.1: EDTA (ethylenediaminetetraacetic acid) is a kind of organic acid that contains four carboxyl (–COOH) group. Its chemical structure is shown as follows.

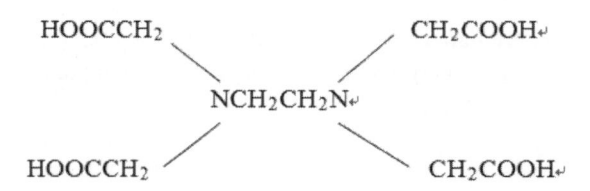

$$\text{HOOCCH}_2 \diagdown \qquad \diagup \text{CH}_2\text{COOH}$$
$$\text{NCH}_2\text{CH}_2\text{N}$$
$$\text{HOOCCH}_2 \diagup \qquad \diagdown \text{CH}_2\text{COOH}$$

A13.2: EDTA has four carboxyl groups, which can form anion $-COO^-$. So, its function is to "sequester" metal ions (Ca^{+2}, Mg^{+2}, Fe^{+2}, Zn^{+2}, Na^{+1}). After being bound by EDTA, metal ions remain in solution. It chelates and protects cations, so that they will not be fixed or absorbed by other matters. EDTA is not a nutrient, but it can improve the availability of cations.

Q14: *Does cobalt sulfate belong to the microelement? What's its function to the plants? Can it be used on the fruit trees?*

A14.1: Cobalt sulfate is a kind of salt compounds ($CoSO_4 \cdot 7H_2O$). Cobalt is a micronutrient of plants, which means that the amount for plants need is small.

A14.2: As for the functions of cobalt to plants, it is needed by the symbiotic nitrogen fixation of the leguminous plants and rhizobia. Its importance to leguminous plants is greater than to any other crops. The cobalt is helpful to the growth of plants when it is applied in the soil (such as sandy soil) which lacks it. It was reported that applying cobalt can improve nutrients of the forage grass and the growth of cotton and cabbage mustard. When the soil doesn't lack cobalt, it is not necessary to apply it to avoid the excess of it.

Q15: *Can alkaline fertilizers be absorbed by plants more easily? What harm will it do to the soil?*

A15.1: Alkaline fertilizer refers to the fertilizer whose value of pH is over 7. Most of the common alkaline fertilizers contain salts of calcium, magnesium, and sodium. Whether the fertilizer can be absorbed by plants depends on crop species, components of the fertilizer, and conditions of the soil. If the alkaline calcium fertilizer is applied to alkaline soils, it is not sure to be effective, for alkaline soils have too much calcium. However, it is better to apply it to acidic soils.

A15.2: Different types of alkaline fertilizers have different degrees of alkalinity. Different amounts of the fertilizer will have different effects. If the fertilizer with high alkalinity and solubility are overused, it will harm roots easily. However, if it is not overused, it may do no harm to roots.

Q16: *Are there any differences among limes, silicate slag and dolomite powder? And, how can they be used?*

A16: Lime is a soil conditioner made from pulverized limestones. It is usually called magnesia limes. The amount of calcium carbonate and magnesium carbonate depends on types of the limestone. Because of different proportion between calcite ($CaCO_3$) and dolomite ($Ca \cdot Mg(CO_3)_2$), limestones have four types: limestone powder with MgO (0 ~ 2.2%), limestone dolomite powder with MgO (2.2 ~ 10.9%), dolomitic limestone with MgO (10.9 ~ 19.5%), and dolomite powder with MgO (19.5 ~ 21.7%). Therefore, the magnesium content of dolomite powder is higher than the limestone powder of common magnesia lime.

The silicate slag is the blast furnace slag of the by-products of the steel industry. It still has the converter slag (usually called "lime slag") and deSulfurization slag (one type of lime slag). In silicate fertilizer, the content of hydrochloric acid soluble silicon oxide is over 10% and the alkalinity is over 30%. The CaO content of common slag is 38 ~ 56% and MgO is 3.4 ~ 8.2%. Therefore, the content of MgO of silicate slag is lower than that of dolomite powder.

The calcareous materials can improve soil acidity and provide nutrients such as calcium and magnesium. But it is not appropriate to be used excessively. Otherwise, the following problems will occur: the acid soil will become alkaline; the organic matters will be over decomposed; and, the soil will get hard. Therefore, calcareous materials are appropriate to improve the soil gradually and apply the organic fertilizer.

10.3 Q & A for How to Improve Fertilization and Management of Soils

Q17: *Are there certain environmental conditions suitable for the absorption and utilization of nutrients in soils? Will the acidity and alkalinity affect them?*

A17: Yes, there are. The environmental conditions, climate, and soil conditions that mainly affect the absorption and utilization of nutrients will be explained as follows:

A17.1: Climate Factors:

The climate factors that mainly affect the absorption of nutrients are temperature, humidity, sunshine, wind, and rain. The climate factors affect transpiration and photosynthesis of plants. The greater the transpiration is, the more favorable it is for the plants to absorb water and nutrients. The greater the photosynthesis rate is, the more abundant photosynthetic products supplied by the plants to roots will be; namely, they are more capable of absorbing nutrients.

A17.2: Soil Factors:

Soil environmental conditions include physical, chemical, and biological properties. They can all affect the absorption and utilization of nutrients. The main factors are soil moisture content, pH value, organic matter, proportion between cations and anions, and microorganisms. Soil moisture content affects soil aeration. Appropriate water is important. If it is too dry, nutrients cannot move or be absorbed, especially for trace elements. If it is too wet, roots of upland crops cannot respire under the condition of hypoxia and they are unable to absorb nutrients.

The pH value affects not only root metabolisms, but also the availability of nutrients. The crop content may lack phosphorus, calcium, magnesium, and molybdenum in the strongly acidic soil (pH < 5.5). If the soil is a strong alkaline soil (pH > 7.5), it may lack iron, zinc, manganese, copper, and boron. The organic soil is helpful to the availability of absorption and utilization of nutrients. As to the proportion between cations and anions, we should not overuse cations, such as calcium, potassium, and magnesium. Cations and anions compete with each other on the absorption of nutrients. For example, the reason for why magnesium is lack is that applying magnesium may not be enough, or that calcium or magnesium fertilizer is overused. If soil microorganism is healthy and root system or rhizosphere has effective microorganisms, it will be helpful for roots to absorb nutrients.

Q18: *How can we choose the most appropriate materials to improve soil pH? Are there any differences between clay soil and sandy soil? How much value can be adjusted at a time?*

A18: Guidelines of choosing materials that can improve over acidity or over alkalinity of soils are that materials should have slow neutralizing capacity, long-term effects, none or fewer side effects, multifunction, and economical benefits. Most of the materials that can improve soil pH come from natural mineral substances or organic matters. For example, the most commonly used material that can improve acidity is calcareous materials, such as magnesium lime, oyster shell powder, silicate slag and dolomite powder. The most commonly used materials that can improve alkalinity are sulfur powder, peat, and so on.

We should notice that the buffering capacity of sandy soil is relatively poor. The neutralizing material cannot be applied in a large amount at one time, in order to avoid side effects or the excessive amount of it. The amount of neutralizer applied to clayey soil is greater than that to sandy soil.

Except peat, materials that can neutralize soil pH will affect physical, chemical, and biological properties more or less. Therefore, the gradual method should be taken instead of overusing it too much at one time. It is appropriate to adjust 0.3 ~ 0.5 of pH value. Because of the great buffering effect of soil, its pH value has changed only about 0.5 with many years application of neutralizer. But the excessive use of calcareous material will decompose soil organic matter in a great amount and harden the soil. Therefore, in order to reduce side effects of the neutralizer, organic materials or organic fertilizers should be used along with it.

Q19: There is a simple method of fertilization, that is, use mixed fertilizer and put it into the source of water irrigation, and then the fertilizer will flow and irrigate the whole plot with water. Is it feasible?

A19: It depends on whether the fertilizer is evenly spread over the whole plot or not. If the mixed fertilizer is soluble or liquid, it is appropriate to fertilize the plot in an extensive way. It is not easy to distribute the fertilizer evenly in the plot, if this simple method is taken. It is feasible when incorporated with waterways or irrigation ditches between plots.

Q20: When is the "soil removed from some other place to improve the local soil (covered soil)" needed? What are the guidelines for doing it?

A20: "Covered soil" means when the soil is not appropriate for the growth of some crops, because of physical, chemical, or biological problems of it, the soil from other places can be removed to the local soil.

A20.1: Common Conditions that Need "Covered Soil":

The Physical Properties of the Soil are Poor:
Many gravel, shallow soil layer, fine texture (such as clay soil), coarse texture (such as sandy soil), and so on.

The Soil Chemical Properties are Poor:
For example, it can be too acidic or too alkaline, have high salinity or polluted land.

The Soil Biological Properties are Poor:
There is high incidence of plant diseases and insects.

A20.2: Guidelines for "Covered Soil":

Complementarity:
For example, if the texture of local soil is too fine or clay, the coarse soil should be used for covered soil. If the soil has many gavels, the fine soil should be used for covered soil; the alkaline soil should be used for acidic soil as covered soil.

Dilution:
To lower the harmful concentration or degree. For example, the removed soil can dilute the salinity or the pollutants for covered soil.

Health:
The removed soil is used to ensure the health of plants and decrease plant diseases and insects.

Q21: Does the groundwater level affect fertilizer utilization?

A21: The groundwater level affects fertilizer utilization. It includes the following conditions:

A21.1: The Absorption Ability of Root System:
The high groundwater level or poor drainage affects the absorption and energy metabolisms of root systems of upland crops, which leads to the slow root growth, poor nutrient absorption or even death, because of suffocation. It is the main factor that affects the fertilizer utilization.

A21.2: The Availability of Nutrients:
Under conditions of the high groundwater level or poor drainage, the soil can create the conditions of chemical reduction. It can increase the reduction of many metal nutrients, thus producing low charge ions which increase the availability of nutrients.

Q22: Will the plastic film covered on plots affect the application rate or availability of fertilizer? Are there any differences when compared with the fertilization in protected cultivation?

A22: The plastic film covered on the plot mainly afects the movement, incoming and outgoing of water in the soil, temperature, aeration, and weed growth. It has significant difference in seasons and at the time of irrigation.

Therefore, it affects the application rate or availability of fertilizer more or less. If it is in the rainy season, the plastic film will decrease fertilizer loss and save it. If it is in the dry season or if the soil lacks irrigation, the fertilizer will not move very well in the soil and the availability of fertilizer will become poor. Therefore, we need irrigation to improve it. In the season with low temperature, it is helpful to improve soil temperature and nutrient absorption of root system. The plastic film can control weeds, so that the fertilizer will not be competed by them.

The difference between plastic film covered on plots and the protected cultivation is that plastic film has not covered the whole plot, but the protected cultivation has covered the whole plot and does not have the rain washing process at all. Therefore, it is not appropriate to overuse the fertilizer in the protected cultivation. Otherwise, there will be much slat accumulation caused by fertilizer application.

Q23: How can we tell if the soil is slightly acidic in terms of pH value?

A23: If the soil pH value is below 7, it is acidic. If the value is close to 7, it is slightly acidic soil. The pH value of weak acid is 6.0 ~ 6.5. If the pH value is below 5.5, it is strongly acidic soil. The classification of the pH value can be seen in the following Table 10-2.

Table 10-2. The Classification of the pH Value of the Soil (1:1 Water Soil)

Classification	pH Value
Ultra Acid	< 3.5
Extreme Acid	3.5 ~ 4.4
Very Strong Acid	4.5 ~ 5.0
Strong Acid	5.1 ~ 5.5
Moderate Acid	5.6 ~ 6.0
Slight Acid	6.1 ~ 6.5
Neutral	6.6 ~ 7.3
Slightly Alkaline	7.4 ~ 7.8
Moderately Alkaline	7.9 ~ 8.4
Strongly Alkaline	8.5 ~ 9.0
Very Strongly Alkaline	> 9.0

Q24: Does the size of granular fertilizer relate to the effectiveness of fertilizer?

A24: The relationship between the size of granular fertilizer and the effectiveness of it depends on the qualities of granules (solubility, components, and so on). If the raw materials of the granules are different, the size of granular relationship with the effectiveness of fertilizer is different. If the raw materials of granules and the amount applied are the same, the small granule will have wider surface with the soil and play its role more easily. In this way, plant roots can touch and absorb the fertilizer easily. If the granules are applied in the orchard on the slopeland, the small granules can be applied and spread relatively wide. However, if the granule is too small, it can be easily washed away while heavy rain.

Q25: What should we do to overcome the problems of continuous cropping?

A25: Overcoming the problems of continuous cropping depends on factors of the problems, so that they can be dealt with properly. Therefore, we should analyze and figure out factors of the continuous cropping problems first. The reasons of continuous cropping problems for different crops and different soils may be different. The problems and solving strategies are listed in the following Table 10-3.

Table 10-3. Problems and Solving Strategy of Continuous Cropping

Problems of Continuous Cropping	Solving Strategy
The increase of plant diseases and insects	Plowing, solarization, soaking, coverage of plastic film, or soil conditioners application.
The dramatically increase or decrease of soil microorganism	Application of soil conditioners, beneficial microorganisms, organic fertilizers or detoxing microorganisms.
Lack of nutrients and decline of organic matters	Supplement of insufficient or uneven nutrients and applying organic matters.
Soil acidification	Application of soil conditioners.
Lack of nitrogen or other trace elements	Supplement of nitrogen fertilizer, or the removal of excessive residues.
Plants are poisoned	Methods such as dilution (soaking and watering), absorption (apply peat), or decomposition (apply organic fertilizer).

Table 10-3. Problems and Solving Strategy of Continuous Cropping (Continue)

Problems of Continuous Cropping	Solving Strategy
High soil electrical conductivity	Soaking, applying organic matters, applying Gypsum (CaCO$_3$) when the rock salt is excessive in alkaline soil, the planting of crops (such as corns) whose absorption capacity of fertilizer is great, or applying the beneficial microorganisms.
High compaction	Plowing in order to loosen the soil.
Existing in the plow pan and compacting	Deep plowing, turning over, or drainage.

Q26: Can the planting or seeding be done right away after plowing residues of vegetables, plants, or green manure into the soil?

A26: Whether the planting or seeding can be done right after plant residues are plowed into the soil depends on the amount of plant residues, the remains of different plants, and the tolerance of the crop. When the residues are plowed into the soil, they will be decomposed and propagate large amounts of microorganisms. The crops compete with microorganisms for nutrients. The lack of nutrients will cause the growth disorders of the crop. If soil nutrients are high or fertilizers are applied, the problem can be overcome with small amount of residues that are plowed into the soil. When the lignification of residues is high, such as sawdust, the problem will occur easily. It is relatively safe to plant crops after several weeks (about 2 ~ 4 weeks) or > 10 days after residues or green manure was plowed into the soil.

Q27: Strawberry is a short-term crop. The organic fertilizer is as the basal fertilizer. If the fast-release organic fertilizer is supplemented during the growth, plants are often hurt. The slow-released fertilizer cannot be supplied in time. So, how can we grasp fertilizer types in order to meet the requirements of plant growth?

A27: Two aspects should be paid attention to in order to supply abundant nutrients for strawberry plants during their growth. One is the ability of continuous nutrient supply of the basal fertilizer. The other is the grasping of the additional fertilizer.

A27.1: Basal Fertilizer:
In order to improve the nutrient supplying ability of the basal fertilizer, we have to pay attention to the qualities of organic fertilizer. The basal fertilizer applied should not be organic matters that are easily decomposed. It should be used together with organic matters with high

ability of preserving fertility and slow decomposition. For example, animal wastes are easily decomposed. Therefore, we should pay attention to the formula of organic matters as the basal fertilizer, so that it can provide nutrients continuously.

A27.2: Additional Fertilizer:

To control the additional fertilizer to provide nutrients without harming the plant growth, we should pay attention to the method of fertilization, as well the amount and types of fertilizers. When the fast-release chemical fertilizer is applied, be aware not to get close to the plants and avoid excessive use. When it is the time of dews or there is still some rain on the leaves, the fertilizer that can easily cause fertilizer injury should not be applied. If it is applied, it is appropriate to spray it with water. The fast-release fertilizer powder applied to plants tend to cause the fertilizer injury. In addition, it cannot be scattered right away easily or insoluble granulated fertilizer will not cause fertilizer injury. Apart from fertilizer application, we should also pay attention to the health of root system during the growth of strawberry. The microbial fertilizer that can promote the growth of root system can be used. It is also important to strengthen the root's absorption ability and scope. Otherwise, if the fertilizer is abundant in the soil, but the root system is poor, plants still cannot absorb nutrients effectively.

Q28: *Should watering be done at once after fertilization?*

A28: It depends on the types of fertilizer and soil water content. When soil water content is not enough, the organic fertilizer will not be decomposed easily and chemical fertilizer will not move easily and will be hard to dissolve. In this situation, the water should be supplied. The fast-release fertilizer applied can be sprayed with water. If soil water content is enough, or if it rains, the water cannot be supplied unless the fertilizer are made to move fast. The water can be irrigated after the fertilization. The fertilizer should not be washed out and ran off. Therefore, water should not be supplied in a large amount or for a long time.

Q29: *The slope land has been developed for twenty years, and it is gravel soil. The peach and orange plants are there at the moment. What can we do about soil improvement or fertilization? Should we focus on the foliar dressing?*

A29: During the twenty-year cultivation of the slope land, the soil has been watered into the lower layer or washed away. The common shortcomings of this soil are too many stones, low organic matter content, low water holding capacity and fertility, and high acidity. Therefore, for the techniques of fertilization, we should pay attention to following things:

A29.1: The organic fertilizer should be incorporated with organic matters that cannot be decomposed easily, such as bark compost and peat.

A29.2: It is not appropriate to apply large amounts of chemical fertilizers in order to prevent the serious soil deterioration and land decline. It can be used together with microbial fertilizers, so as to strengthen the health and absorption ability of the root system and promote the availability of nutrients.

A29.3: The foliar dressing is regarded as an assisting method. In the key periods of growth (the younger fruit period and middle-size fruit period), the foliar dressing is applied to supplement some parts. Because the soil improvement is the basis of management, the absorption of leaves in the foliar dressing is limited.

Q30: *After continuous heavy rain, what can be done to supplement the lost nutrients? What needs our attention?*

A30: Different soils will lose different amounts of fertilizer after continuous rain. Therefore, the soil fertility of different soils will not be the same after rain. For example, when soil organic matter content in the sandy soil is low, the nitrogen is washed away severely. So, nitrogen fertilizers should be supplemented. However, when soil organic matter content is high and the amount of lost nitrogen is great, it is not necessary to supplement nitrogen fertilizers. The most effective way to supplement the nitrogen fertilizer is to supply the chemical nitrogen fertilizer (such as urea and ammonia sulfate) or other organic nitrogen fertilizers with small molecules.

Q31: *Does continuous cropping problems mean that the same crop cannot be planted continuously or even the crop from the same family cannot? How often does it occur? The peppers have been planted in the same plot for two seasons. Is it necessary to prevent continuous cropping problems? In what ways can we avoid them?*

A31: Problems of continuous cropping refers to problems caused by continuous planting the same crop. If the next crop is from the same family, the problems will be caused the same as continuous cropping. For example, tomatoes and eggplants of Solanaceae, or cabbages, Chinese cabbages and turnip of Cruciferae have the problems of continuous cropping easily. Different soils and ways of management can affect the occurrence of problems of continuous cropping. In order to prevent problems of continuous cropping, it is best to rotate crops that are from different families. If it is inevitable and peppers have been planted for two seasons, it is appropriate to rotate crops from other families. The prevention of problems of continuous cropping depends on conditions of the soil. For example, lime matters can be used in acidic soil. The high-quality organic matters can be used. The soil can be exposed to the sun before planting. The beneficial microbial fertilizer can be applied and plant residues in the field can be cleaned. If there are nematodes, the soil can be plowed and exposed to the sun after several weeks soaking.

Q32: How can we increase the sweetness of fruits?

A32: The ways that can increase the sweetness of fruits are listed as below:

A32.1: The Nitrogen Fertilizer should Not be Overused:
The excessive nitrogen fertilizer make crops grow fast or the leaves become big and thin. Then photosynthetic products are distributed to the growth of the new leaves, thus the ripe and the sweetness of fruits are decreased. If it is serious, it may even cause fruits to fall.

A32.2: The Supplement of Phosphorus and Potassium Fertilizers after the Middle Fruit Period should be Paid Attention to:
If the phosphorus and potassium are not enough or the proportions of them are low, the phosphorus fertilizer and the potassium fertilizer should be added.

A32.3: Applying the Organic Matters or the Microbial Fertilizers should be Paid Attention to:
The organic matters and microbial fertilizer should be applied during the time of the basal fertilization. The quality of organic fertilizer should be emphasized and the excessive use of e chemical nitrogen fertilizer should be prohibited.

A32.4: The Foliar Dressing should be Paid Attention to:
Under special conditions, foliar dressing with chemical or organic compounds having small molecules can improve needs of the crops immediately.

Q33: If continuous cropping problems are dealt with by using black fertilizer (nitrolime), how much should this fertilizer be used in soil so that it can be effective?

A33: This book can be referred to in terms of the ways that can decrease the continuous cropping problems. The black fertilizer is nitrolime or calcium cyanamide (Ca^{2+}) ($^-N = C = N^-$). The content of nitrogen is about 20 ~ 35%. The black fertilizer contains cyanogens (CN^-), which are toxic and can kill some soil microorganisms, such as nitrofiers. But there are also microorganisms that can make use of cyanide of nitrolime. They can be decomposed greatly after about three days of the fertilization and change into nitrogen fertilizer. The amount of nitrolime that can be applied in paddy field is usually about 200 ~ 400 kg/ha. The amount of the fertilizer applied in uplands can be increased, according to the fertilizer formula.

Q34: If the irrigation water contains high salinity (4 dS/m) and pH value (pH 8.5), can sulfuric acid and gypsum be used to lower the pH value and the amount of salts?

A34: The problem of irrigation water is that it is difficult to treat the quality of irrigation water with high salinity and pH value. The removal of salts requires much work and money. It is not economical to apply desalination methods to irrigation water, especially when water is in a great amount. If EC value is above 4 dS/m, the water cannot be used for irrigation. The addition of sulfuric acid can only reduce pH value, but not high salinity. The crops should be chosen, but their tolerance to salts is different. The addition of citric acid can lower the pH value and combine small amount of salts, but it cannot improve the quality of water on the whole. Water with low pH value can be improved with the help of alkaline lime, but the problem of high salinity still exists. It is good and effective to use clean water to dilute water with high salinity. It is important that EC value (salinity) should not be high. As to other nutrients, it is accepted as long as they are not excessive (such as standards of content limitations of irrigation water). The fertilizer should be applied on the basis of nutrient contents of soil, unless it is applied on the hydroponics vegetable (depending on the qualities of hydroponics vegetable).

Q35: Why do the margins of old leaves in the grape plants get yellow, burned, and even withered after a rain or a long-time rain?

A35: The reasons may be listed as follows:

A35.1: The Water Damage of Root System:
The water saturation will lead to the hypoxia of soil and damage of root system caused by suffocation. When it is sunny after rain, and water loss from leaves is much more than that of the water absorption from roots, namely, the leaves lose water and become soft, the symptoms of leaves turn out.

A35.2: The Leaves or Roots are Suffering from Disease:
If the leaves or roots get any disease, in case of rain, they would go rotten, followed by a shortage of water absorbed which would in turn lead to the problem above.

A35.3: The Root System is Injured by Fertilizer:
The usage of fertilizer with high water-solubility may easily lead to damages to roots when the rain water sinks into the soil.

A35.4: Acid Rain:
Foliar injury resulted from strong acidic.

A35.5: Damage from Agricultural Chemicals:
It happens when the chemical is absorbed by plants.

Items listed above are all possible reasons, so it must be analyzed and diagnosed according to the practical situation. If the accumulation of potassium fertilizer in the soil brings about a huge amount of salts, and if the rain persists when there are too much amount of organic fertilizer in the soil, the root system is liable to rot. This is an integration of the 1, 2, 3 items listed above.

A35.6: Remedies and Matters to be Attended to:
Choose from the remedied below according to reasons listed above:

- Spraying liquid paraffin or water retention agents: the use on leaves can protect plants from problems caused by dissipation of water.
- Spraying pesticides: this is for the prevention of diseases.
- If the root is injured by water or fertilizers, it is recommended that the soil or the root is not to be applied fertilizers again, and applicable act is to supply the nutrients through foliar dressing.

Q36: Will organic materials in the soil speed up their decomposition when the wet season is over? If that is the case, is there any need for a reinforcement of nutrients or improvements?

A36: It is true that rain water can promote the decomposition of organic materials and release many nutrients. However, when it rains a lot, nutrient loss is prone to happen. Nevertheless, a reasonable amount of rain is beneficial to the availability of various nutrients. When there is any over-fertilization, the problems of over-application of the nitrogen fertilizer will easily occur, which will lead to vain-type over growth of the plant. Therefore, when the wet season ends, it is not advisable to apply too much nitrogen fertilizer in case of serious vain-type growth or reduction in fruit quality or sweetness. At the same time, attention should be paid to the application of phosphorus, potassium, and calcium fertilizers. Whether the soil is in need of a reinforcement of nutrient depends on the basal fertilizer and the situation of the plant. If the basal fertilizer is enough for the whole process of yielding, there will be no need for additional fertilizers; if not, a dressing is certainly needed.

Q37: Is there any difference in soil management between facility cultivation and the open filed cultivation of grape? Is there any phenomenon to be particularly guarded against?

A37: Soil management in condition of facility cultivation is slightly different from that in the condition of open field. The main difference lies in the amount and proportion of fertilizers used. For example, the fertilizer amount should not be excessive in the facility cultivation, where appropriate fertilization is more important. This is for the prevention of salt accumulation in the soil of facilities culture. The amount of nutrients absorbed from the soil by the crop in the facility cultivation is less than the crop in the open field. If over-fertilization occurs, it is prone to have vain-type growth, which is a disadvantage to fruit quality. Particularly, the amount of nitrogen fertilizer should be reduced.

In the circumstance of growing grape in facility cultivation, the covering plastic film could be removed for the rain to leach into the soil for natural adjustment. Drip irrigation of nutrients can be used for saving water. On the other hand, sprinkling irrigation or flooding irrigation can be employed for an overall adjustment of the soil in facility cultivation.

Soil in facility cultivation needs a strengthened management physically, chemically, and biologically. Attention should be paid to soil in facility cultivation as for pH value, and electric conductivity, and watch out whether there is something worsening.

Q38: *The method of cultivation carried out in mango orchard is restricting the growth of roots in order to dwarf the tree shape. Is a change of soil management necessary?*

A38: The preferable method for dwarfing the tree shape is ordinary pruning or layering. Yet, vain growth can be reduced by using low nitrogen fertilizer, high concentrations of phosphoric, and potassium fertilizer.

Q39: *The EC value in soil should not reach too high. It is much appreciated if a detailed explanation for how to deal with this is given.*

A39: EC value is the index of the salt amount in soil. To prevent the increase of EC value, excessive use of fertilizer should be avoided, because most fertilizers are salts. Moreover, the salinity of irrigation water is another issue of concern: it is safe under 0.05 dS/m, and once the amount surpasses 2.25 dS/m, there will be extremely serious accumulation of salts. The most effective way is to leach and soak the soil, thus the salts can be washed out. Meanwhile, growing aquatic crops is a good alternative. It can be costy digging or using underground pipes drainage for salt elimination. Growing crops with high salt resistance can check the deterioration.

10.4 Q & A for Organic Matters and Organic Fertilizers

Q40: *What's the difference between odors of well microbial fermentation and bacterial decay?*

A40: Fermentation and aerobic respiration are terms in biological metabolisms. They can differentiate anaerobic effects from aerobic effects, respectively. Fermentation is to supply and receive electron during the process of microbial metabolism. They are all organic compounds without stink, such as alcohols, acids, and so on. Specifically, ethanol, lactic acid, and acetic acid are commonly seen. If the microorganism gives out the odor of decay, obviously, the case is that the bacteria produce decay odor such as organic nitrogen, hydrogen sulfide, and so on by means of organic matters.

Q41: *When making farmyard manure or liquid fertilizer, how to speed up the fermentation and avoid decaying?*

A41: It can be worked out from several aspects such as adjustment of materials, addition of microorganisms, and improvement of production environment. Briefs are listed as follows:

A41.1: Adjusting Materials:
> The materials used for fermentation are of various kinds. In the earlier period, the supply of organic matters and nutrients which can be easily used must be ensured; that is, the fermentation effect is promoted. It offers benefits to adjust the balance of acid and base by using calcium carbonate.

A41.2: Adding Microorganisms:
> The addition of beneficial decomposer can immediately make an increase in the amount of decomposer, which would do good to a rapid fermentation, and, at the same time, it would also prevent the undesirable bacteria from going stunk.

A41.3: Controlling Environment:
> Apart from the above factors, the speed of fermentation is highly influenced by temperature, water, and oxygen. Generally speaking, it will speed up when the temperature is high. In winter or high mountains, it will slow down. If water is sufficient, the speed of fermentation is accelerated. Too little water would lead to a difficulty in fermentation; yet, too much water would cause oxygen deficiency, which would further lead the bacteria to produce odor. When supply oxygen to the fermentation, redox potential will increase and the odor production will be reduced. Therefore, the air supply to the fermentation can reduce the odor production.

Q42: When applying organic fertilizers, take farmyard manure for example, how to make the choice? Poultry dung, mushroom residue, or anything else. Is there any principle to follow?

A42: The choice of proper farmyard manure should be made according to the different soils and crops. First of all, the principles concerning the materials of various farmyard manure, nutrient content, maturity, and economic benefit of the farmyard manure must be known. Through materials, the grade of decomposition and quality can be figured out. Through the nutrient status of the farmyard manure, the kind of crop it is applied to can be known; the maturity of the farmyard manure, a key factor, can decide the supply quantity, time, and economic benefit. The manure does not contain pathogenic bacteria or the speed of decomposition should not be too fast to produce any pollution. When using mushroom residues to make farmyard manure, make sure they are thoroughly decomposed and devoid of any pathogens or toxic phenolic compounds. Every organic material has its advantages and disadvantages, so the best way is to make use of its advantages and minimize its disadvantages.

Q43: When applying organic fertilizer along with chemical fertilizer, what kind of question should be paid attention to? Whether it is advisable to add magnesia lime at the same time?

A43: When applying organic fertilizer along with chemical fertilizer, things needed to be noted include the application or mixing, the proportion among different nutrients, and whether there is a loss of chemical fertilizer. On account of the difference of various organic fertilizers in their nutrient content, even, difference between high nitrogen and low nitrogen, so, it must be clear that which kind of nutrient should be mixed with the organic fertilizer. Therefore, pay attention to the application, amount of mixture, and the proportion of nutrients.

Magnesia limes should be applied to the soil before fertilization; the best timing is after rain or sprinkler irrigation. The effect can be better seen if the magnesia limes are neutralized with the soil before fertilizer application. If magnesia limes contact with phosphorus or ammonium nitrogen in the soil, the effect of phosphorus may be reduced or may cause ammonia evaporation. However, small quantity of magnesia lime (about 1%) can be used to adjust the acidity.

Q44: What is the standard of organic matter content in soils? Which level can be deemed as favorable?

A44: The standard of organic matter content must be adjusted according to the form of organic matter and situation of the land (dry or paddy). Generally, it is best when organic matter content reaches beyond 2%. The less content is worse. Yet, too much content (above 10%) may also cause problems, especially when the organic matter is mainly consisted of fresh residues or immature compost -- it is a disadvantage because many soil problems are prone to occur during the decomposition of fresh residues. As a common rule, the upland need more care in the content of soil organic matter compared with paddy field, because its water holding capacity and the perseverance of soil fertility would directly influence crop production. This is because the amount of organic matter content is a key index of soil fertility and closely related to quantity and quality of crops.

Q45: How to make a judgment on whether it is necessary to add organic matter every year?

A45: If most of organic matters in soils consist of fresh residues or easily decomposed organic fertilizer, more than a half or 80 ~ 90% of them may be decomposed in one year. Therefore, it depends on the conditions of the organic matter tested. Generally, it is uneasy to make the content of soil organic matter reach too high (above 6%), since organic fertilizer has a rather positive effect on the soil, as well as crop quantity and quality. If the amount of organic matter content is above 6%, a small amount of organic fertilizer can be applied. If the amount reaches beyond 8%, then it is unnecessary to add organic matter every year.

Q46: Since organic fertilizers abound in kinds in the market, how to testify the reliability of their contents? How to calculate the content of nitrogen, phosphorus, and potassium? Is there any principle to follow?

A46: Organic fertilizers in the market differ from each other, because of various material kinds and the mixture proportion. It is difficult for farmers to make judgment from the appearance, since the nutrients in plants such as nitrogen, phosphorus, potassium, calcium, and magnesium cannot be seen by our eyes. It can only be known through laboratory test. So, when purchasing organic fertilizer, pay attention to whether it has a brand, whether it is lawfully registered, and whether it has a reliable manual or direction of contents. Only through these can the amount of application be known; otherwise, it may result in a blind application which would exert influence on the soil environment and crop quality. Especially when there is a high quality request on fruit and vegetable plants, the controlling technique of the amount of fertilization must be guaranteed.

Organic materials include residues of animals, plants and microorganisms, or wastes of agriculture and livestock, which provide materials to produce organic fertilizers. Animal material (such as fish meals) has a high content of nitrogen, the content of leguminous meals are next. Plant material has a larger proportion of potassium, while manure materials differ with contents.

Q47: When applying humic acid, how to control the time and quantity of its application?

A47.1: Although the application period has a great range, every season is suitable. However, the best time of applying humic acid comes when the crop growth is at its beginning of a year or a period, specifically, before the germination of plant, seedling stage, or before blooming or grafting. The main purpose is to promote the growth of root system and soil improvement.

A47.2: Humic acid must be diluted before application to the root. The more the amount of diluted humic acid is applied, the better the response of crops.

Remarkable effects can only be reached when 20% of the root system is influenced after application.

Q48: Is it accepted to apply the organic fertilizer to fruit trees by hole-application or scattering on the soil surface?

A48: When applying organic fertilizer to fruit trees, the choice between hole-application and scattering is decided by the different materials and purposes. All kinds of organic fertilizers are mainly applied in holes or buried into the soil, which is easily for the absorption of fertilizer and the decreasing loss of fertilizer. It is a common practice for farmers to apply organic fertilizers by scattering or tiling. It is because of the high cost of labor and the unawareness of the disadvantage of scattering. Special purpose, such as reduce the water loss and prevent weed spreading, can be fulfilled by covering organic matters (such as rough tree bark and bagasse) on the soil surface.

Q49: When making farmyard manure, is a complete covering of plastic film necessary? What are the concerns when decomposing personally? How to make the judgment on whether it is maturity or wholly fermented?

A49: Whether a complete covering of plastic film is necessary or not depends on the season and organic matter.

A49.1: Season:
It is unnecessary to cover in dry seasons, while necessary in rainy seasons and low temperature seasons. Covering is beneficial to a speedy decomposition.

A49.2: Organic Matter:
If the manure is smelly or easily breed flies, covering is necessary in order not to pollute the environment. If the organic matter contains too much water and aeration that can speed up the decomposition, covering is harmful to aeration and water release.

The process of decomposition must be conducted with a caution against rain and an awareness of providing proper aeration. The compost must be prevented occasionally in case that it may stink seriously. Materials of organic matter must be mixed by adding limes and chemical fertilizers (the amount is decided by the raw material), and avoid using single certain material only. The judgment of maturity should be made according to different materials. Some materials can be used, in advance or mixed, even not being decomposed. Actually, the complete decomposition of organic matter definitely refers to mineralization, that is, transform from organic to inorganic. Thus, it is like ash which cannot be called organic fertilizer. Therefore, maturity represents the extent of nutrient availability and utilization after applied to the soil. There are many methods to judge whether it is decomposed or not. Different farmyard manures and their additions can lead to different judging methods. For example, the measuring of C/N ratio is unavailable as for farmyard manure which has a great addition of nitrogen fertilizers. Generally, composting farmyard manure can be initially judged from the composting time, the temperature change, the degree of color change; otherwise, it can only be measured by chemical methods such as measuring soluble carbon content, C/N ratio, and soluble nutrients.

Q50: Should different brands of organic fertilizers be used every year?

A50: The brand of organic fertilizer is not necessarily changed each year. Quality organic fertilizer is already a diversified and multi-kind formulation, so it is not necessary to be changed every year. If the organic fertilizer is made of only single material, mixture of organic materials is needed, so that a diversity of nutrients can prevent any bad tendency.

Q51: When should soluble fish extracts be applied to fruit trees (i.e., litchi plant)? How many times a year? It is said that soluble fish extracts contain a large amount of salts, is it true that fruit trees will die with a long time application?

A51.1: Soluble fish extracts is a kind of organic fertilizer that contains a lot of nitrogen. When it is applied to fruit trees such as litchi, the amount of application and chemical fertilizers (such as phosphorus, potassium, calcium, and magnesium) should be paid attention to; otherwise, it may result in excessive vain growth of branches and leaves, which would lead to reduction, dropping of flower, and fruit dropping. It becomes most serious when it is rainy. During the time of application, it is usually the case that organic fertilizer is used as basal fertilizer, before the rainy season or during the fruit growing period. One or three times each year is suitable. It depends on the soil condition. The soil with high nitrogen content should be applied with a little amount of soluble fish extracts; soil lacking in nitrogen should be applied with small amount and high frequency.

A51.2: There are many resources of soluble fish extracts. Fish production made by different means differ in their salt content. Production with high salt content should not be used with high concentration, but rather be used when diluted properly. Otherwise, the root may be damaged by fertilizer because of high salinity. So, the hilly land in the place with much rain has a great amount of salt leaching. Under the condition of protected cultivation without salt leaching, a large amount of application is not suitable. But it can be used after high-ratio dilution in order to avoid accumulation of salts. Otherwise, there is a possible result of flower and fruit dropping.

Q52: What kind of question should be attended to when applying organic fertilizer? How to distinguish between the true and the false?

A52: When applying organic fertilizer, the problems to be cared are as follows:

A52.1: Choose Suitable Organic Fertilizer:
The kinds of organic fertilizer are numerous and the nutrient content is different. Because of the different purposes when applying to plant growth with long-term and short-term phase, both conditions should be taken into consideration regarding the capability of releasing large amount of nutrients or the capability of releasing nutrients in a short period. When the purpose is soil improvement, it is necessary to choose organic material which is not easily decomposed. And, it can be added up by other material whether easily decomposed or not. Besides, a mixture of various materials designed in advance is another choice. All in all, only single material should be avoided.

A52.2: Proper Time of Application:
Based on the need of crop, it can be applied in the whole process or when the crop remains dormant. Make sure that the organic fertilizer provides nutrients for crop in a suitable time.

A52.3: Proper Amount of Application:
Either too little or too much amount of application is unsuitable. Too much application may lead to loss or excessive absorption of nutrients; too little application may result in insignificant effect. The amount of application should be coordinate with the chemical and microbial fertilizer. At the same time, economic benefit also needs consideration.

A52.4: Proper Method of Application:
The preferred effect can be reached when the organic fertilizer is completely mixed with buried soil. On the hilly land, it is not advisable

to apply on the surface soil, because it may lead to loss and thus cause pollution to the environment. It should be buried or covered by the soil in order not to breed flies which would endanger the sanitation of environment.

Issues concerning the differentiation of true organic fertilizers from false ones: any residues of animal, plant and microorganism can be used as the material of organic fertilizer, as long as they do not contain toxic matter, pathogens or pests, weed seeds, mixed soils or odor pollution. Therefore, if a fertilizer is made from organic matters, or the product is regulated by the governmental rules on fertilizers, it is organic fertilizer. The quality of organic fertilizers should be estimated according to economic benefit, organic matter content, nutrient content, water content, impurity content, the speed of providing nutrients, and the capability of promoting crop growth.

Q53: (1) What are the functions of fat in organic fertilizer? (2) Whether it is true: the more fat, the better? (3) Is oilcake the best resource of fat? Is the residue of sesame seed usable? (4) It is heard that there is a kind of organic fertilizer added with fat, is it true?

A53: Generally, the amounts of fat in organic fertilizers differ widely. The fat content is under 6% in oilcakes, and lower than 1% in solvent extracted product from oil seeds, while the figure goes even lower in manures (beyond 1%). Farmyard manure with plant residues as its material also has a low fat content.

A53.1: Fat consists of carbon, hydrogen, and oxygen, and is unimportant in fertilizer, because it lacks plant nutrients such as nitrogen, phosphorus, potassium, and calcium. Its function is merely for utilization by microorganisms, and the product of decomposition is no other than carbon dioxide and water.

A53.2: It is not necessarily true that the more fat, the better. For example, meals, which have a high fat content, are relatively low in protein. However, amino acid and ammonium (nitrogen fertilizers) will be produced during the decomposition of protein.

A53.3: Oilcakes are made by mechanic pressing rather than by solvent exacting, and the fat contend is of high level. However, because meal flour is made by solvent extracting, there is a great loss in potassium. Both of them can be used.

A53.4: Fat oil could be added to organic fertilizer for speeding up the extrusion process to produce grain organic fertilizer, it only serves as lubricating agent and is not necessarily the only choice.

Q54: *As organic fertilizers have been advocating in the past few years, how to choose the proper ones? Is there any soil management in net house cultivation? When growing papaya, how to make the suitable cooperation of chemical fertilizers?*

A54: Choose proper organic fertilizer based on the species of crop and the condition of the soil. Because of the fact that organic fertilizers abound in kinds and materials, differences abound in the nutrients, their proportion, price, maturity and rate of decomposition. Therefore, to choose proper organic fertilizer, the characteristics of certain organic fertilizer must be known. Along with the supply of other nutrients, the highest returns may be achieved. The good quality of organic fertilizers must satisfy the following requirements:

A54.1: Rich in Nutrients:
Because nutrient content varies in different organic materials, the more organic matter rich in nutrients is used, the better the quality. On account of the fact that different material has its own advantage in nutrients and contents, the mixture of different organic materials would fulfill the function of complementation.

A54.2: Rich in Organic Matter:
Because many of the compound organic fertilizers have an addition of chemical fertilizers, the organic matter content is decreased. A great amount of organic matter can ensure more accumulation of organic matter content in the soil.

A54.3: A Nutrient Ratio Suitable to the Crop:
The useful parts of crops are different, some crops take leaves and stems, some take flowers and fruits, and others take roots and stems. As a result, the formula of nutrients in organic fertilizers varies greatly.

A54.4: Decomposability and Nutrients Decomposed Can Satisfy the Need of Crop:
The decomposition of organic matter may be fast or slow, and either has its own advantage. The decomposition of organic fertilizer should not be too fast, when it comes to crop for long term growth. A compounding of

materials that can be quickly decomposed and materials that is slow in decomposing are necessary. The amount and proportion of decomposed nutrients should coordinate with the crop. The nutrients provided to fruits should not contain too much nitrogen fertilizer; the situation of leaf crop is different from this. Therefore, quality of organic fertilizers needs assortment.

A54.5: Promoting the Health and Growth of Crop:
The decomposition of organic matter is resulted from the function of microorganism reproduced. The products of decomposition or secretion of microorganism contain beneficial matter which can promote the growth of root system. They vary with different organic matters.

A54.6: Devoid of Pests, Pathogens and Toxic Matters:
The resource of organic matter is different, so there should not has any pests, pathogens, flyblow or toxic matters, nor should there be anything eliciting pests or diseases.

Soil management in net house cultivation is about the same with that in the open field. In the net house cultivation, though the rain would drop into the field, shade is provided, so nitrogen can be applied moderately to avoid vain-type growth. The application of all other fertilizers needs full awareness of adequate supply of nutrients in the early period.

Q55: *Is it necessary to make compost for some organic materials before application? How to decide whether it is the best time of application and when it is suitable to add beneficial microorganisms? What's the function?*

A55: Whether composting for some organic materials depends on the species of crops, time of application, amount of application, and the organic matter. Explanations are made respectively as follows:

A55.1: The Species of Crops:
In order to fast supply nutrients, it is necessary to compost the organic matter for short growing crops (such as vegetables and grasses). On the other hand, it is unnecessary to compost growing perennial crops (such as fruit trees). It can be applied directly during the dormancy stage of perennial crops or after fruit harvesting, with the exception of some castoffs, such as sawdust from mushroom residues, which must be made into compost before used.

A55.2: The Time of Fertilizer Application:
> It must be applied after composting, when the crop is in immediate need for large amount of nutrients. There would be not enough time for the release of nutrients, if it is not composted in advance.

A55.3: The Amount of Application:
> If it is organic matter that has not been composting, a small amount of application is recommended. And, large amount of application is not advisable.

A55.4: Organic Materials:
> If there contains any pests, pathogens, and weed seeds in the organic materials, composting is suggested to avoid these problems. High temperature during fermentation would minimize these problems. If no pests, pathogens or weed seeds are involved, it can be applied when it is suitable to the crop, proper in time and quantity.
>
> The higher the maturity of compost is, the more the nutrients can be quickly used. Beneficial microorganisms can be added at the beginning of composting to boost the decomposition, shorten the time needed, and even deodorize and improve the quality of the compost.

10.5 Q & A for Soil Microorganism

Q56: How to verify the vitality and effectiveness of microbiological products?

A56: It can be verified through the plate count to measure the ability of propagation. By means of medium culture to count the reproduction, the amount of viable organism, or active specific index such as the activity of enzyme, it can be worked out.

Q57: Would climatic factors and soil condition influence the functioning of microbial fertilizers?

A57: Different kinds of microorganisms are influenced by environmental factors to different extents. Microorganisms with lesser influences can better exert their function. Therefore, the superior microbial selection is an important work.

Q58: What are the microbial species that have been developed into microbial fertilizers and used in agriculture recently?

A58: There are many types of microbial fertilizers as following examples in Table 10-4.

Table 10-4. Types of Microbial Fertilizers and Their Main Functions

Microbial Fertilizers	Main Functions
N_2-fixing microorganisms (symbiotic or free-living)	Fix nitrogen and provide the crop with nitrogen fertilizer.
Phosphate-solubilizing bacteria	Dissolve insoluble phosphates from soils.
Mycorrhizal fungi	Promote the root absorption of phosphorous, resist pathogens and drought.
Decomposing bacteria	Decompose organic matter and provide nutrients.
Secretion bacteria of polymer compounds	Secrete polysaccharide and promote the soil aggregation.
Synthetic bacteria of growth hormone	Secrete growth hormone and promote the growth of crops.

Q59: What is the meaning of "expiration date" of microbial fertilizers? Is it valid beyond the "expiration date"?

A59: Microorganisms can be influenced by the store time, thus causing the decrease of bacteria populations or activeness. So, "expiration date" does exist. Over the "expiration date," the quality and effects of microbial fertilizers will decrease.

Q60: How to preserve microbial fertilizers?

A60: It is best preserved in place of low temperature and shade.

Q61: How do microorganisms keep their activeness and resist the unfavorable environment?

A61: Under favorable conditions, microorganisms keep their activeness by normal activity and metabolism. Under unfavorable circumstances, microorganisms would resist by dormancy or spore formation so that its population will decrease.

Q62: Does dormancy exist in microorganisms?

A62: In order to survive, microorganisms will change or diminish metabolisms, enhance the resistibility of their structures by the stress, and keep alive by a state of dormancy.

Q63: Why bottle-swelling sometimes happen to liquid microbial products?

A63: If liquid microbial products are polluted by other microorganisms from the environment, the polluted microorganisms could produce gases such as H_2S, CO_2, and CH_4. At such moment, bottle-swelling may occur in liquid microbial products.

Q64: Is it possible to culture different microorganisms at the same time in a bioreactor?

A64: It is possible for microorganisms to be compatible with each other.

Q65: Are the conditions the same when culturing different microorganisms?

A65: It depends. The conditions needed are different according to characters of the microbial species.

Q66: Is it true that the more microbial species contained in microbial fertilizers, the better? What is the highest density?

A66.1: It depends, because the more species contained in a given volume, the less the individual microorganism there is. It is difficult for less individual microorganism to exert their capability. The amount of application must be increased; otherwise, the effects may be nullified. Many species in microbial product must be cared whether there is any antagonism among microorganisms. One kind or few kinds of microorganisms can give a full play to their specific function.

A66.2: The higher the density of culture in microbial product, the more effects there are. The common average level of microorganisms is as much as 10^8 per milliliter. Long-time storage and transportation may decrease the microbial population.

Q67: Can different species of microorganisms be used at the same time? Will it have any influence on certain function?

A67: There is no definite answer to this question. Whether different species of microorganisms can be used at the same time depends on whether there is antagonism between different species. Generally, if there is not antagonistic or suppression between them, mixture is suitable. Whether there will be influence on certain function also depends on the function. If the functions are the same, possibilities are that it may be enhancement or adjustment. As for microorganisms of different products, it is suggested that the effects be testified thus the chance of supplementation can be controlled.

Q68: How to distinguish bacteria, fungi, and actinomycetes by our eyes? Can we see the microbial moving through the microscope?

A68: Bacteria, fungi, and actinomycetes can be distinguished only on growing plate in the certain special substrate of medium by our eyes. For example, if different strains grow on different solid medium, the bacteria colony can be roughly differentiated, for the bacteria individual is very small (about 1 μm). Actinomycetes is a filamentous and branching hyphae and shares the similar appearance with the fungi hyphae, with the diameter of 0.5 ~ 1.2 μm, which is the same with the bacteria. On the other hand, the fungi grow as filamentous hyphae. With the liquid medium, bacteria are in turbit, while actinomycetes and fungi are not, but with massive appearances.

Q69: How to identify whether the farm land need microbial fertilizer?

A69: The microbial fertilizer is used to reduce the side effect of the chemical fertilizer, assist the crops to absorb nutrients, and increase the availability of the nutrients. So, any crops and farm lands can use the beneficial microbial fertilizer. It is not necessary to decide whether to use it or not, but to decide which kind to use for the best economic benefit. For example, if phosphorus fertilizer is important for the land, then the phosphorus solubilizing bacteria should be used.

Q70: How to maximize the function of the microbial fertilizer?

A70: The best function of any kind of fertilizer depends on the matchup between the soil and the fertilizer. So does the microbial fertilizer. Pay attention to the followings.

A70.1: The soil cannot be too acid or alkaline. Too acid (pH < 5.0) and too alkaline (pH > 7.8) soils can affect both the absorbing of the nutrients and their availability for plants. The function of the microbial fertilizer would be limited in the strong acidic or alkaline soils. Highly acid soils can be neutralized by lime materials (such as magnesia lime, oyster shell, silicate lime, and dolomite lime); highly alkaline soils can be neutralized by acid materials, such as sulfur powder or acidic peat.

A70.2: The environment and materials have to match up the microbial propagation. The microbial fertilizer is viable, so they need to propagate and survive. The best place for them is the rhizospere, so putting the fertilizer around the root system can gain the best and direct effect. If a few humic acid, molasses, or nutrients are put in the dilution of the microbial fertilizer, it would be helpful for the survival and propagation of microorganisms.

A70.3: It is important to match up the crop requirement. The different crops within different growing periods need different microbial fertilizers. It is the earlier the better for crops to have microbial fertilizer, especially in the time of younger seedling. For example, perennial fruit trees need microorganisms that enhance the absorbing nitrogen and phosphorus in the growing period, and need the microorganisms that enhance the function of the phosphorus in the middle of fruiting.

A70.4: The quality of the microbial fertilizer: the microbial fertilizer is viable, so the inoculant needs to be with high microbial count and high degree of activity. The microbial fertilizer needs to adapt to the environment and there has to be few contaminated microorganisms.

Q71: After growing vegetables for several times, the soils become brick-red on the soil surface, what's the reason?

A71: In order to find the reason of the brick-red soil, the soils need to be analyzed by chemical methods. Usually, there are two reasons. Firstly, it may be a kind of red microorganism when feces organic matter are used over application, which causes the propagation of the red microorganisms, inhibits the other microorganism to grow, and causes the unbalance. The other possible reason is that it is only inorganic matter, mainly iron oxide. It is more possible for the soils that contain high degree of iron element, or the water of irrigation with the high iron element.

Q72: Which microbial product is better, the ones made in the local or those imported? What are the differences between the two and the matters needing attention?

A72: There are many factors that would affect the function of the microbial cultivation. The quality of the cultivation depends on the ability of the strains to adapt to soils, crops, and weather environment; and also the compatible ability to the crops, the viable count, the activity degree, the amount of contaminated microorganisms, the type of cultivation, the timing for using, the ways, and so on. When the above mentioned factors are positive, then it is a good product. For the protection of the environment, there are rules for the import of the microbial products to avoid the bad consequences. So, the development of the microbial products in the local area is very important, for problems that cannot be seen by our eyes need to be studied.

References

1. The Crop Fertilization Handbook. 1987. Council of Agriculture (COA) and Taiwan Provincial Department of Agriculture and Forestry (Department of Agriculture and Forestry and Fertilizer Technology Group, Lo, Chiu-Shyoung penned, Su Nan-Rong and Huang Po-En Reviewed). (in Chinese)

2. Fertilizer Diagnostic Techniques for Crops. 1981. Taiwan Agricultural Research Institute (Lin, Chia Fen Ed.). (in Chinese)

3. Taiwan Provincial Department of Agriculture and Forestry, 71, 72, 73, 74, 75, 76 Summary Reports for Experimental Study. (in Chinese)

4. Lan, Shen (translation). 1998. National Certification of Organic Farming-Draft Standards (The United States, Dec. 16, 1997). Chinese Society of Soil and Fertilizer. (in Chinese)

5. Council of Agriculture Committee, 2001. Biodiversity Promotion Plan. (in Chinese)

6. Feng, Fong-Long, 1997. Environment Impact Assessment of Taiwan Agricultural Development. Journal of Soil and Water Conservation 29 (3):220-236. (in Chinese)

7. Young, C. C., and C. C. Chen, 1989. Reseach and Planning on the Improvement of Taiwan Soil Fertility. The Society of Soil and Fertilizer of ROC., pp. 1-38. (in Chinese)

8. Young, C. C., W. L. Choa, C. C. Lou, S. N. Haung, S. S. Zun, and W. F. Chu, 2003. Collection and Application of Microorganisms in Taiwan. Special Research Report No.7, Chung Cheng Agriculture Science & Social Welfare Foundation, Taipei., pp. 1-125. (in Chinese)

9. Young, C. C., 2005. Characteristics and Exploitation of Biological Fertilizer. A Quarterly Publication of Agricultural Biotechnology, 4:18-22. (in Chinese)

10. Young, C.C., and F. T. Shen, 2007. Agricultural Application in Soil Microbial Diversity. Council of Agriculture, Executive Yuan and Chinese Society of Sustainable Agriculture, 26:3-9. (in Chinese)

11. Young, C. C., and F. T. Shen, 2008. Characteristics and Benefits of Phosphate Solubilizing Biofertilizers. Agricultural Biotechnology Industry Quarterly, 16:57-59. (in Chinese)

12. Xie, Zhao-Shen, and Ming-Guo Wang, 1991. Main Soil Maps of Taiwan. Experimental Center of Soil Survey, National Chung Hing University, Taichung, Taiwan. (in Chinese)

13. Alexandor, M, 1977. Introduction to Soil Microbiology. 2nd Ed. Wiley, New York.

14. Atlas, R. M., and R. Bartha, 1998. Microbial Ecology: Fundamentals and Applications. 4th Ed., Beajamin /Cummings Pub. Co., Inc., California.

15. Curl, E. A., and B. Truelove, 1986. The Rhizosphere. Springer Verlag, Berlin.

16. Frinck, A. 1982. Fertilizers and Fertilization: Introduction and Practical Guide to Crop Fertilization. Verlag Chemie GmbH, Wein. Heim.

17. Hausenbuiller, R. L, 1985. Soil Science: Principles & Practices. Wm. C. Brown Pub., Dubuque, Iowa.

18. Liebhardt, W., and R. Harwood, 1985. Organic Farming in Technology Public Policy, and the Changing Structure of American Agriculture. (Vol II-Background Papers, No. 21) Office of Technology Assessment, Congress of the United States.

19. MOA, 1996. MOA Nature Farming Guidelines. MOA International Foundation of Nature Ecology.

20. Mengel, K. and E. A. Kirkby, 1982. Principles of Plant Nutrition. International Potash Institute, Worblaufen-Bern.

21. Metcalfe, D. S., and D. M. Elkins, 1980. Crop Production: Principles and Practices. Macmillan, New York.

22. Potash & Phosphate Institute, 1979. Soil Fertility Manual, PPI, Altanta.

23. Powell, C.L., and D. J. Bagyaraj, 1984. VA Mycorrhiza, CRC Press, Boca Raton, Forida.

24. Russell, E.W, 1973. Soil Conditions and Plant Growth. Longman, London.

25. Tortora, G. J., B .R. Funke, and C. L. Case, 2001. Microbiology: An Introduction. Addison Wesley Longmen, Inc., N.Y.

26. Young, C. C., 2007. Development and Application of Biofertilizers in the Republic of China. In Business Potential for Agricultural Biotechnology Products. The Asia Productivity Organization, Tokyo, pp. 51-57.

27. Young, C.C., 1996. Biological N_2 Fixation and P-Solubilizing Bacteria. International Training Workshop on Microbial Fertilizers and Composting, pp. 1-14.

28. Young, C.C., 1994. Selection and Application of Biofertilizer in Taiwan Agriculture. Rural Development Administration, and Food and Fertilizer Technology Center, pp. 1-19.

SOIL and FERTILIZER
Concepts and Practices

作　　　者／	Chiu-Chung Young（楊秋忠）
總 編 輯／	官大智
責任編輯／	黃俊升、廖敬華、葉菀婷
美術編輯／	林玫秀
封面圖片／	楊秋忠
版面編排／	劉紋伶

發 行 人／	李德財
總 經 理／	鄭學淵
經　　理／	范雅竹
發　　行／	楊子朋
出　　版／	國立中興大學

　　　　　　　地　　址：402 臺中市南區國光路 250 號
　　　　　　　電　　話：(04)2284-0291　傳真：(04)2287-3454
　　　　　　　服務信箱：press@nchu.edu.tw

　　　　　　　華藝學術出版社（Airiti Press Inc.）
　　　　　　　地　　址：234 新北市永和區成功路一段 80 號 18 樓
　　　　　　　電　　話：(02)2926-6006　傳真：(02)2923-5151
　　　　　　　服務信箱：press@airiti.com

發　　行／	華藝數位股份有限公司

　　　　　　　戶　　名（郵局／銀行）：華藝數位股份有限公司
　　　　　　　郵政劃撥帳號：50027465
　　　　　　　銀行匯款帳號：045039022102（國泰世華銀行　中和分行）

法律顧問／立暘法律事務所　歐宇倫律師
ISBN ／ 978-986-5663-01-8
GPN ／ 1010301290
DOI ／ 10.6140/AP.9789865663018
出版日期／ 2014 年 8 月
定　　價／新台幣 680 元

　　　　　　　　版權所有・翻印必究　　Printed in Taiwan
　　　　　　　　（如有缺頁或破損，請寄回本社更換，謝謝）